ENDOCRINE, METABOLIC AND IMMUNOLOGIC FUNCTIONS OF KERATINOCYTES

ANNALS OF THE NEW YORK ACADEMY OF SCIENCES
Volume 548

ENDOCRINE, METABOLIC AND IMMUNOLOGIC FUNCTIONS OF KERATINOCYTES

Edited by Leonard M. Milstone and Richard L. Edelson

The New York Academy of Sciences
New York, New York
1988

Cover (paper edition): *Schematic drawing of the epidermis.*

Library of Congress Cataloging-in-Publication Data

Endocrine, metabolic, and immunologic functions of
   keratinocytes.

   (Annals of the New York Academy of Sciences,
ISSN 0077-8923; v. 548)
   Papers presented at a conference held by the New
York Academy of Sciences in New York City, Feb. 17–19,
1988.
   Includes bibliographies and index.
      1. Keratinocytes—Congresses.   I. Milstone,
Leonard M.   II. Edelson, Richard L.   III. Series.
[DNLM: 1. Epidermis—cytology—congresses.   2. Keratin—
immunology—congresses.   3. Keratin—physiology—
congresses.   W1 AN626YL v. 548 / WR 101 E56 1988]
Q11.N5   vol. 548   500s   89-2922
[QP88.5]   [599′.01858]
ISBN 0-89766-504-X (alk. paper)
ISBN 0-89766-505-8 (pbk. : alk. paper)

SP
Printed in the United States of America
ISBN 0-89766-504-X (cloth)
ISBN 0-89766-505-8 (paper)
ISSN 0077-8923

ANNALS OF THE NEW YORK ACADEMY OF SCIENCES

Volume 548
December 31, 1988

ENDOCRINE, METABOLIC AND IMMUNOLOGIC FUNCTIONS OF KERATINOCYTES[a]

Editors
LEONARD M. MILSTONE and RICHARD L. EDELSON

CONTENTS

[a]The papers in this volume were presented at a conference entitled Endocrine, Metabolic and
Immunologic Functions of Keratinocytes, which was held by the New York Academy of Sciences
in New York City on February 17–19, 1988.

Part III. Peptides with Paracrine/Endocrine Actions

Part IV. Peptides with Autocrine Actions

Part V. Peptides Which Modulate the Immune System

Major funding for this conference was provided by:
- THE NATIONAL INSTITUTE OF ARTHRITIS AND MUSCULOSKELETAL AND SKIN DISEASES, NATIONAL INSTITUTES OF HEALTH

Other financial assistance was received from:
- CHEMEX PHARMACEUTICALS
- COLLAGEN CORPORATION
- CONVATEC (A SQUIBB COMPANY)
- ELDER PHARMACEUTICALS, INC.
- HERBERT LABORATORIES
- HOECHST-ROUSSEL PHARMACEUTICALS, INC.
- HOFFMANN-LA ROCHE, INC.
- ICN PHARMACEUTICALS, INC.
- JANSSEN PHARMACEUTICALS
- JOHNSON AND JOHNSON
- MERCK SHARPE AND DOHME RESEARCH LABORATORIES
- MILES PHARMACEUTICAL
- ORTHO PHARMACEUTICAL CORPORATION
- OWEN/ALLERCREME PHARMACEUTICALS
- PFIZER CENTRAL RESEARCH
- SCHERING LABORATORIES
- SMITH KLINE AND FRENCH LABORATORIES
- STIEFEL LABORATORIES
- SYNTEX LABORATORIES, INC.
- THE UPJOHN COMPANY
- WESTWOOD PHARMACEUTICALS, INC.

Preface

Few would argue with the statement that the epidermal keratinocyte is not the most important or interesting cell in the body. Yet, there also would be no argument that the number of investigators contributing to the science of keratinocyte biology has increased enormously during the past ten years. In 1985, Richard Edelson and I had several lengthy conversations about our own work, the investigative strengths of the dermatology department at Yale, and the directions that research in skin would be taking over the next decade. It was clear that he, a cellular immunologist, and I, an epidermal biologist interested in cell structure and renewal, had become peripherally involved with actions of keratinocytes that raised more questions than they answered. Our experiences were by no means unique, and they seemed to follow a pattern. When methods for cultivating large numbers of keratinocytes—developed largely in the laboratories of Yuspa, Ham and Green—were introduced to a fertile investigative community, unexpected observations were made. Ready accessibility of the culture methods often engendered these studies and accounts for their obvious geographic clustering. This certainly had been true in New Haven, where I began growing keratinocytes in 1977. The more the culture systems were probed, the more pieces of provocative data were found. The more data accumulated, the more interesting and important the keratinocyte seemed, but the more difficult it became to understand those data from a functionally balanced perspective.

The time seemed appropriate to collect and organize the existing data. The hope was to raise the consciousness of investigators who, like ourselves, had inadvertently placed a small piece in a puzzle whose overall form was only beginning to emerge. The result was a meeting held in New York City during February 17–19, 1988.

We are grateful to the New York Academy of Sciences for sponsoring the meeting and to the National Institute of Arthritis and Musculoskeletal and Skin Diseases for providing the major financial support. Geraldine Busacco, Cynthia Wishengrad and Ellen Marks at the Academy office were especially helpful with the local planning for the meeting. Bill Boland and Cook Kimball were a great help in assembling and editing the manuscripts. Janet Fears' administrative assistance in my office was indispensable.

I am indebted to my colleagues, whose work made the program possible and whose participation made the meeting a success. Lorne Taichman, Stuart Yuspa and Peter Elias deserve added thanks for thoughtful consultation during all phases of planning for the meeting. Just as the meeting was enlightening and useful to the participants, it is my hope that this volume will facilitate the entry of others into this new area of keratinocyte biology.

Leonard M. Milstone

Effector Functions of Epidermal Keratinocytes

LEONARD M. MILSTONE[a]

Veterans Administration Medical Center
West Haven, Connecticut
and
Yale University
New Haven, Connecticut

Until ten years ago, most biologists considered the epidermis to be little more than an elegant form of Saran wrap: it kept water in and toxins and microbial agents out, and it imparted a certain distinctive character to the individual. Research in epidermal biology was concerned mainly with structural proteins, barrier formation, and the regenerative capacity of keratinocytes, the cells responsible for keeping the structure intact.

More recent experimental data reveal the epidermis as a remarkably versatile organ, possessing previously unrecognized metabolic, endocrinologic, and immunologic capabilities. Moreover, by virtue of its total mass, its contact with the environment, and its proximity to the rich supply of underlying dermal vessels and nerves, the epidermis is in a unique position to act as the signalling interface between the rest of the body and its surrounding environment. Within the epidermis, the keratinocyte has emerged as the cell responsible for many of these newly described functions, and this information has become available largely through recent advances in keratinocyte cell culture technology, improvement in protein isolation techniques, and development of sensitive bioassays.

This conference was organized not only to compile, analyze, and evaluate these newly appreciated keratinocyte activities, but also to consider to what extent we should view the epidermis in a new light—as a tissue with effector functions. By effector function I mean an activity of one cell that alters the behavior of another cell. It can result from *de novo* synthesis and secretion or from biochemical modification of molecules that ultimately have a distant action.

The concept of the keratinocyte as an effector cell may seem somewhat surprising and even contrary to classical teachings, which present the keratinocyte as a responsive but not an evocative cell. Thus, we learn that keratinocytes respond to inductive signals from underlying mesenchyme by differentiating into hair or nails; they respond to external chemical, physical or microbial stimuli by altering their rate of proliferation or turnover; they respond to genetic and metabolic deficiencies and to the inevitable immunologic battles which occur in the skin by making the wide variety of scales and blisters which delight the diagnostic eyes of so many of us who are clinicians. With the exception of the long-established role of the keratinocyte in vitamin D synthesis, the epidermis was formerly ignored as having no significance to the rest of the organism apart from its barrier function. However, now that several keratinocyte activities with

[a]Address for correspondence: Leonard M. Milstone, M.D., Department of Dermatology/185, Veterans Administration Medical Center, West Spring Street, West Haven, CT 06516.

potentially distant functions have been identified, we should take a moment to reflect on the clues that were always, unmistakably there.

One clue that the epidermal keratinocyte has the capacity to be an effector cell comes from comparative anatomy and embryology. During embryogenesis, some keratinocytes from some species differentiate into specialized, glandular epithelia that synthesize and secrete all manner of interesting molecules, whose chemical structures range from alkaloids to zymogens and whose effector functions range from toxins to pheromones. Curiously, these glands mostly evolved to overcome the barrier function of the epidermis and deliver effector molecules to the external environment. Rarely would such effector molecules have an internal function. Nonetheless, such examples are clear indications that keratinocytes have the capacity to synthesize or metabolically modify and secrete molecules intended to have distant actions.

A second clue is found in the way in which terrestrial mammals produce and maintain the epidermal barrier. Two well-known properties of mammalian epidermis are its ability to rapidly regenerate following trauma and its normal, ongoing steady-state of continuous renewal of the epidermal keratinocytes. Both these properties have long been thought to be regulated, to a certain degree, by keratinocyte-derived effector signals. Cybernetic theory predicts the existence of feedback messengers that regulate the steady-state process of continuous renewal. Some of these messengers could be stimulatory, other inhibitory. Some could be made by differentiated cells and act on proliferating cells. Others could be made by proliferating cells and act on stem cells. The result of such a signalling network is that cells do not excessively accumulate in one compartment nor do they become depleted by the renewal process. The important issue for this meeting is not the validity of any specific scheme, but the logical expectation that epidermal keratinocytes should be secreting effector molecules. Furthermore, the keratinocyte itself should be considered as a possible target for any of the newly discovered effector functions of keratinocytes.

There are three major goals for this conference. The first is to compile a current catalogue of these potential effector functions of keratinocytes. In so doing, we should be asking each other questions about production, characterization and function. What regulates the activity or the production of the putative function? Is it restricted to a specific compartment within the epidermis? Is it produced constitutively, during all phases of embryonic and adult life, or is it a response to certain inductive stimuli? Is it found *in vivo*, under normal circumstances, or is it an artifact of our experimental systems? What is the molecular nature of the effector and is it structurally related to other effectors isolated from the keratinocyte? Are the keratinocyte molecules unique or related to molecules made elsewhere in the body? Aside from the effector functions readily identified by our bioassays, what is the real physiologic role of each effector function?

The second goal is to educate each other about which are the really important questions to be asking, and how, technically to go about answering them.

The third goal is to encourage additional serious thought about the role of the keratinocyte as an effector cell. I venture to say that if each of the keratinocyte activities described here were viewed as an isolated fact, it would be considered little more than a curiosity. In the aggregate, however, the two dozen or so activities begin to make a strong statement about the potential of the keratinocyte as an effector cell.

The papers are organized under six headings:

- the role of keratinocytes in vitamin and hormone biogenesis
- distant effects of keratinocyte metabolism
- peptides with paracrine/endocrine actions

- peptides with autocrine actions
- peptides which modulate the immune system
- multifunctional peptides from keratinocytes

These groupings are admittedly arbitrary, since they more often reflect the assay system we use to identify and characterize each molecule rather than an understanding of its true function. They do, nonetheless, have the heuristic value of emphasizing categories of potential effector function.

Interspersed with these topics are several more general lectures, included to anticipate important issues that will arise when considering the epidermis as an effector tissue. What are the natural barriers and gradients in the epidermis? How can we quantitate the contribution of an epidermal activity that may be only a part of the entire systemic activity? How can the peptide's structure be modified either naturally or artificially to alter its activity? What are the important signal transduction systems in the epidermal keratinocyte? How do we begin to identify the true function of a molecule with several potential functions or targets?

Let me acknowledge that this program is incomplete. Practical and organizational choices had to be made, sometimes arbitrarily. All topics on the program have been described in mammalian keratinocytes and have either established or potential distant importance. The restriction to mammals was particularly troublesome because of the rich literature of effector molecules—ranging from newly described antimicrobial peptides to the centuries-old toxins used on poison darts—found in amphibian epidermis. We must remember, however, that most of these keratinocyte activities probably evolved within the context of the basic barrier-protective function of the epidermis. Thus, an organism's evolutionary history and environmental niche should be significant in determining which of these activities would be retained and expressed. Consequently, while amphibians' epidermis provides inspiring examples of the versatility of the keratinocyte, it may not be relevant to understanding the potential of human keratinocytes.

What implications will this conference have for future investigation? At the very least, the compilation of newly appreciated keratinocyte capabilities should facilitate a reassessment of a variety of issues about which we are ignorant, and, perhaps, open entire new areas of investigation. What is the role of the keratinocyte in regulating morphogenesis during development? How is the steady-state in the epidermis controlled? Do these newly appreciated effector functions have a role in the etiology, pathogenesis or therapy of disease? Can the keratinocyte be used, either through genetic engineering or selective stimulation, as an effector peptide factor?

I think you will find the following presentations enlightening and provocative. They should provide you with a totally new view of the epidermal keratinocyte.

Lipid-Related Barriers and Gradients in the Epidermis[a]

PETER M. ELIAS[b] AND KENNETH R. FEINGOLD

*Department of Dermatology
and
Department of Medicine
Veterans Administration Medical Center
and
University of California School of Medicine
San Francisco, California*

EVOLUTION OF CURRENT CONCEPTS OF STRATUM CORNEUM STRUCTURE

The old image of mammalian stratum corneum, based on routine histologic preparations, is one of a loosely-bound layer in various stages of disorganization and shedding (TABLE 1). This misleading image, refuted by physical-chemical,[1] frozen-section,[2-4] and freeze-fracture[5-7] studies (see below), led to the long-held view that the barrier resides at the stratum granulosum (SG)-stratum corneum (SC) interface.[8] New appreciation for the integrity of the SC led initially to the concept that the full thickness of the SC is functionally competent,[1,9] a view that analogized the stratum corneum to a homogenous film.[9] The "plastic wrap" hypothesis held that the SC consisted of keratin fibers embedded in a lipid-enriched matrix, a view supported by X-ray diffraction studies that demonstrated highly ordered lipid structures in SC.[10] The homogenous film analogue views percutaneous transport as occurring transcellularly, without regard to membrane, intercellular, or intracellular compartments.[9,11] Finally, this model holds that the SC is devitalized, *i.e.*, metabolically inert, and participating in the permeability phenomenon only as a passive membrane.[9]

In 1968, Middleton first reported certain straight-forward yet elegant studies that tended to refute the homogenous film concept (TABLE 2).[11] Whereas Irvin Blank had shown years earlier that organic-solvent extraction destroyed the water-holding capacity of SC,[12] Middleton showed that pulverization was as effective as solvent-extraction in this respect, thereby providing evidence that SC lipids might be organized as osmotically active membranes within the SC. Shortly thereafter the existence of separate lipophilic vs. hydrophilic pathways of percutaneous absorption was suggested by workers at Alza Corporation.[13]

The destruction of stratum corneum during processing for routine histologic and ultrastructural preparations obscured further advances until the early 1970s, when the cells of the SC were shown by frozen sectioning to comprise tightly arrayed, polyhedral structures in vertical, interlocking columns.[2-4] The initial awareness of lipid-protein

[a]Supported by National Institutes of Health Grant AM 19098 and by the Medical Research Service, Veterans Administration.

[b]Address correspondence to Peter M. Elias, M.D., Dermatology Service (190), Veterans Administration Medical Center, 4150 Clement Street, San Francisco, California 94121.

4

segregation to specific tissue compartments came with freeze-fracture replication, which revealed for the first time the presence of multiple, broad lamellations in the interstices of several types of mammalian keratinizing epithelia.[5-7] Several other lines of indirect evidence also supported such structural heterogeneity, including lipid histochemistry of frozen sections,[6] as well as the ability of SC to be dispersed by certain organic solvents.[14,15] Definitive evidence for the compartmentalization of lipids came with the isolation of SC membrane "sandwiches", containing trapped intercellular lipids. These preparations:[16] a) comprised about 50% lipid by weight, and accounted for over 80% of SC lipid; b) displayed the same lipid profile as whole SC; c) contained the same broad lamellae on freeze-fracture as are present in the interstices of whole SC; and d) generated the same ordered X-ray diffraction pattern previously ascribed to the "interfilamentous lipid matrix".[17] More recently, the colocalization of lipid catabolic enzymes to SC membrane domains, both by ultrastructural cytochemistry and by enzyme biochemistry, can be considered further evidence of the structural heterogeneity of mammalian SC.[18-20]

The localization of the barrier continued to be debated (TABLE 3): Is the entire SC functionally competent, or does the principal barrier reside in the lower layers? Tracer perfusion studies, as early as 1969, showed that water-soluble molecules injected into the dermis do not reach the SC.[21,22] Outward percolation halted in the outer SG, at intercellular sites engorged with discharged lamellar body contents.[5] These studies,

TABLE 1. Stages in the Understanding of the Stratum Corneum

Concepts	Approx. Dates
1. Disorganized, nonfunctional	up to 1950
2. Homogenous film	up to the 1970s
3. Lipid-protein compartmentalization	1975–present
4. Metabolically active	1984–present

while admittedly employing tracers considerably larger than the water molecule itself, pointed to the presence of a barrier in the outer stratum granulosum. Direct evidence of the barrier capabilities of different layers of the SC came with the recent isolation of intact sheets of porcine stratum compactum after prior enzymatic stripping of the stratum disjunctum. Whereas these studies did not address the capacity of the stratum disjunctum to contribute to barrier function, they did demonstrate the functional integrity of the stratum compactum.[23]

LIPID BIOSYNTHETIC GRADIENTS IN THE EPIDERMIS

The fact that phospholipids disappear from stratum corneum, while neutral lipids and sphingolipids (primarily ceramides) increase in relative abundance, is now common knowledge. A very similar change in lipid profile seems to occur in human,[24] hairless mouse,[25] neonatal mouse,[26] and porcine epidermis.[27] Certain striking features of the modulations suggest that these alterations are important for barrier function. Sphingolipids, in particular, have been pinpointed as the critical molecule for barrier function, because: a) they possess a long-chain, hydrophobic base;[28,29] b) they have very long-chain, highly saturated, N-acylated fatty acids;[27-31] c) some species contain an additional ester-linked, terminal linoleate residue;[30-32] and finally d) oleate-linoleate

TABLE 2. Evidence for Lipid-Protein Compartmentalization in the Stratum Corneum (SC)

1. Pulverization destroys water-holding capacity of SC.[11]
2. Hydrophilic + lipophilic substances cross separate SC pathways.[13]
3. Freeze-fracture reveals lipid lamellae in SC interstices.[5–7]
4. Frozen sections display neutral lipids solely in SC interstices.[6]
5. SC can be dispersed into individual cells with organic solvents.[14,15]
6. Isolated SC membrane "sandwiches" account for most SC lipids.[16]
7. X-ray diffraction shows ordered lipids in isolated SC membranes.[17]
8. Catabolic enzymes colocalize with lipids in SC interstices.[18–20]

substitution occurs in association with defective barrier function in essential fatty acid deficiency (EFAD).[33] Together, these results have been used to assert the primacy of sphingolipids for barrier function in mammalian SC.[34] This lipid analytic approach is obviously indirect, however, and leaves unanswered the role, if any, of neutral lipids, particularly free fatty acids, free sterols, and the lesser quantities of alkanes, sterol esters, triglycerides, and cholesterol sulfate in barrier function.

To establish more definitively the role of lipids for the permeability barrier, a more physiological approach is necessary. In the early 1980s, one of us (KRF) noted that the skin of both rodents and primates synthesized abundant cholesterol and other nonsaponifiable lipids (NSL).[35,36] In fact, the synthetic activity of the skin rivalled that of the liver and gastrointestinal (GI) tract, the two major putative sites of sterologenesis.[35,38] Moreover, in contrast to hepatic and GI synthesis, cutaneous sterologenesis was not influenced by circulating sterol levels,[37] suggesting an unusual degree of independence from systemic regulation. The basis for this autonomy appears to reside in the absence of plasma membrane receptors for low-density lipoproteins (LDL) on differentiating keratinocytes, both *in vitro* and *in vivo*.[38–40] However, since LDL receptors are present on the membranes of preconfluent, cultured human keratinocytes,[39] and on both human and murine basal cells *in vivo*,[40] these studies have not ruled out the possibility that regulation of cholesterol synthesis via LDL-receptor-mediated mechanism may occur in the proliferating pool of the epidermis.

In order to link lipid species to epidermal function, we recently have eschewed purely analytical studies in favor of a lipid metabolic approach (TABLE 4). The basic strategy is, first, to perturb epidermal barrier function, typically with an organic solvent such as acetone, and then to correlate barrier status with both an assessment of lipid replenishment in oil red O- and nile-red-stained frozen sections, and with rates of lipid biosynthesis (from ^3H_2O) in the same samples. These results have shown: a) that the epidermis is a major site of sterol synthesis, accounting for about 30% of total cutaneous sterologenesis;[37] and b) that both epidermal sterol and fatty acid synthesis are stimulated by perturbation of the permeability barrier.[41,42] Moreover, such stimulation is localized to treated sites, is limited to the epidermis,[41,42] corrects as barrier function returns to normal (FIG. 1),[41,42] and correlates first with removal and

TABLE 3. Location of the Barrier-Evolving Concepts

Concepts	Dates
1. Stratum granulosum-stratum corneum interface	through the 1950s
2. Full thickness of stratum corneum	1960s—may still be valid
3. Limited to stratum compactum	1985—may still be valid
4. Initially formed in stratum granulosum	1968—may still be valid.

TABLE 4. Dynamic vs. Static Approach to the Delineation
of Lipid Functions in the Epidermis

	Static	Dynamic
Advantages	Can determine precise chemical structures	Can directly infer metabolic regulation
	Methodology straight-forward	Can express data in meaningful terms, *i.e.,* specific activity
Disadvantages	Cannot directly infer function	Radioactivity interferes with some analytical techniques
	Proper denominator unknown	Potential problems with choice of precursor

subsequently with repletion of lipid in the stratum corneum.[41] Furthermore, c) epidermal sterologenesis is stimulated in essential fatty acid deficiency in relation to the defect in barrier function; *i.e.,* lipogenesis rates normalize with occlusion despite persistence of the underlying deficiency state.[43] Finally, d) synthesis is normalized when an impermeable membrane is applied to the perturbed skin.[41–43] Extracutaneous sterols and fatty acids do not appear to contribute significantly to the pool of newly synthesized lipids.[41,42] Also, the relationship of epidermal lipogenesis to barrier

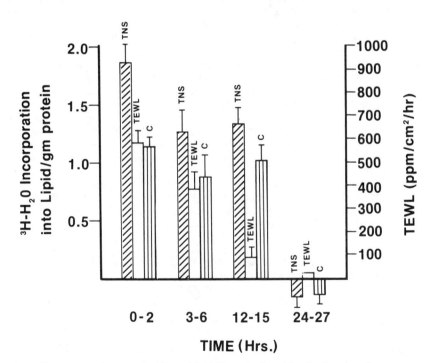

FIGURE 1. The time course of epidermal lipid biosynthesis and barrier function after treatment with acetone. Note that transepidermal water loss (TEWL), total nonsaponifiable lipid biosynthesis (TNS), and cholesterol (C) synthesis exhibit a parallel return toward normal over 24 hours.[41]

TABLE 5. Evidence That Barrier Function (Water Loss)
Regulates Epidermal Lipogenesis

---

1. Dietary or solvent-induced barrier disruption stimulates epidermal lipogenesis.[41-43]
2. Extent of lipid biosynthetic rates parallels severity of barrier defect.[41,42]
3. Normalization of lipid biosynthetic rates parallels barrier recovery.[41,42]
4. Occlusion with impermeable, but not vapor-permeable membranes after barrier disruption blocks acceleration of lipid biosynthesis.[41-44]

---

function is underscored further by the lack of modulations in cutaneous sterologenesis in response to either vitamin D deficiency or excess,[44] despite the known importance of the cutaneous free sterol, 7-dehydrocholesterol, for vitamin D synthesis.

In more recent studies, the role of the water molecule itself as the regulatory signal has been explored in greater detail. (TABLE 5).[45] Whereas occlusive membranes blunted the expected burst in lipid biosynthesis that follows barrier disruption, application of semipermeable membranes neither blocked synthesis, retarded the rate of return of normal barrier function, nor impeded the return of stainable lipid in the stratum corneum. These studies lend strong support to the concept that the rate of water loss itself may be the regulatory signal for epidermal lipogenesis.

Although the above body of evidence demonstrates the capacity of the epidermis to synthesize lipids in response to barrier requirements, they leave undetermined the sites of synthesis in the epidermis, and the relative importance of specific SC lipids for barrier function. The rapid rate of return of barrier function to normal, i.e., by 15–20 hrs, even after exhaustive delipidization accompanied by TEWL rates >1,000 ppm/cm^2/hr, suggests that lipid synthesis may not be limited to the proliferating compartment (the basal layer) but may extend to, or even be accelerated in the outer layers of the viable epidermis. Indeed, both in vivo and in organ culture, the rates of epidermal lipogenesis are higher in the stratum granulosum (SG) of neonatal mouse epidermis than in the subjacent basal/spinous layers, even under basal conditions (TABLE 6),[46] i.e., in the absence of barrier perturbation. The importance of this observation is underscored further by the simultaneous, precipitous decrease in protein, DNA, and CO_2 generation in the same layer (TABLE VI). Whether the SG is the site of accelerated synthesis following barrier perturbation is currently under investigation.

Certainly, the above studies establish that nonsaponifiable lipids, namely sterols and possibly hydrocarbons, are important for barrier function. Although the role of sphingolipids has not been assessed utilizing the metabolic approach, the observation that fatty acid synthesis is also regulated by barrier requirements[42] is consistent with a

TABLE 6. Comparison of Lipid, Protein, and DNA Biosynthesis Rates
in Neonatal Mouse Epidermal Cell Layers[a]

Fraction	Number of Animals	Incorporation		Ratio SG/SB + SS
		SS/SB	SG	
Total lipids[b]	16	40.9	85.7	2.1
Total protein[c]	7	56.1	12.7	0.226
Total DNA[d]	7	9.96	0.23	0.024

[a]Adapted from REFERENCE 46.
[b]nmoles of ^3H_2O incorporated/mg protein/hr.
[c]fmoles ^{3}H-leucine incorporated/mg protein/hr.
[d]fmoles of ^{3}H-thymidine incorporated/mg protein/hr.

role for acylated lipids, including sphingolipids, in barrier function. Recently, we assessed the importance of relatively polar SC lipid species—sphingolipids and free sterols vs. nonpolar species (free fatty acids, sterol esters, and hydrocarbons)—for barrier function.[47] Using a highly nonpolar organic solvent, petroleum ether, vs. the more bipolar solvent, acetone, we found that removal of highly nonpolar species alone produced a significant break in the barrier, but TEWL rates never exceeded 150 mg/cm^2/hr. Whereas petroleum ether removed large quantities of nonpolar lipids, it left the polar species in place.[47] In contrast, acetone treatment caused profound barrier defects, as more and more lipid was removed. Moreover, whereas petroleum ether removed only nonpolar lipids, acetone removed a much greater proportion of polar species. These experiments show: a) the linear relationship between lipid content and barrier function; b) that nonpolar lipids, in the absence of sphingolipids, provide a

TABLE 7. Stratification of Lipid Catabolic Enzyme Activities in Neonatal Mouse and Human Epidermis[a]

	Specific Activities		
Layer	Acid Lipase[b] (nmol/min/mg)	Sphingomyelinase[b] (pmol/min/mg)	Steroid Sulfatase[c] (cpm/mg/hr × 10^3)
SC	1.38 ± 0.08 (n = 10)	51.6 ± 1.3 (n = 12)	37.3 (n = 5)
SG	3.07 ± 0.11 (n = 10)	49.2 ± 2.0 (n = 10)	97.8 (n = 5)
SS/SB	ND[d] (n = 8)	14.3 ± 0.6 (n = 8)	10.5 (n = 5)

[a]Adapted from REFERENCES 18, 19.
[b] = neonatal mouse.
[c] = human.
[d]ND = none detectable.

"first-line" of barrier function; and c) that relatively polar lipids appear to provide a more profound level of barrier integrity and cohesion.[a]

LIPID CATABOLIC GRADIENTS IN THE EPIDERMIS

That the outer layers of the epidermis are sites of intense metabolic recycling is well known. During terminal differentiation, nucleotides, certain amino acids, trace metals, and certain ions are salvaged and apparently reutilized. A zone of catabolic enzyme activity, including nucleotidase, proteases, esterases, and phosphatases, apparently mediates this recycling process.[48–50] Recently, a detailed picture of lipid catabolic events in the epidermis has resulted from a combination of cytochemical and

[a]Although this paper deals only with the role of lipids in the epidermal permeability barrier, several lines of evidence point to the function of SC lipids in cohesion and desquamation. Therefore, it is perhaps relevant that acetone treatment, after prior petroleum ether applications, caused the SC to disintegrate into individual cells and aggregates (unpublished observations). This suggests that in addition to providing for barrier function, sphingolipids and free sterols may underlie desquamation as well.

TABLE 8. Hydrolytic Enzyme Content of Epidermal Lamellar Bodies[a]

	Present	Absent or Not Increased
Lipid catabolic	acid lipase phospholipase A sphingomyelinase glycosidases	steroid sulfase
Others	acid phosphatase cathepsins carboxypeptidase	aryl sulfatase A & B beta- glucuronidase
extracellular protease		plasminogen activator

[a]Adapted from REFERENCES 51, 52.

biochemical studies. A variety of lipases, including phospholipase, acid lipase, sphingomyelinase, and steroid sulfatase, all were found to be (TABLE 7): a) concentrated in the SG and SC,[18–20] with much less activity in the basal/spinous layers;[18] and b) localized to epidermal lamellar bodies in the SG (except steroid sulfatase which is not present in lamellar bodies).[51,52] This organelle is the putative precursor of the lipids in the SC interstices. All were also found to be c) concentrated in membrane domains in the SC.[18–20] Likewise, glycosidases,[52] acid phosphatase,[51,52] and certain proteases[51] appear in the outer epidermis and within lamellar bodies (TABLE 8). The localization of these enzymes to the outer epidermis, as well as their colocalization within sites of lipid deposition in intercellular domains, suggests that they mediate changes in extracellular lipids (and other materials) that might first result in the formation of a hydrophobic barrier and subsequently mediate normal desquamation (FIG. 2, TABLE 9). This model is supported by recent biochemical and morphological studies that demonstrate progressive loss of phospholipids, glycosphingolipids, and cholesterol sulfate during movement from the SG to the stratum compactum,[53] and finally the stratum disjunctum, where only small amounts of cholesterol sulfate still remain (FIG. 3).[20,53] Moreover, accompanying these compositional changes are dramatic changes in the morphology of secreted lamellar body disks, as they first elongate by end-to-end fusion, then transform from elongated discs to the broad stacked lamellae found in the mid-to-outer SC (FIG. 4).[36,53] Still unresolved is the fate of these lamellations during the first stages of shedding—is membrane bilayer break-up a prerequisite for shedding

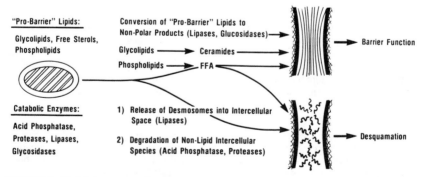

FIGURE 2. Model that summarizes available data about lipid modulations that result from the secretion of lamellar body lipids and hydrolytic enzymes.

TABLE 9. Possible Relationships of Biochemical Modulations and Observed Changes in Stratum Corneum Membrane Structure[a]

Step	Membrane Event	Responsible Enzyme	Biochemical Alteration
1	Fusion/elongation of lamellar body discs	phospholipases	diacylphospholipids → lysolecithin
		sphingomyelinase	sphingomyelin → ceramides
		glycosidases	glycolipids → ceramides
2	Transformation of elongated discs to broad lamellae	phospholipases	degradation of residual polar lipids (*e.g.,* lysolecithin → FFA)
		sphingomyelinase	
		glycosidases	
3	Break-up of membrane bilayers → desquamation	steroid sulfatase	cholesterol sulfate → cholesterol
		acid lipase	triglycerides → FFA
		? ceramidase	ceramides → sphingosine base + FFA

[a]Adapted from REFERENCE 20.

(TABLE 9)? Do further changes in composition and/or the physical-chemical properties of these bilayers lead to desquamation? Indeed, this is likely to be the case since abnormal lamellar body contents and bilayer structures occur in certain inherited and acquired disorders of cornification.[54]

SUMMARY

The stratum corneum, once regarded as a degenerate, inconsequential tissue, is now respected as a structurally heterogeneous and metabolically active tissue. The segregation of lipids into intercellular domains, and the shift in composition from a mixture of polar lipids and neutral lipids to sphingolipids and neutral lipids have important implications for both barrier function and desquamation. Metabolic studies demonstrate the capacity of epidermis to synthesize lipids, the relative autonomy of such synthesis from extracutaneous influences, and the regulation of epidermal lipogenesis by local barrier requirements. Abundant lipid biosynthesis appears to occur in the stratum granulosum, consistent with the rapid recovery of barrier function

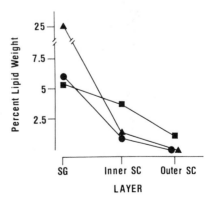

FIGURE 3. Changes in lipid composition that occur during epidermal terminal differentiation. Note that small amounts of phospholipids (▲), glycosphingolipids (●), and cholesterol sulfate (■) remain in the inner stratum corneum (SC) (stratum compactum), while only cholesterol sulfate persists in the outer SC. SG = stratum granulosum. (Adapted from REFS. 20, 26, 53).

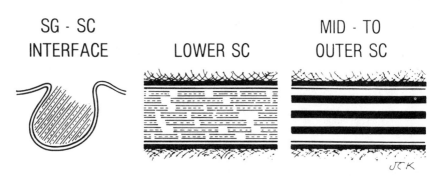

FIGURE 4. Under the influence of hydrolytic enzymes present in the intercellular spaces (cf. FIG. 2 and TABLE 9), the morphology of lamellar body-derived lipids changes from discs to broad sheets. SG = stratum granulosum; SC = stratum corneum. (From Elias et al.[20] Reprinted by permission from the *Journal of Investigative Dermatology*.)

following solvent treatment. Whereas nonpolar lipids, including sterol esters and hydrocarbons, provide a superficial barrier, sphingolipids and free sterols provide a more profound barrier. In parallel with the synthesis of lipids for barrier function, hydrolysis of phospholipids, glycolipids, and cholesterol sulfate occurs in the outer epidermis leading to a more hydrophobic lipid mixture with the evolution of broad membrane bilayers that may regulate both transcutaneous water loss and desquamation.

ACKNOWLEDGMENTS

We appreciate the critical suggestions of Dr. Mary L. Williams during the writing of this paper, and the typing assistance of Mr. Bill Chapman and Ms. Sally Michael.

REFERENCES

1. KLIGMAN, A. M. & E. CHRISTOPHERS. 1964. Arch. Dermatol. **88:** 702–705.
2. CHRISTOPHERS, E. 1971. J. Invest. Dermatol. **56:** 165–169.
3. MACKENZIE, I. C. 1969. Nature **222:** 881–882.
4. MENTON, D. N. & A. Z. EISEN. 1971. J. Ultrastruct. Res. **35:** 247–264.
5. ELIAS, P. M. & D. S. FRIEND. 1975. J. Cell Biol. **65:** 180–191.
6. ELIAS, P. M., J. GOERKE & D. FRIEND. 1977. J. Invest. Dermatol. **69:** 535–546.
7. ELIAS, P. M., N. S. MCNUTT & D. FRIEND. 1977. Anat. Rec. **189:** 577–593.
8. STUPEL, H. & A. SZAKALL. 1957. Die Wirkung von Waschmitteln auf die Haut. Huthig. Heidelberg.
9. SCHEUPLEIN, R. J. & I. H. BLANK, 1971. Physiol. Rev. **51:** 702–747.
10. SWANBECK, G. 1962. Acta Dermatovener. **42:** 445–457.
11. MIDDLETON, J. D. 1968. Br. J. Dermatol **80:** 437–450.
12. BLANK, I. H. 1953. J. Invest. Dermatol. **21:** 259–271.
13. MICHAELS, A. S., S. K. CHANDRASEKARAN & J. E. SHAW. 1975. J. Am. Inst. Chem. Eng. **21:** 985–996.
14. ELIAS, P. M. 1981. Int. J. Dermatol. **20:** 1–19.
15. Smith, W. P., M. S. CHRISTENSEN, S. NACHT & E. H. GANS. 1982. J. Invest. Dermatol. **78:** 7–10.

16. GRAYSON, S. & P. M. ELIAS. 1982. J. Invest. Dermatol. **78:** 128–135, 1982.
17. ELIAS, P. M., L. BONAR, S. GRAYSON & H. P. BADEN. 1983. J. Invest Dermatol. **80:** 213–214.
18. ELIAS, P. M., M. L. WILLIAMS, M. E. MALONEY, J. A. BONIFAS, B. E. BROWN, S. GRAYSON & E. H. EPSTEIN, JR. 1984. J. Clin. Invest. **74:** 1414–1421.
19. MENON, G. K., S. GRAYSON & P. M. ELIAS. 1986. J. Invest. Dermatol. **86:** 591–597.
20. ELIAS, P. M., G. K. MENON, S. GRAYSON & B. E. BROWN. 1988. J. Invest. Dermatol. In press.
21. SCHREINER, E. & K. WOLFF. 1969. Arch. Klin. Exp. Dermatol. **235:** 78–88.
22. SQUIER, C. A. 1973. J. Ultrastruct. Res. **43:** 160–177.
23. BOWSER, D. A. & R. J. WHITE. 1985. Br. J. Dermatol. **112:** 1–14.
24. LAMPE, M. A., A. L. BURLINGAME, J. WHITNEY, M. L. WILLIAMS, B. E. BROWN, E. ROITMAN & P. M. ELIAS. 1983. J. Lipid Res. **24:** 120–130.
25. ELIAS, P. M., M. A. LAMPE, J.-C. CHUNG & M. L. WILLIAMS. 1983. Lab. Invest. **80:** 44–49.
26. ELIAS, P. M., B. E. BROWN, P. O. FRITSCH, R. J. GOERKE, G. M. GRAY & R. J. WHITE. 1979. J. Invest. Dermatol. **73:** 339–348.
27. GRAY, G. M. & H. J. YARDLEY. 1975. J. Lipid Res. **16:** 441–447.
28. GRAY, G. M. & R. J. WHITE. 1978. J. Invest. Dermatol. **70:** 336–341.
29. WERTZ, P. W. & D. T. DOWNING. 1983. J. Lipid Res. **24:** 759–765.
30. GRAY, G. M., R. J. WHITE & J. R. MAJER. 1978. Biochim. Biophys. Acta **528:** 122–137.
31. WERTZ, P. W. & D. T. DOWNING. 1982. J. Lipid Res. **24:** 753–758.
32. BOWSER, P. A., D. H. NGUTEREN, R. J. WHITE, U. M. T. HOUTSMULLER & C. PROTTEY. 1985. Biochim. Biophys. Acta **834:** 419–428.
33. WERTZ, P. W., E. J. CHO & D.T. DOWNING. 1983. Biochim. Biophys. Acta **753:** 350–355.
34. WERTZ, P. W. & D.T. DOWNING. 1982. Science **217:** 1261–1262.
35. FEINGOLD, K. R., M. H. WILEY, G. MACRAE, S. R. LEAR, A. H. MOSER, G. ZSIGMOND & M. D. SIPERSTEIN. 1983. Metabolism **32:** 75–81.
36. FEINGOLD, K. R., M. H. WILEY, A. H. MOSER, S. R. LEAR & M. D. SIPERSTEIN. 1982. J. Lab. Clin. Med. **100:** 405–410.
37. FEINGOLD, K. R., B. E. BROWN, S. R. LEAR, A. H. MOSER & P. M. ELIAS. 1983. J. Invest. Dermatol. **81:** 365–369.
38. PONEC, M., L. HAVEKES, J. KEMPENAAR & B. J. VERMEER. 1983. J. Invest. Dermatol. **81:** 125–130.
39. WILLIAMS, M. L., S. L. RUTHERFORD, A.-M. MOMMAAS-KEINHUIS, S. GRAYSON, B. J. VERMEER & P. M. ELIAS. 1987. J. Cell Physiol. **132:** 428–440,
40. MOMMAAS-KEINHUIS, A.-M., S. GRAYSON, M. C. WIJSMAN, B. J. VERMEER & P. M. ELIAS. 1987. J. Invest. Dermatol. **89:** 513–517.
41. MENON, G. K., K. R. FEINGOLD, A. H. MOSER, B. E. BROWN & P. M. ELIAS. 1985. J. Lipid Res. **26:** 418–427.
42. GRUBAUER, G., K. R. FEINGOLD & P. M. ELIAS. 1987. J. Lipid Res. **28:** 746–752.
43. FEINGOLD, K. R., B. E. BROWN, S. R. LEAR, A. H. MOSER & P. M. ELIAS. 1986. J. Invest. Dermatol. **87:** 588–591.
44. FEINGOLD, K. R., M. L. WILLIAMS, S. PILLAI, G. K. MENON, B. P. HALLORAN, D. D. BIKLE & P. M. ELIAS. 1987. Biochim. Biophys. Acta **930:** 193–200.
45. GRUBAUER, G., P. M. ELIAS & K. R. FEINGOLD. 1988. Clin. Res. In press.
46. MONGER, D. J., M. L. WILLIAMS, K. R. FEINGOLD, B. E. BROWN, & P. M. ELIAS. 1988. J. Lipid Res. In press.
47. GRUBAUER, G., K. R. FEINGOLD & P. M. ELIAS. 1987. J. Invest. Dermatol. **88:** 492 (abstract).
48. GRAY, G. M. 1981. Front. Matrix. Biol. **9:** 83–101.
49. HOPSU-HAVU, V. K. & C. J. JANSEN. 1969. Acta Dermatovener. **49:** 525–535.
50. MIER, P. & J. VAN DEN HURK. 1975. Br. J. Dermatol. **93:** 509–517.
51. GRAYSON, S., A. D. JOHNSON-WINEGAR & P. M. ELIAS. 1983. Science **221:** 962–964.
52. FREINKEL, R. K. & T. N. TRACZYK. 1985. J. Invest. Dermatol. **85:** 295–298.
53. LAMPE, M. A., M. L. WILLIAMS & P. M. ELIAS. 1983. J. Lipid Res. **24:** 131–140.
54. ELIAS, P. M. & M. L. WILLIAMS. 1985. Arch. Dermatol. **121:** 1000–1008.

Skin: Site of the Synthesis of Vitamin D and a Target Tissue for the Active Form, 1,25-Dihydroxyvitamin D$_3$

MICHAEL F. HOLICK

Boston University School of Medicine M-1013
Vitamin D, Skin and Bone Research Laboratory
80 East Concord Street
Boston, Massachusetts 02118

Photobiology of Vitamin D$_3$

In 1822, Snaidecki reported that exposure to sunlight was important for the prevention and cure of rickets.[1] However, another century would pass before Hess and Unger[2] found that exposure to natural sunlight could heal this dreaded bone-deforming disease. It is now known that when human skin is exposed to sunlight the high energy ultraviolet photons with energies between 290 and 315 nm, penetrate into the epidermis and photolyze cytoplasmic pools of 7-dehydrocholesterol (provitamin D$_3$) to previtamin D$_3$ (FIG. 1).[3,4] Although half of the provitamin D content in human skin is found in the dermis while the other half is found in the epidermis, during exposure to sunlight the high-energy photons are absorbed in the epidermis and greater than 90% of the previtamin D$_3$ synthesis occurs in this layer. Once formed previtamin D$_3$, a thermally labile compound, undergoes rearrangement of its double bonds to form a more stable isomer known as vitamin D$_3$ (FIG. 1). This process takes about 3 days to reach completion. Vitamin D$_3$ is then translocated from the epidermis into the dermal capillary bed by the circulating vitamin D-binding protein.

It has been suggested that the major factor that limits the production of vitamin D$_3$ in the skin during excessive exposure to sunlight is melanin pigmentation.[5] Although, melanin is an effective filter which can absorb the same radiation that is responsible for converting provitamin D$_3$ to previtamin D$_3$, it is not the most important factor for regulating the epidermal production of vitamin D$_3$. Once previtamin D$_3$ is formed in the skin it begins to isomerize to vitamin D$_3$. Because this reaction takes approximately three days to reach completion, previtamin D$_3$ is susceptible to photodegradation by sunlight. The major photoproducts that result from this photodegradation reaction are biologically inert isomers, lumisterol and tachysterol (FIG. 1).[6] Similarly, if vitamin D$_3$ does not escape into the circulation it too can be photolyzed to 5,6-transvitamin D$_3$ suprasterol 1 and suprasterol 2.[7] Therefore, sunlight appears to be the major factor that limits the cutaneous production of vitamin D$_3$.

Metabolism of Vitamin D$_3$

Once vitamin D$_3$ enters the circulation, it is transported to the liver where it is hydroxylated to 25-hydroxyvitamin D$_3$ (25-(OH)D$_3$).[8,9] 25-Hydroxyvitamin D$_3$ is the major circulating form of vitamin D$_3$, and its determination is often used to measure the vitamin D status of a patient. 25-Hydroxyvitamin D$_3$, however, is biologically inert and requires an additional hydroxylation in the kidney to form 1,25-dihydroxyvitamin D$_3$ (1,25-(OH)$_2$D$_3$), the hormonal and biological active form of vitamin D$_3$.[8,9] The major biological functions of 1,25-$(OH)_2D_3$ are the enhancement of intestinal calcium absorption and the mobilization of calcium stores from bone (FIG. 2).[8] When ionized calcium concentrations in the circulation decrease, the parathyroid glands increase the secretion of parathyroid hormone. Parathyroid hormone travels to the kidney where it increases the tubular reabsorption of calcium and excretion of phosphate. In addition, parathyroid hormone through its action on phosphate metabolism enhances the kidney's production of 1,25-$(OH)_2D_3$. Once formed 1,25-$(OH)_2D_3$ travels to the intestine to increase the efficiency of absorption of dietary calcium and in concert with parathyroid hormone to mobilize calcium stores from bone (FIG. 2).

1,25-Dihydroxyvitamin D_3 is a steroidlike molecule. It interacts with a nuclear receptor in its target tissues and causes the transcription of several specific messenger

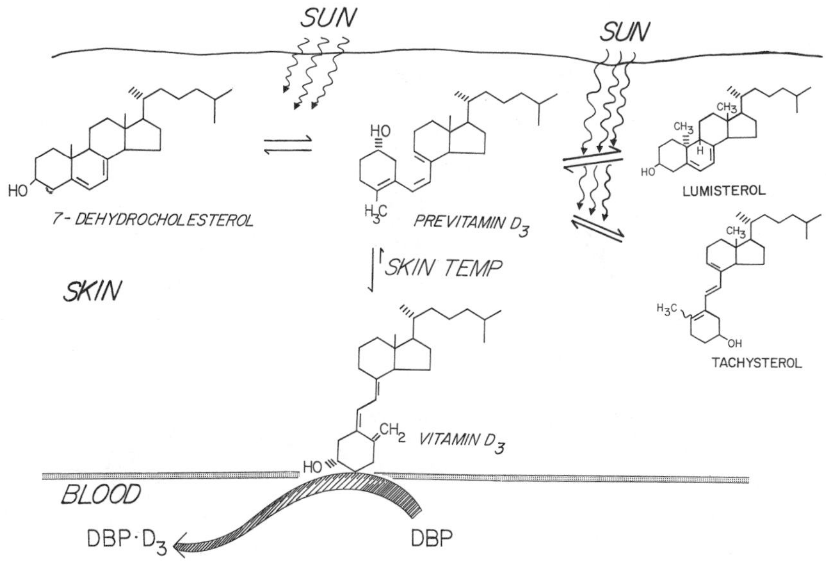

FIGURE 1. Cutaneous vitamin D_3 synthesis. Previtamin D_3 formation in skin occurs during exposure to the sun, and it is then thermally isomerized to vitamin D_3, which, in turn, is translocated into the circulation by vitamin D-binding protein. Further sun exposure to previtamin D_3 allows for photoisomerization to lumisterol and tachysterol. (From Holick et al.[6] Reprinted by permission from the American Association for the Advancement of Science.

RNAs (FIG. 3).[8-10] In the early 1970s specific high-affinity, low-capacity nuclear receptors for 1,25-$(OH)_2D_3$ were identified in cells from the intestine and bone. By the end of 1970 a variety of other tissues were also identified as having nuclear receptors for 1,25-$(OH)_2D_3$. These included such diverse tissues and cells as the gonads, central nervous system, parathyroid glands, pituitary glands, pancreas, stomach, skin, and activated T and B lymphocytes.[8-14] In addition, a variety of tumor cell lines including malignant melanoma, HL-60 human promyelocytic leukemic cells and breast cancer cells possess nuclear receptors for 1,25-$(OH)_2D_3$.[8,15,16] 1,25-Dihydroxyvitamin D_3 inhibits the proliferation of these receptor-positive tumor cells in culture. In addition, this hormone will induce differentiation of receptor-positive leukemic cells into

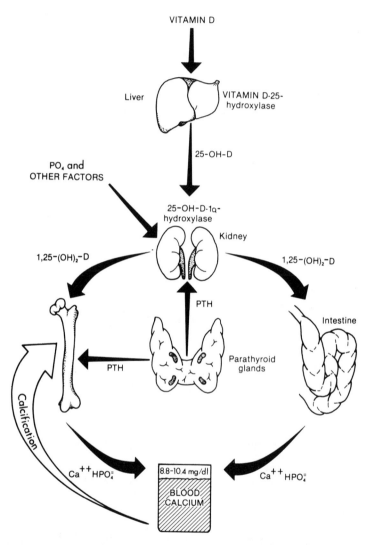

FIGURE 2. Schematic representation of the hormonal control loop for vitamin D metabolism and function. A reduction in the serum calcium below approximately 8.8 mg/100 ml serum prompts a proportional secretion of parathyroid hormone that acts to mobilize calcium stores from the bone. Parathyroid hormone also promotes the synthesis of 1,25-$(OH)_2$D in kidney, which, in turn, stimulates the mobilization of calcium from bone and intestine. (From Holick.[8] Reprint by permission from *Kidney International*.)

biochemically functional macrophages (FIG. 4).[8,15] However, the physiologic relevance of 1,25-$(OH)_2D_3$ receptors in a variety of tissues and cells that are not responsible for maintaining calcium homeostasis remains to be determined. It has been shown *in vitro* that 1,25-$(OH)_2D_3$ will have such diverse effects as stimulating insulin secretion, enhancing thyroid-stimulating hormone synthesis and secretion, inhibiting interleukin 2 production by activated T lymphocytes, inhibiting parathyroid hormone production, and inhibiting immunoglobulin synthesis by activated B lymphocytes.[8,10,12] These interesting *in vitro* observations do not necessarily mean that 1,25-$(OH)_2D_3$ is a dominant regulator for many of these biochemical functions. We now know that patients with a rare hereditary disorder known as vitamin D-dependent type II (these patients either lack or have a defective receptor for 1,25-$(OH)_2D_3$) have normal endocrine function, and there are no obvious defects in any of the other tissues in the body other than an abnormality in calcium homeostasis and severe bone disease.[8,17]

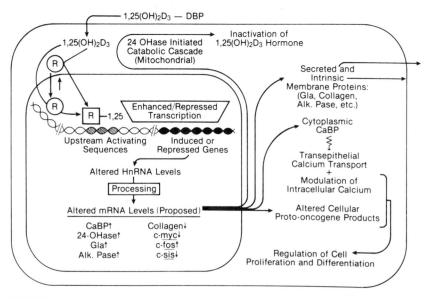

FIGURE 3. Proposed mechanism of action of 1,25-$(OH)_2D_3$ in target cells resulting in a variety of biologic responses. (From Haussler *et al.*[10] Reprinted by permission from Walter de Gruyter.)

However, there is mounting evidence that 1,25-$(OH)_2D_3$ may play a fundamental role in the mobilization of monocytic stem cells in the peripheral circulation and bone marrow to become mature osteoclasts[10] (FIG. 4).

The Physiologic Function of 1,25-Dihydroxyvitamin D_3 in the Skin

Autoradiographic analysis of rat skin has demonstrated localization of radiolabeled 1,25-$(OH)_2D_3$ in the stratum Malpighium and stratum granulosum of the epidermis and in the outer root sheath of hair.[11,21] Receptors with a high affinity and low capacity for the hormone have been found in mouse, rat, and human skin, cultured

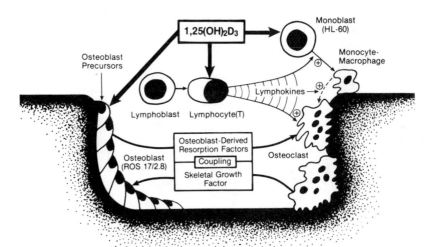

FIGURE 4. Proposed function of 1,25-(OH)₂D₃ and its receptor in bone remodeling and immunomodulation. (From Haussler et al.[10] Reprinted by permission from Walter de Gruyter.)

human dermal fibroblasts, and epidermal keratinocytes, murine keratinocytes, squamous cell carcinoma, and melanoma cells.[16-20] When cultured melanoma cells are exposed to 1,25-(OH)₂D₃ this hormone in a dose-dependent manner increases melanogenesis.[22] The biologic action of 1,25-(OH)₂D₃ on melanogenesis in malignant melanoma cells may not be physiologically relevant. When cultured human melanocytes were exposed to 1,25-(OH)₂D₃ there was no increase in melanogenesis.[24] This observation is consistent with clinical observations that have shown no increase in melanogenesis in patients with chronic renal failure who are taking high doses of 1,25-(OH)₂D₃.[8]

The growth of normal receptor-positive cultured human dermal fibroblasts is inhibited in a dose-dependent fashion by 1,25-(OH)₂D₃ (FIG. 5).[19] The effect of this hormone is probably a receptor-mediated event, inasmuch as, unlike normal fibroblasts there was no growth inhibition of vitamin D-dependent rickets type II dermal fibroblasts after incubation with 1,25-(OH)₂D₃.[19]

When receptor-positive cultured murine keratinocytes were exposed to 1,25-(OH)₂D₃ this hormone inhibited the proliferation and stimulated differentiation of these cells.[26] Similarily, 1,25-(OH)₂D₃ inhibited the proliferation and induced terminal differentiation of cultured human keratinocytes that were maintained in a serum-free medium.[25] After incubation with 1,25-(OH)₂D₃, there was a dose-dependent increase in the proportion of differentiated squamous cells and terminally differentiated desquamated cells in human keratinocyte cultures (FIG. 6).[25] Incubation with 10^{-10} or 10^{-8} M of this hormone for 2 weeks resulted in a significantly greater percentage of attached squamous cells (20.0 ± 1.0%; 27.7 ± 0.6%, respectively) than in cultures incubated with vehicle alone (12.2 ± 0.7%).[19] The percentage of terminally differentiated desquamated floater cells in control cultures (6.9 ± 0.1%) was less than in cultures incubated with 10^{-10} M (9.8 ± 1.3%) or 10^{-8} M (24.6 ± 0.8%) of 1,25-(OH)₂D₃.

During the epidermal terminal differentiation process there was a shift to lower cellular density due to the loss of nuclei and other cellular components. In human keratinocyte cultures treated with 1,25-(OH)₂D₃ there was a shift to less dense cells.[25]

In the least dense fraction (differentiated cells) there were greater than a 4-fold increase in the number of cells in 1,25-$(OH)_2D_3$-treated cultures than in control cultures.[27]

During the epidermal terminal differentiation process, cornified envelopes are formed beneath the plasma membranes of keratinocytes. 1,25-Dihydroxyvitamin D_3 increased in a time- and dose-dependent manner cornified envelope formation.[25]

Transglutaminase is an enzyme that causes the cross-linking of proteins to form the cornified envelope. An evaluation of the biological effect of 1,25-$(OH)_2D_3$ on tranglutaminase activity revealed that this hormone stimulated in a dose-dependent fashion

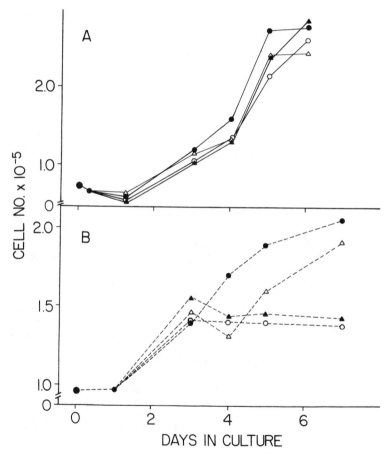

FIGURE 5. Effect of 1,25-$(OH)_2D_3$ on the growth of cultured dermal fibroblasts obtained from a patient with DDR-II (A) and normal fibroblasts (B) obtained from breast skin. Cells were plated at 0.75×10^5 (DDR II) or 0.9×10^5 (normal) in 35-mm dishes in DMEM containing 5% NBS. After attachment, cells from two dishes were counted, and medium was removed and replaced with fresh medium alone (●) or medium containing 1,25-$(OH)_2D_3$ at a final concentration of 10^{-10} (△), 10^{-8} (▲), or 10^{-6} M (○). At intervals thereafter, replicate plates were trypsinized, and cells were counted using a Coulter counter. (From Clemens *et al.*[38] Reprinted by permission from the *Journal of Clinical Endocrinology and Metabolism.*)

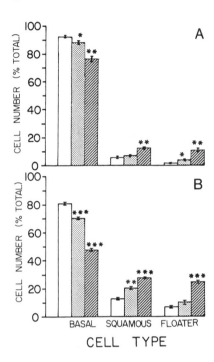

FIGURE 6. Effect of $1\alpha,25\text{-}(OH)_2D_3$ on the morphological differentiation of cultured human keratinocytes. The proportion of different keratinocyte cell types after 1 wk (**A**) or 2 wk (**B**) of incubation with vehicle alone (open bar); 1,25-$(OH)_2D_3$ at 10^{-10} M (dotted bar); or 1,25-$(OH)_2D_3$ at 10^{-8} M (striped bar). Each bar represents the mean of triplicate determinations $\pm$ SEM. Student t test was used to assess level of significance (*, $p < 0.05$; **, $p < 0.01$; ***, $p < 0.001$). (From Smith et al.[25] Reprinted by permission from the *Journal of Investigative Dermatology*.)

the activity of this enzyme. This effect was specific for 1,25-$(OH)_2D_3$, as the biologically inert structural analog of 1,25-$(OH)_2D_3$, $1\beta,25$-dihydroxyvitamin D_3, did not stimulate epidermal transglutaminase activity (FIG. 7).[25,27] Recently, it was demonstrated that there are two forms of the transglutaminase, a soluble form which is found in a variety of cells and whose biologic function is unknown, and a particulate form that is believed to be responsible for the production of the cornified envelope.[28,29] Because it is known that retinoic acid will increase transglutaminase activity without affecting cornified envelope development it was of interest to determine the biologic effect of retinoic acid and 1,25-$(OH)_2D_3$ on the two forms of transglutaminase in cultured human keratinocytes. 1,25-Dihydroxyvitamin D_3 (10^{-8} M) and retinoic acid (10^{-6} M) increased soluble transglutaminase activity by approximately 20%. When the particulate transglutaminase activity was evaluated in the same cells, it was found that 1,25-$(OH)_2D_3$ increased its activity by 40% while retinoic acid inhibited its activity by approximately 30%. These observations clearly show that 1,25-$(OH)_2D_3$ has a different biologic effect on cultured human epidermal cells than retinoic acid and provides an explanation for why retinoic acid does not increase cornified envelope development whereas 1,25-$(OH)_2D_3$ does.[27]

Effect of 1,25-(OH)₂D₃ on the Regulation of Intracellular Calcium Concentrations

It is well known that calcium has a major effect on the proliferative activity of cultured human keratinocytes.[30] Cells incubated in a low calcium medium will proliferate, whereas normal and high calcium will inhibit proliferation and induce terminal differentiation. Recently, it was demonstrated that 1,25-$(OH)_2D_3$ can alter intracellular calcium concentrations in hepatocytes and HL-60 cells.[31,32] Because calcium plays such a critical role in epidermal differentiation, the effect of 1,25-$(OH)_2D_3$ on free cytosolic calcium concentrations was determined. By means of a Quin 2 fluorescent dye indicator free cytosolic calcium determinations were made in the presence of 1,25-$(OH)_2D_3$ (10 ng/ml). Immediately upon its addition there was a significant increase in free cytosolic calcium that was maintained over 5 minutes when compared to cells treated with vehicle alone (FIG. 8).[27]

The Use of 1,25-Dihydroxyvitamin D_3 for the Treatment of Psoriasis

The realization that 1,25-$(OH)_2D_3$ was a potent inhibitor of proliferation of cultured keratinocytes prompted an investigation into the possible therapeutic use of 1,25-$(OH)_2D_3$ for the treatment of the hypoproliferative skin disorder, psoriasis. Before launching into a clinical trial with this hormone, skin biopsies from patients with psoriasis were obtained and fibroblasts and keratinocytes were cultured in order to determine whether skin cells from psoriatic patients were responsive to this hormone. Cultured human psoriatic fibroblasts and keratinocytes possess normal

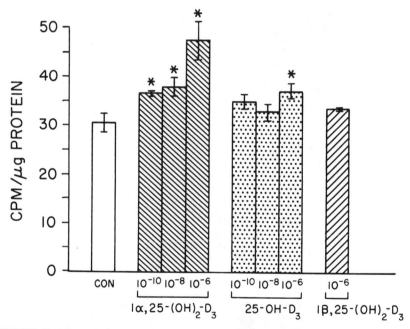

FIGURE 7. Effect of 1,25-$(OH)_2D_3$ on the transglutaminase activity of cultured human keratinocytes. After 1 wk in culture, cells were incubated for 24 h with medium containing vehicle alone; $1\alpha,25$-$(OH)_2D_3$ (10^{-10}, 10^{-8}, or 10^{-6} M); 25-$(OH)D_3$ (10^{-10}, 10^{-8}, or 10^{-6} M); or $1\beta,25$-$(OH)_2D_3$ (10^{-6} M). Cultures were extracted for transglutaminase, and the enzyme activity was indexed against protein concentration. Each point represents the mean of triplicate determinations ± SEM. There was a significant increase in transglutaminase activity (*, $p < 0.05$) after treatment with 1,25-$(OH)_2D_3$ (10^{-10} to 10^{-6} M) or 25-$(OH)D_3$ (10^{-6} M). (From Smith et al.[25] Reprinted by permission from the *Journal of Investigative Dermatology*.)

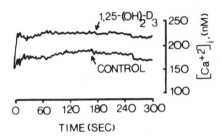

FIGURE 8. Effect of 1,25-$(OH)_2D_3$ on free cytosolic calcium concentration of cultured human keratinocytes. Cells were loaded with Quin 2/AM and then treated with vehicle alone or 1,25-$(OH)_2D_3$ (10^{-8} M). The fluorescence was measured at 495 nm over the 5-min time-course. The calcium concentration was calculated according to the following equation: $[Ca^{+2}]_i = 115$ nM $(F - F_{min})/(F_{max} - F)$. (From Smith and Holick.[27] Reprinted by permission from *Steroids*.)

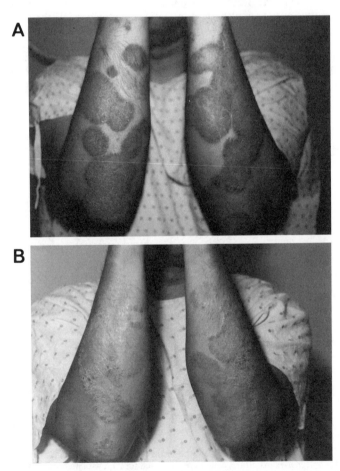

FIGURE 9. Effect of oral administration of 1,25-$(OH)_2D_3$ for a psoriatic patient. This 43-year-old male has had psoriasis for 4 years. Photographs were taken of his forearms prior to treatment (**A**) and after 2½ months of receiving a daily oral dose of 0.5 µg of 1,25-$(OH)_2D_3$ (**B**). There was a substantial decrease in erythema and central clearing of lesions with residual hyperpigmentation after treatment with 1,25-$(OH)_2D_3$. (From Smith and Holick.[27] Reprinted by permission from *Steroids*.)

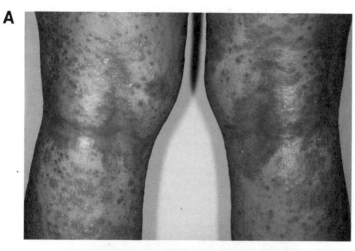

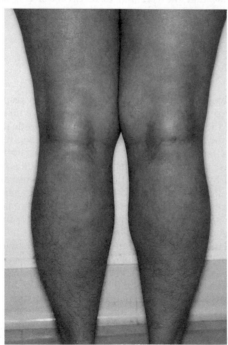

FIGURE 10. Effect of oral administration of 1,25-(OH)$_2$D$_3$ on psoriatic patient. This female has had psoriasis for 4 years. Photographs were taken of her legs prior to treatment (**A**) and after 3 months of receiving a daily oral dose of 2.0 μg of 1,25-(OH)$_2$D$_3$ (**B**). There was a substantial decrease in erythema and clearing of lesions after treatment with 1,25-(OH)$_2$D$_3$. (From Holick.[8] Reprinted by permission from *Kidney International*.)

receptors for $1,25\text{-}(OH)_2D_3$.[18,33,34] Initially, it was found that dermal fibroblasts from patients with psoriasis had a partial resistance to the growth-inhibition effect of $1,25\text{-}(OH)_2D_3$.[33] However, a more extended evaluation revealed that approximately 20% of patients with psoriasis have a partial resistance to $1,25\text{-}(OH)_2D_3$, while the majority have a normal response to the hormone.[18,27,34]

Based on these encouraging *in vitro* results, a clinical trial was launched to evaluate the therapeutic efficacy of orally or topically administered $1,25\text{-}(OH)_2D_3$. Three patients received topical $1,25\text{-}(OH)_2D_3$ (1 ml containing 3 μg of $1,25\text{-}(OH)_2D_3$ was applied over a 4 cm^2 area and then occluded with a saran wrap). All three patients responded within 2 weeks and 1 patient showed clearing of her lesions after a 6-week period.[34] Greater than two thirds of the 16 patients treated with orally administered $1,25\text{-}(OH)_2D_3$ had moderate to complete resolution of their lesions as shown in FIGURES 9 and 10.

These results are consistent with other observations that have demonstrated that orally administered 1-hydroxyvitamin D_3 or $1,25\text{-}(OH)_2D_3$ or topically applied $1,25\text{-}(OH)_2D_3$ or 1,24-dihydroxyvitamin D_3, was effective for clearing psoriatic lesions.[35-37] There is concern, however, about the biologic effect of $1,25\text{-}(OH)_2D_3$ and its analogs on calcium metabolism. Although it is true that $1,25\text{-}(OH)_2D_3$ will enhance intestinal calcium absorption, it was recently reported that if the drug is given at night as a single dose before bedtime that the potential toxic effects of this hormone on enhancing calcium in the urine and blood is minimized.[34]

CONCLUSION

It is remarkable that the skin is not only the organ responsible for vitamin D synthesis but is also a potential target tissue for its active metabolite $1,25\text{-}(OH)_2D_3$. Although many of the *in vitro* results have demonstrated wide ranging biologic actions for this hormone, they need to be put into some perspective. It is known for example, that patients with vitamin D-dependent rickets type II have normal skin. However, it is intriguing that these patients often suffer from alopecia. It has been suggested that the alopecia is caused by the inability of the hair follicle to respond to $1,25\text{-}(OH)_2D_3$.[17] However, there is no evidence that $1,25\text{-}(OH)_2D_3$ regulates hair growth. Patients that are unable to make $1,25\text{-}(OH)_2D_3$, such as those with chronic renal failure, do not experience baldness, and patients that receive $1,25\text{-}(OH)_2D_3$ do not report an increase in hair growth and development. Similarly, there is no evidence that $1,25\text{-}(OH)_2D_3$ will stimulate melanogenesis in a normal person or in our psoriatic patients who are taking up to 2 μg of $1,25\text{-}(OH)_2D_3$ a day.

Although $1,25\text{-}(OH)_2D_3$ has a marked effect on inhibiting proliferation of cultured human fibroblasts and keratinocytes and inducing terminal differentiation of cultured keratinocytes, there is no evidence that a deficiency in this hormone will increase proliferative activity in human skin. Thus, it would appear that $1,25\text{-}(OH)_2D_3$ is not a major regulator of dermal and epidermal proliferation and differentiation.

Despite the fact that $1,25\text{-}(OH)_2D_3$ is not a major participant in the regulation of epidermal differentiation, the observations that $1,25\text{-}(OH)_2D_3$ can inhibit the hyperproliferative activity of epidermal cells in patients with psoriasis is intriguing. It should be noted that patients with psoriasis have a normal metabolism of vitamin D to 25-hydroxyvitamin D and $1,25\text{-}(OH)_2D_3$.[24] Therefore, $1,25\text{-}(OH)_2D_3$ and its analogs hold promise as unique pharmacologic agents for treating hyperproliferative disorders of the skin such as psoriasis.

REFERENCES

1. SNIADECKI, J. 1840. Cited by W. MOZOLOWSKI. 1939. Jedrzej Sniadecki (1768–1833) on the cure of rickets. Nature **143:** 121.
2. HESS, A. F. & L. J. UNGER. 1921. J. Am. Med. Soc. **77:** 39.
3. MACLAUGHLIN, J. A., R. R. ANDERSON & M. F. HOLICK. 1982. Science **216:** 1001–1003.
4. HOLICK, M. F., J. A. MACLAUGHLIN, M. B. CLARK, S. A. HOLICK, J. T. POTTS, JR., R. R. ANDERSON, I. H. BLANK, J. A. PARRISH & P. ELAIS. 1980. Science **210:** 203–205.
5. LOOMIS, F. 1967. Science 157:501–506.
6. HOLICK, M. F., J. A. MACLAUGHLIN & S. H. DOPPELT. 1981. Science **211:** 590–593.
7. WEBB, A. R., D. DE COSTA & M. F. HOLICK. 1986. Photochem. Photobiol. **43s:** 116S.
8. HOLICK, M. F. 1987. Kidney Int. **32:** 912–929.
9. DELUCA, H. F. 1984. The metabolism, physiology, and function of vitamin D. *In* Vitamin D, Basic and Clinical Aspects. R. Kumar, Ed. 1–68. Nijhoff. Boston, MA.
10. HAUSSLER, M. R., C. A. DONALDSON, M. A. KELLY, D. J. MANGELSDORF, S. L. MARION & J. W. PIKE. 1985. Functions and mechanism of action of the 1,25-dihydroxyvitamin D_3 receptor. *In* Vitamin D: Chemical, Biochemical and Clinical Update. A. W. Norman, K. Schaefer, H.-G. Grigoleit & D. von Herrath, Eds. 83–92. Walter de Gruyter. New York, NY.
11. STUMPF, W. E., M. SAR, F. A. REID, Y. TANAKA & H. F. DELUCA. 1979. Science **206:** 1188–1190.
12. REICHEL, H., H. P. KOEFFLER, R. BARBERS, R. MUNKER & A. W. NORMAN. 1985. 1,25-Dihydroxyvitamin D_3 and the hematopoietic system. *In* Vitamin D: Chemical, Biochemical and Clinical Update. A. W. Norman, K. Schaefer, H.-G. Grigoleit & D. von Herrath, Eds. 167–176. Walter de Gruyter. New York, NY.
13. AMENTO, E. P., A. K. BHALLA, J. T. KURNICK, R. L. KRADIN, T. L. CLEMENS, S. A. HOLICK, M. F. HOLICK & S. M. KRANE. 1984. J. Clin. Invest. **73:** 731–739.
14. PROVVEDINE, D. M., C. D. TSOUKAS, L. J. DEFTOS & S. C. MANOLAGAS. 1983. Science **221:** 1181–1182.
15. ABE, E., C. MIYAURA, H. SAKAGAMI, M. TAKEDA, K. KANNO, T. YAMAZAKI, S. YOSHIKI & T. SUDA. 1981. Proc. Natl. Acad. Sci. USA **78:** 4990–4994.
16. COLSTON, K., M. J. COLSTON & D. FELDMAN. 1981. Endocrinology **108:** 1083–1086.
17. MARX, S. J. 1984. Resistance to vitamin D. *In* Vitamin D, Basic and Clinical Aspects. R. Kumar, Ed. 721–745. Nijhoff. Boston, MA.
18. HOLICK, M. F., E. SMITH & S. PINCUS. 1987. Arch. Dermatol. **123:** 1677a–1683a.
19. CLEMENS, T. L., N. HORIUCHI, M. NGUYEN & M. F. HOLICK. 1981. FEBS Lett. **134:** 203–206.
20. FELDMAN, D., T. CHEN, M. HIRST, K. COLSTON, M. KARASEK & C. CONE. 1980. J. Clin. Endocrinol. Metab. **51:** 1463–1465.
21. STUMPF, W. E., S. A. CLARK, M. SAR & H. F. DELUCA. 1984. Cell Tissue Res. **238:** 489–496.
22. SIMPSON, R. U. & H. F. DELUCA. 1980. Proc. Natl. Acad. Sci. USA **77:** 5822–5826.
23. HOSOI, J., E. ABE, T. SUDA & T. KUROKI. 1985. Cancer Res. **45:** 1474–1478.
24. MANSUR, C. P., P. GORDON, S. RAY, M. F. HOLICK & B. A. GILCHREST. 1988. Vitamin D, its precursors, and metabolites do not affect melanization of cultured human melanocytes. J. Invest. Dermatol. **91:** 16–20.
25. SMITH, E. L., N. C. WALWORTH & M. F. HOLICK. 1986. J. Invest. Dermatol. **86:** 709–714.
26. HOSOMI, J., J. HOSOI, E. ABE, T. SUDA & T. KUROKI. 1983. Endocrinology **113:** 1950–1957.
27. SMITH, E. L. & M. F. HOLICK. 1987. The skin: the site of vitamin D_3 synthesis and a target tissue for its metabolite 1,25-dihydroxyvitamin D_3. Steroids. **49:** 103–131.
28. LICHI, U., T. BEN & A. H. YUSPA. 1985. J. Biol. Chem. **260:** 1422–1426.
29. GOLDSMITH, L. A. 1983. *In* The Epidermal Cell Periphesy in Biochemistry and Physiology of the Skin. L. A. Goldsmith, Ed. 184–196. Oxford University Press. New York, NY.
30. HENNINGS, H., P. STEINERT & M. M. BUXMAN. 1981. Biochem. Biophys. Res. Commun. **102:** 739–745.

31. BARAN, D. T. & M. L. MILNE. 1986. J. Clin. Invest. **77:** 1622–1626.
32. DESAI, S. S., M. C. APPEL & D. T. BARAN. 1986. J. Bone Min. Res. **1:** 497–501.
33. MACLAUGHLIN, J. A., W. GANGE, D. TAYLOR, E. SMITH & M. F. HOLICK. 1985. Proc. Natl. Acad. Sci. USA **82:** 5409–5412.
34. SMITH, E. L., S. PINCUS, L. DONOVAN & M. F. HOLICK. A novel approach for the evaluation and treatment of psoriasis: oral or topical use of 1,25-dihydroxyvitamin D_3 can be a safe and effective therapy for psoriasis. J. Am. Acad. Dermatol. In press.
35. KATO, T., M. ROKUGO, T. TERUI & H. TAGAMI. 1986. Br. J. Dermatol. **115:** 431–433.
36. MORIMOTO, S., T. ONISHI, S. IMANAKA, H. YUKAWA, T. KOZUKA, Y. KITANO, K. YOSHIKAWA & Y. KUMAHARA. 1986. Calcif. Tissue Int. **38:** 119–122.
37. MORIMOTO, S., K. YOSHIKAWA, T. KOZUKA, Y. KITANO, S. IMANAKA, K. FUKUO, E. KOH & Y. KUMAHARA. 1986. Br. J. Dermatol. **115:** 421–429.
38. CLEMENS, T. L., J. S. ADAMS, N. HORIUCHI, B. A. GILCHREST, H. CHO, Y. TSUCHIYA, N. MATSUO, T. SUDA & M. F. HOLICK. 1983. J. Clin. Endocrinol. Metab. **56:** 824–830.

Vitamin D Metabolite Production and Function in Keratinocytes

DANIEL D. BIKLE AND SREEKUMAR PILLAI

Endocrine Unit
Veterans Administration Medical Center
University of California
4150 Clement Street (111N)
San Francisco, California 94121

INTRODUCTION

The ability of the skin to synthesize vitamin D is well established. Vitamin D has little or no intrinsic biologic activity but must be further metabolized. The first required step occurs in the liver where vitamin D is converted to 25-hydroxyvitamin D (25-(OH)D). This step is not tightly regulated. 25-(OH)D is the major circulating form of the vitamin, and its measurement provides a useful indicator of the vitamin D status of the individual. However, 25-(OH)D has only limited biologic activity; rather 25-(OH)D must be further metabolized to achieve full biologic potency. Of the numerous metabolites formed from 25-(OH)D, 1,25-dihydroxyvitamin D (1,25-$(OH)_2$D) is the most potent in terms of stimulating intestinal calcium transport, stimulating bone resorption, and raising serum calcium. It is the only metabolite of vitamin D which has a specific, high affinity intracellular receptor. Other metabolites such as 24,25-dihydroxyvitamin D (24,25-$(OH)_2$D) may have biologic roles different from that of 1,25-$(OH)_2$D, but such roles are not yet clearly defined.

Both 1,25-$(OH)_2$D and 24,25-$(OH)_2$D are produced from 25-(OH)D primarily in the mitochondria of the renal proximal tubules. However, the kidney is not unique in its ability to form either of these metabolites. The placenta[1-3] and bone cells[4,5] produce 1,25-$(OH)_2$D and 24,25-$(OH)_2$D. Likewise, lymphoid tissue[6] and pulmonary alveolar macrophages[7] from patients with sarcoidosis, melanoma cells,[8] and HTLV-1 infected lymphocytes[9,10] convert 25-(OH)D to metabolites which appear to be 1,25-$(OH)_2$D and 24,25-$(OH)_2$D. Keratinocytes can now be added to the list.[11,12]

The significance of extrarenal production of 1,25-$(OH)_2$D *in vivo* was brought into question by studies that were unable to demonstrate the appearance of ^{3}H-1,25-$(OH)_2$D in rats acutely nephrectomized and given a single dose of ^{3}H-25-(OH)D.[13,14] In contrast to these negative studies in acutely nephrectomized rats, 1,25-$(OH)_2$D has been detected in both anephric patients and pigs, and the levels are increased with vitamin D or 25-(OH)D administration.[15,16] Thus it appears that at least in humans and pigs, 1,25-$(OH)_2$D production is not limited to the kidney, although in these species as well as in rats, the kidney is the major source of 1,25-$(OH)_2$D. Our data to be reviewed here indicate that the keratinocyte is one such alternative source for 1,25-$(OH)_2$D production in the human.

Why should the skin make 1,25-$(OH)_2$D? The skin contains receptors for 1,25-$(OH)_2$D,[17,18] and 1,25-$(OH)_2$D induces morphologic changes in cultured keratinocytes consistent with a role for this metabolite in epidermal differentiation.[19,20] Thus, the skin appears to be a target tissue for 1,25-$(OH)_2$D. As we will demonstrate, the production of 1,25-$(OH)_2$D by the keratinocyte changes with differentiation in a manner consistent with a role for 1,25-$(OH)_2$D production in the differentiation

27

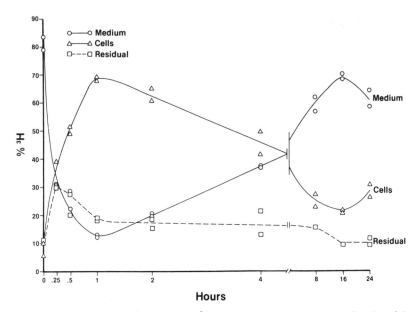

FIGURE 1. The distribution of radioactivity (^{3}H) between cells and medium as a function of time after the administration of ^{3}H-25-(OH)D$_3$ to the cultured keratinocytes. In this experiment, 18 cultures dishes received equal aliquots of ^{3}H-25-(OH)D$_3$. The medium was removed from the duplicate dishes at each of the indicated times after the addition of ^{3}H-25-(OH)D$_3$ and counted for radioactivity. The cells were gently washed once with 1 ml fresh medium (serum free); little radioactivity was present in the wash, so it was discarded. The cells were then scraped into an additional 1 ml of fresh medium, removed from the plate, and counted for radioactivity. The residual radioactivity is that which was extracted by methanol rinses of the dishes after the medium and cells were removed.

process.[21] Conceivably, the skin makes 1,25-(OH)$_2$D for its own endogenous use to serve as a modulator of its ongoing processes of proliferation and differentiation. This review will emphasize our data supporting this hypothesis.

Culture Conditions

The keratinocytes employed in our studies are obtained from human neonatal foreskins. They are isolated after an overnight digestion of the foreskins with 0.25% trypsin at 4°C. After growing the cells to confluence in DMEM/20%FCS on a feeder cell layer of mitomycin C-treated 3T3M cells,[22] we can passage the cells using essentially the same culture conditions (DMEM/5%FCS with 3T3M feeder cells) or using a serum free system employing Keratinocyte Growth Medium (KGM) from Clonetics Corporation (San Diego, CA). This latter system is completely defined except for the addition of bovine pituitary extract, and does not require 3T3M feeder cells.[23] The use of KGM is preferable for our studies because it eliminates the uncertain contribution to the results of undefined serum constituents and 3T3M cell metabolism. All studies to be described employed first or second passage keratinocytes. If 3T3M feeder cells were used, they were eliminated prior to study with 0.1% EDTA. The keratinocytes obtained by these methods have few (<5%) or no contaminating cells.

25-(OH)D Metabolism by Human Keratinocytes: Kinetics and Structural Identification of Metabolites

The addition of ^{3}H-25-(OH)D to keratinocytes results in a rapid uptake of the radiolabel by the cells reaching a maximum by 1 h. Beyond this point the radiolabel leaves the cell and reappears in the medium (FIG. 1). However, the radiolabel reappearing in the medium after 1 h is not 25-(OH)D, as seen in FIGURE 2. Rather, the 25-(OH)D is rapidly converted to a variety of metabolites, some of which are shown in the HPLC profile depicted in FIGURE 3. Most of these metabolites appear in both the

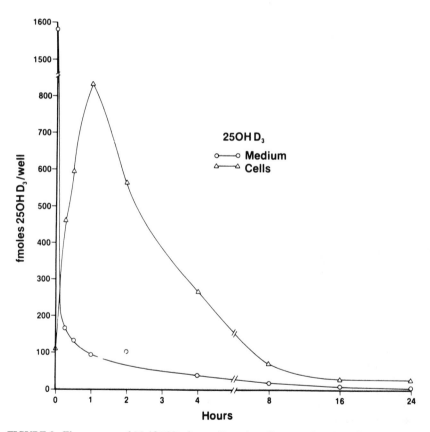

FIGURE 2. The recovery of 25-(OH)D$_3$ from cells and medium as a function of time following the addition of 1,600 fmol ^{3}H-25-(OH)D$_3$ to the culture medium (as in FIG. 1). The medium was removed, and the cells were rinsed once and scraped off into fresh medium. Both cells and medium were extracted separately and analyzed for 25-(OH)D$_3$ by HPLC. The mean value of determinations from duplicate wells is shown.

cells and medium. 1,25-$(OH)_2$D is an important exception in that it stays within the cells. The levels of these metabolites change with time of incubation, each metabolite having its own optimal incubation period. The principal metabolite produced within the first few hours is 1,25-$(OH)_2$D which was identified in the following manner.[11]

Keratinocytes were incubated with 5×10^{-7} M ^{3}H-25-(OH)D, specific activity 2 Ci/mmol. After 2 h the cells and medium were extracted with chloroform:methanol, and the chloroform extract was chromatographed through a μ-Porasil column eluted with hexane:isopropanol (90:10). The presumptive 1,25-$(OH)_2$D peak was further purified using three additional HPLC systems: C18 μ-Bondapak column eluted with

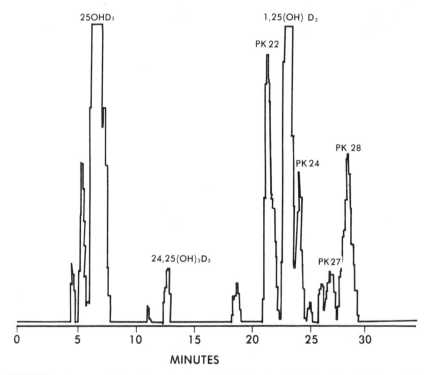

FIGURE 3. A representative chromatogram of the lipid extract of a keratinocyte culture (cells plus medium) incubated with ^{3}H-25-(OH)D$_3$. This sample was incubated with 1.5×10^{-8} M 25-(OH)D$_3$ for 4 h. The peaks eluting in the positions of the standards 25-(OH)D$_3$, 24,25-$(OH)_2$D$_3$ and 1,25-$(OH)_2$D$_3$ (peak 23) are labeled as such.

methanol:water (75:25), Zorbax-Sil column eluted with hexane:isopropanol (97:3– 90:10 gradient), and Zorbax-Sil column eluted with dichloromethane:isopropanol (96:4). The co-chromatography of the presumptive 1,25-$(OH)_2$D from keratinocytes with authentic 1,25-$(OH)_2$D on two of these systems is shown in FIGURE 4.

After purifying the presumptive 1,25-$(OH)_2$D we tested its biologic potency via its ability to displace authentic ^{3}H-1,25-$(OH)_2$D from the intestinal cytosol receptor for 1,25-$(OH)_2$D. By knowing the specific activity of the presumptive 1,25-$(OH)_2$D purified we were able to compare a known amount of the purified 1,25-$(OH)_2$D from keratinocytes with chemically synthesized 1,25-$(OH)_2$D obtained as a gift from Dr.

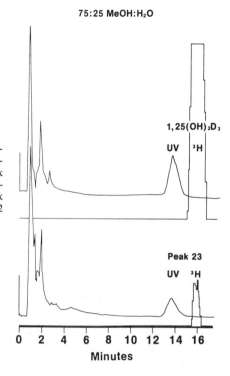

75:25 MeOH:H₂O

FIGURE 4A. The elution pattern of radioactive and nonradioactive 1,25-(OH)₂D₃ standards (*top panel*) compared with purified peak 23 (presumptive 1,25-(OH)₂D₃) (*bottom panel*) using a 3.9 mm × 30 cm C18 μ-Bondapak column eluted with MeOH:H₂O (75:25) at 2 ml/min.

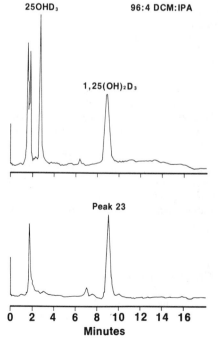

FIGURE 4B. The elution patterns of 25-(OH)D₃ and 1,25-(OH)₂D₃ standards (*top panel*) compared with purified peak 23 (presumptive 1,25-(OH)₂D₃) (*bottom panel*) using a 4.6 mm × 25 cm Zorbax Sil column eluted with dichloromethane:isopropanol (96:4) at 2 ml/min. Only the UV monitor trace is shown since dichloromethane quenched the radioactivity below the limits of detection under these conditions.

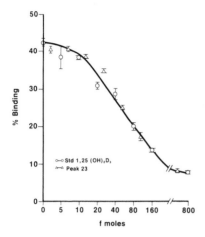

FIGURE 5. A comparison of the ability of peak 23 and authentic 1,25-$(OH)_2D_3$ to displace ^3H-1,25-$(OH)_2D_3$ from the chick intestinal cytosol receptor. The amount of peak 23 that was added was determined by the specific activity of the initial substrate (25-$(OH)D_3$) and the radioactivity in the final purified product (peak 23). Peak 23 and authentic 1,25-$(OH)_2D_3$ have comparable affinity for the intestinal 1,25-$(OH)_2D_3$ receptor. *Error bars* enclose mean and range of duplicate determinations.

M. R. Uskokovic (Hoffmann-LaRoche). The results (FIG. 5) show that the presumptive 1,25-$(OH)_2D$ produced by keratinocytes has the same binding affinity to the 1,25-$(OH)_2D$ receptor as authentic 1,25-$(OH)_2D$.

We then subjected the purified presumptive 1,25-$(OH)_2D$ to mass spectral analysis, and compared the spectrum to that obtained with authentic 1,25-$(OH)_2D$. The results (FIG. 6) show the equivalence of the presumptive 1,25-$(OH)_2D$ produced by keratinocytes with that of chemically synthesized, authentic 1,25-$(OH)_2D$. In particular, the molecular ion is seen at m/z 416; fragment ions are observed at m/z 398, 380, and 362 representing losses of 1, 2, or 3 molecules of water; other fragment ions at 383, 365, and 347 indicate the further loss of CH_3; the ions at m/z 287, 269, and 251 indicate the loss of the 8-membered side chain and additional losses of 1 or 2 molecules of water; and the ions at 152 and 134 indicate fission between C_7 and C_8 plus the loss of 1 water molecule, respectively.

Identification of the other metabolites of 25-$(OH)D$ produced by the keratinocyte has been less extensive. A periodate-sensitive peak which elutes in the position of authentic 24,25-$(OH)_2D$ in several HPLC systems is produced by the keratinocytes especially after they have been incubated with 1,25-$(OH)_2D$ under conditions expected to induce the 25-$(OH)D$ 24-hydroxylase (the enzyme that produces 24,25-$(OH)_2D$). We refer to this metabolite as 24,25-$(OH)_2D$. Likewise, a periodate-sensitive metabolite that elutes in the position of 1,24,25-$(OH)_3D$ in several HPLC systems has been tentatively identified as 1,24,25-$(OH)_3D$. Two prominent metabolites (called peak 24 and peak 28 in FIG. 3) are produced from both 1,25-$(OH)_2D$ and 25-$(OH)D$. Their elution positions in an HPLC system comparable to ours and their sensitivity to periodate suggest that they may be 24-keto 1,25-$(OH)_2D$ and 24-keto 1,23,25-$(OH)_3D$, respectively.[24,25] Of all the metabolites produced by the keratinocyte, only 1,25-$(OH)_2D$ has been shown to influence keratinocyte function. However, the others have not been adequately evaluated.

Regulation of 1,25-(OH)₂D Production

Parathyroid hormone (PTH), from 2 to 200 ng/ml, stimulates 1,25-$(OH)_2D$ production by human foreskin keratinocytes (FIG. 7). This effect is best seen if the keratinocytes are exposed to PTH for 4 h prior to adding the ^{3}H-25-(OH)D for an additional 1 h incubation. The phosphodiesterase inhibitor isobutylmethylxanthine (IBMX) also leads to an accumulation of 1,25-(OH)$_2$D after the addition of 25-(OH)D, but in this case IBMX blocks the further metabolism of 1,25-(OH)$_2$D (to peak 28) rather than stimulates 1,25-(OH)$_2$D production (TABLE 1). The reduced production of 24,25-(OH)$_2$D from 25-(OH)D in keratinocytes treated with IBMX (TABLE 2) suggests that IBMX inhibits 24-hydroxylase, a major pathway for the metabolism of 1,25-(OH)$_2$D. Maximum accumulation of 1,25-(OH)$_2$D in the keratinocytes occurs in

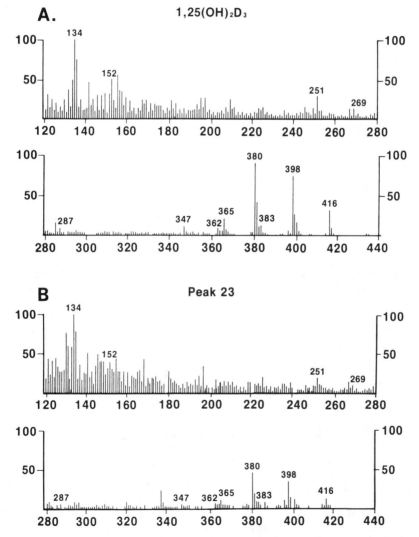

FIGURE 6. Mass spectrum determination of authentic 1,25-(OH)$_2$D$_3$ (**A**) and purified peak 23 (**B**). Fragmentations characteristic of 1,25-(OH)$_2$D$_3$ are indicated and described in the text.

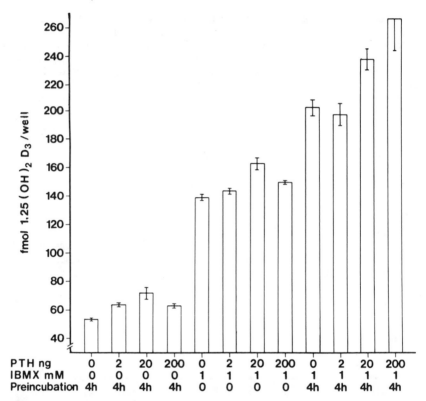

PTH ng	0	2	20	200	0	2	20	200	0	2	20	200
IBMX mM	0	0	0	0	1	1	1	1	1	1	1	1
Preincubation	4h	4h	4h	4h	0	0	0	0	4h	4h	4h	4h

FIGURE 7. The effect of IBMX and PTH on $1,25\text{-(OH)}_2D_3$ levels. Three experiments are shown. *To the left* the cells were incubated for 4 h with the indicated concentrations of PTH in the absence of IBMX prior to the addition of 625 fmol $^3H\text{-25-(OH)}D_3$ for an additional 4 h incubation. *In the middle* the cells were placed into medium containing 1 mM IBMX and the indicated concentrations of PTH at the beginning of the 4 h incubation with 625 fmol of $^3H\text{-25-(OH)}D_3$. *To the right* the cells were preincubated for 4 h with IBMX and the indicated PTH concentration before the 4 h incubation with 625 fmol $^3H\text{-25-(OH)}D_3$. In all cases, duplicate dishes were run and the data are expressed as mean ± range of duplicate determinations. In each of the three experiments, analysis by ANOVA showed that PTH increased $1,25\text{-(OH)}_2D_3$ significantly ($p = 0.017$, $p < 0.001$, and $p = 0.017$ for the three experiments, respectively). For the combined data, IBMX had a significant ($p = 0.001$) effect on $1,25\text{-(OH)}_2D_3$ levels as did preincubation with PTH and IBMX ($p = 0.001$).

the presence of PTH and IBMX, conditions in which production is stimulated (by PTH) and further metabolism is inhibited (by IBMX) (Fig. 7, Table 2). $3'5'cAMP$ and its more permeable derivatives cannot substitute for IBMX or PTH, and have no effect on $1,25\text{-(OH)}_2D$ production by keratinocytes. Therefore, PTH and IBMX appear to alter $25\text{-(OH)}D$ metabolism by the keratinocyte through non-cAMP-dependent mechanisms.

$1,25\text{-(OH)}_2D$ inhibits its own production and induces $24,25\text{-(OH)}_2D$ and $1,24,25\text{-(OH)}_3D$ production (Fig. 8). The production of peaks 24 and 28 (tentatively identified as 24-keto $1,25\text{-(OH)}_2D$ and 24-keto $1,23,25\text{-(OH)}_3D$, respectively) are also increased by exogenous $1,25\text{-(OH)}_2D$ (data not shown). The concentration of $1,25\text{-(OH)}_2D$

required for 50% inhibition of 1,25-$(OH)_2$D production and 50% stimulation of 24,25-$(OH)_2$D production (ED50) is 10^{-11} M. 1,24,25-$(OH)_3$D production is increased initially in parallel with 24,25-$(OH)_2$D production, but falls at concentrations of 1,25-$(OH)_2$D greater than 10^{-10} M. Similar observations were made for peaks 24 and 28. This increase, then decrease in 1,24,25-$(OH)_3$D, peak 24, and peak 28 production is consistent with the need for both the 24-hydroxylase, which is stimulated by 1,25-$(OH)_2$D, and the 1α-hydroxylase which is inhibited by 1,25-$(OH)_2$D, in the production of these metabolites.

1,25-$(OH)_2$D must be incubated with the keratinocytes for at least 4 h to observe the inhibition of 1,25-$(OH)_2$D production and longer (we routinely use 16 h) to observe the stimulation of 24,25-$(OH)_2$D production. Actinomycin D (2 μg/ml) blocks these effects of exogenous 1,25-$(OH)_2$D on 25-(OH)D metabolism indicating that these effects of 1,25-$(OH)_2$D on 25-(OH)D metabolism are mediated through genomic mechanisms.

The half maximally effective concentrations (ED50) of 1,25-$(OH)_2$D for the regulation of 1,25-$(OH)_2$D production in serum free medium is 10^{-11} M. When the percent free 1,25-$(OH)_2$D is reduced in the medium with the addition of human serum albumin (0.1 to 10%) or serum (0.1 to 10%), the apparent ED50 is shifted to increasingly higher concentrations of total 1,25-$(OH)_2$D. However, when the free concentration of 1,25-$(OH)_2$D (determined directly by centrifugal ultrafiltration)[26] is plotted against the biologic response (decreased 1,25-$(OH)_2$D production), the ED50 remains approximately 10^{-11} M (TABLE 3). These results indicate that the free concentration and not the total concentration of 1,25-$(OH)_2$D is the important variable in regulating 25-(OH)D metabolism by the cell. This suggests that 1,25-$(OH)_2$D does not need a serum carrier such as the vitamin D-binding protein to enter the cell.

Role of 1,25-$(OH)_2$D Production in Keratinocyte Differentiation

The capacity of the keratinocyte to produce 1,25-$(OH)_2$D changes as the keratinocytes differentiate *in vitro* (FIG. 9). In early preconfluent cultures, little 1,25-$(OH)_2$D is made. As the cells reach confluence, 1,25-$(OH)_2$D production increases to a maximum only to decline as the keratinocytes begin to terminally differentiate. Terminal differentiation is associated with an increase in 24,25-$(OH)_2$D production.

TABLE 1. Effect of PTH and IBMX on 1,25-$(OH)_2D_3$ metabolism[a]

		% Recovery			
PTH	IBMX	1,25-$(OH)_2D_3$	Peak 28	1,24,25-$(OH)_3D_3$	Aqueous
0	0	28.9 ± 1.5	30.7 ± 1.2	21.8 ± 0.8	8.5 ± 0.5
20 ng/ml	0	28.5 ± 4.0	28.2 ± 1.0	21.2 ± 0.7	7.3 ± 2.9
0	1 mM	46.9 ± 3.3	10.5 ± 0.9	21.0 ± 4.7	4.6 ± 0.3
20 ng/ml	1 mM	58.5 ± 8.0	6.6 ± 0.4	14.4 ± 1.0	4.6 ± 0.3

[a]In this experiment the cells were incubated with the indicated concentration of PTH and/or IBMX for 4 h at which point 0.05 μCi ^{3}H-1,25-$(OH)_2D_3$ was added for an additional 4 h incubation. The data are expressed as percent of the label recovered in the indicated peak following HPLC of the chloroform extract or in the aqueous extract. Each incubation was performed in duplicate, and the mean ± range of duplicates is shown. Multivariate ANOVA analysis showed that only IBMX (not PTH) significantly increased the levels of 1,25-$(OH)_2$D ($p = 0.008$) while decreasing the levels of peak 28 ($p < 0.001$) and aqueous radioactivity ($p < 0.05$).

TABLE 2. Effects of PTH and IBMX on 25-(OH)D Metabolism (4 Hour Incubation)[a]

| PTH | IBMX | 24,25-(OH)$_2$D$_3$ | 1,25-(OH)$_2$D$_3$ | fmol Recovered | | |
				Peak 24	Peak 28	1,24,25-(OH)$_3$D$_3$
ng/ml	mM					
0	0	28 ± 0.6	54 ± 0.6	36 ± 3	113 ± 6	89 ± 1
20	0	23 ± 6	79 ± 8	34 ± 6	102 ± 0.3	110 ± 2
0	1	2 ± 1	231 ± 5	10 ± 0.9	13 ± 5	46 ± 12
20	1	1.3 ± 0.6	274 ± 9	12 ± 1	12 ± 1	50 ± 3

[a]In this experiment the cells were incubated with the indicated concentration of PTH and/or IBMX for 4 h prior to the addition of 625 fmol ³H-25-(OH)D$_3$. Each incubation condition was performed in duplicate, and the results are expressed as fmol/well ± range of duplicates.

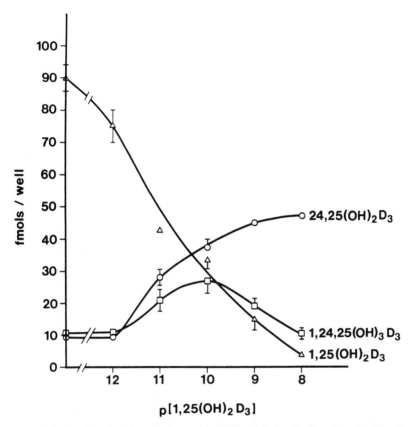

FIGURE 8. The effect of 1,25-(OH)$_2$D on 25-(OH)D metabolism by keratinocytes. Keratinocytes were incubated with nonradioactively labeled 1,25-(OH)$_2$D$_3$ at the indicated concentrations (p[1,25-(OH)$_2$D$_3$] is the negative log of [1,25-(OH)$_2$D$_3$]) for 16 h. The medium was removed, and the cells were washed once with fresh medium and then incubated with ^{3}H-25-(OH)D for 1 h. The production of 1,25-(OH)$_2$D$_3$ ($\triangle$), 24,25-(OH)$_2$D$_3$ ($\bigcirc$), and 1,24,25-(OH)$_3$D$_3$ ($\square$) is shown. Each point is the mean value ± range of duplicate determinations. ANOVA analysis showed that 1,25-(OH)$_2$D significantly decreased 1,25-(OH)$_2$D levels ($p < 0.001$), significantly increased 24,25-(OH)$_2$D levels ($p < 0.001$), and significantly increased, 1,24,25-(OH)$_3$D$_3$ levels ($p = 0.029$).

When expressed on a mg protein basis, the rise and fall of 1,25-(OH)$_2$D production precedes the rise and fall of transglutaminase activity, whereas the appearance of cornified envelopes precedes the increase in 24,25-(OH)$_2$D production (FIG. 10).

The question next addressed is whether the production of 1,25-(OH)$_2$D plays an important role in keratinocyte differentiation or whether it serves only as a marker of keratinocyte differentiation. We first examined whether the receptors for 1,25-(OH)$_2$D within the cell changed with differentiation. The data in TABLE 4 show that although the receptor number appears to fall as the cells terminally differentiate (from 27 to 19 fmol/mg protein), the affinity of the receptor for 1,25-(OH)$_2$D remains constant (Kd ≈ 250 pM). The ability of 1,25-(OH)$_2$D to inhibit its own production

TABLE 3. Comparison of the ED50 for Total and Free 1,25-(OH)$_2$D in Regulating Its Own Production[a]

Medium	% Free[c]	ED50[b]	
		Total	Free
0.1% HSA	34%	2.5×10^{-11} M	0.84×10^{-11} M
1.1% HSA	3.3%	2.7×10^{-10} M	0.89×10^{-11} M
10.1% HSA	0.53%	1.6×10^{-9} M	0.84×10^{-11} M
0.1% serum[d]	30%	3.0×10^{-11} M	0.90×10^{-11} M
1.0% serum	15%	6.0×10^{-11} M	0.90×10^{-11} M
10.0% serum	3.4%	2.8×10^{-10} M	0.95×10^{-11} M

[a]In this experiment, keratinocytes were incubated with 0, 10^{-12} to 10^{-8} M 1,25-(OH)$_2$D in media containing 0.1 to 10% human serum albumin (HSA) or human serum (serum). After a 4 h incubation, the media were removed, and the cells washed, and then incubated in serum-free KGM to which 0.1 μCi ^{3}H-25-(OH)D was added for an additional 1 h. The ^{3}H-1,25-(OH)$_2$D produced during the 1 h incubation was quantitated by HPLC equipped with a radioactivity detector. The free concentration of 1,25-(OH)$_2$D was determined in each medium used for the initial 4 h incubation by centrifugal ultrafiltration. The amount of ^{3}H-1,25-(OH)$_2$D produced was plotted against the total 1,25-(OH)$_2$D in the initial 4 h incubation and against the free 1,25-(OH)$_2$D in the initial 4 h incubation. The concentration of total and free 1,25-(OH)$_2$D causing half maximal inhibition of ^{3}H-1,25-(OH)$_2$D production is shown as the ED50.
[b]ED50 is the concentration of total or free 1,25-(OH)$_2$D required to inhibit by 50% the endogenous production of 1,25-(OH)$_2$D by the keratinocytes.
[c]The percentage of free 1,25-(OH)$_2$D was directly measured by centrifugal ultrafiltration.
[d]The media containing human serum also contained 0.1% human serum albumin (HSA) which was the vehicle for the added 1,25-(OH)$_2$D.

remains intact as the cells differentiate (TABLE 5) even though 1,25-(OH)$_2$D production is much reduced in the terminally differentiated cells. These data indicate that even the terminally differentiated cells are capable of responding to 1,25-(OH)$_2$D.

To more directly determine whether endogenously produced 1,25-(OH)$_2$D could alter keratinocyte differentiation we grew keratinocytes to confluence in serum free KGM supplemented with vehicle (0.1% human serum albumin), 10^{-9} M 25-(OH)D, or 10^{-10} M 1,25-(OH)$_2$D. At the time of confluence the cells were assessed for total DNA content, ^{3}H-thymidine incorporation, and ^{35}S-methionine incorporation into total protein and cornified envelopes. The data are summarized in TABLE 6. Neither

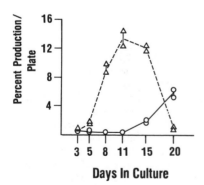

FIGURE 9. Expression of 1α- and 24-hydroxylase activity in keratinocytes at different stages of differentiation. Duplicate dishes at each time point were evaluated for their ability to produce ^{3}H-1,25-(OH)$_2$D$_3$ and ^{3}H-24,25-(OH)$_2$D$_3$ from ^{3}H25-(OH)D$_3$. Production of 1,25-(OH)$_2$D$_3$ ($\triangle$) and 24,25-(OH)$_2$D$_3$ (O) is expressed as percent conversion of the total radioactivity of ^{3}H-25-(OH)D$_3$ added to the culture dish.

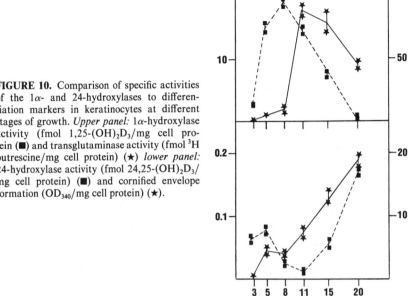

FIGURE 10. Comparison of specific activities of the 1α- and 24-hydroxylases to differentiation markers in keratinocytes at different stages of growth. *Upper panel:* 1α-hydroxylase activity (fmol $1,25\text{-}(OH)_2D_3$/mg cell protein (■) and transglutaminase activity (fmol ^3H putrescine/mg cell protein) (★) *lower panel:* 24-hydroxylase activity (fmol $24,25\text{-}(OH)_2D_3$/mg cell protein) (■) and cornified envelope formation (OD_{340}/mg cell protein) (★).

25-(OH)D nor $1,25\text{-}(OH)_2D$ had a consistent effect on total DNA content, DNA synthesis, or protein synthesis. However, both 25-(OH)D and $1,25\text{-}(OH)_2D$ stimulated cornified envelope formation 20–30-fold. Since the $1,25\text{-}(OH)_2D$ receptor has an affinity for 25-(OH)D that is three orders of magnitude less than that for $1,25\text{-}(OH)_2D$, it seems unlikely that 10^{-9} M 25-(OH)D has a direct effect on cornified envelope formation, although this possibility is not excluded by these results. We prefer to interpret the data as suggesting that 25-(OH)D stimulates cornified envelope formation by first being converted to $1,25\text{-}(OH)_2D$. Subsequent experimentation will be required to establish this hypothesis. Regardless, the observation that both 25-(OH)D and $1,25\text{-}(OH)_2D$ stimulate cornified envelope formation suggests a direct role for these metabolites in keratinocyte differentiation.

TABLE 4. Cytosolic $1,25\text{-}(OH)_2D_3$ Receptor Concentrations in Keratinocytes at Different Stages of Growth[a]

Days in Culture	Receptor Concentration (fmol/mg Protein)	Equilibrium Dissociation Constant (pM)
8 days (70–80% confluent)	26 ± 2.5	278 ± 49
10 days (100% confluent)	27 ± 2.0	222 ± 31
21 days (2 week post-confluent)	19 ± 1.5	237 ± 39

[a]These values (± SE) result from a computer-assisted analysis of binding data plotted by the method of Scatchard.

TABLE 5. Regulation of ^{3}H-1,25-(OH)$_2$D$_3$ Production by Exogenous 1,25-(OH)$_2$D$_3$[a]

	^{3}H-1,25-(OH)$_2$D$_3$ Production		
	0	10^{-11} M	10^{-9} M
Pre-confluent	41.9 ± 7.10	14.4 ± 1.67	6.32 ± 2.89
Confluent	55.4 ± 1.14	39.78 ± 3.38	3.83 ± 1.00
Post-confluent	2.22 ± 0.30	0.85 ± 0.29	0

[a]Data are expressed as fmol ^{3}H-1,25-(OH)$_2$D$_3$ produced per mg cell protein ± standard deviation of triplicate dishes. When analyzed by analysis of variance (ANOVA), 1,25-(OH)$_2$D$_3$ significantly inhibited its own production at all stages of confluence ($p < 0.001$).

Production of 1,25-(OH)$_2$D by Squamous Cell Carcinomas

To provide additional evidence supporting the role for 1,25-(OH)$_2$D production in keratinocyte differentiation we examined the ability of squamous cell carcinoma (SCC) lines for their ability to make and respond to 1,25-(OH)$_2$D. These lines were chosen to provide a broad spectrum in terms of their ability to express various features of normally differentiated keratinocytes *in vitro*.[27-30] The results shown in TABLE 7 show that the line most capable of differentiation, SCC 12F2, produces the most 1,25-(OH)$_2$D, whereas the lines least capable of differentiation, SCC 15 and A431, make the least amount of 1,25-(OH)$_2$D. However, all lines respond to exogenous 1,25-(OH)$_2$D with a fall in 1,25-(OH)$_2$D production. Such data are consistent with a role for 1,25-(OH)$_2$D production in keratinocyte differentiation, although other interpretations are possible.

Model for the Role of 1,25-(OH)$_2$D Production in Keratinocyte Differentiation

Our current model for the role that 1,25-(OH)$_2$D might play in keratinocyte differentiation is depicted in FIGURE 11. The essential role of calcium in this process has been well established by Hennings *et al.*[31,32] Within the keratinocyte are a number of calcium sensitive enzymes which appear to play an important role in keratinocyte differentiation. These enzymes include the membrane bound transglutaminase which is essential for the formation of the cornified envelope,[33] protein kinase C which when activated by phorbol esters leads to an increase in transglutaminase activity[34] and cornified envelope formation,[35] and phospholipase C which cleaves phosphoinositol 4,5-bis phosphate to diacyl glycerol, an activator of protein kinase C, and inositol 1,4,5-triphosphate, a stimulator of calcium release from intracellular stores. In

TABLE 6. Effects of 25-(OH)D and 1,25-(OH)$_2$D on Keratinocytes[a]

	Vehicle	10^{-9}M 25-(OH)D	10^{-10}M 1,25-(OH)$_2$D
DNA-μg/dish	17.1 ± 2	14.7 ± 0.2	18.7 ± 1.7
3H-thymidine, cpm/μg DNA	9,688 ± 2,700	9,187 ± 1,566	10,161 ± 975
^{35}S-met, total protein	8,713 ± 2,212	8,713 ± 2,212	11,518 ± 2,680
^{35}S-met, cornified envelopes	182 ± 67	5,400 ± 1,956	4,053 ± 1,829

[a]Data are presented as mean ± SD of triplicate determinations.

TABLE 7. SCC Production of and Response to 1,25-(OH)$_2$D^a

	Added 1,25-(OH)$_2$D	
Cell Line	0	10^{-8} M
KIP^b	1,541 ± 66	93 ± 31
SCC12F2	2,457 ± 274	93 ± 67
SCC12B2	1,034 ± 26	34 ± 14
SCC25	1,122 ± 40	13 ± 13
SCC 15	306 ± 37	33 ± 18
A431	255 ± 10	6 ± 6

^{a}Data are expressed as fmol 1,25-(OH)$_2$D/mg protein ± range of duplicates.
^{b}KIP are first passage normal keratinocytes.

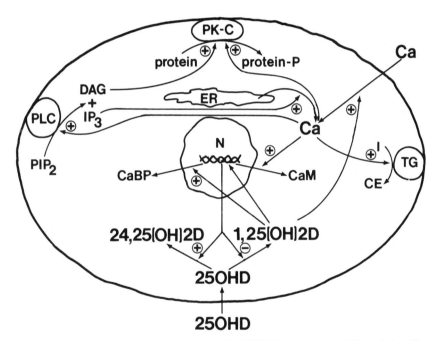

FIGURE 11. Model of the role of calcium and 1,25-(OH)$_2$D in keratinocyte differentiation. The intracellular free concentration of calcium is critical to the differentiation of the keratinocyte. Calcium enters the cell by a poorly understood process shown here as regulated by 1,25-(OH)$_2$D. Inside the cell calcium may trigger differentiation by stimulating cornified envelope formation (CE) from precursors such as involucrin (I) by activating membrane-bound transglutaminase activity (TG). In addition calcium may stimulate protein kinase C activity (PK-C), leading to the phosphorylation of critical proteins important in the differentiation process. Calcium may also activate phospholipase C (PLC), leading to the hydrolysis of phosphoinositide 4,5-bisphosphate (PIP$_2$) to form diacylglycerol (DAG) and inositol 1,4,5-triphosphate (IP$_3$). DAG stimulates PK-C, and IP$_3$ stimulates the release of calcium from intracellular stores thus providing amplification of the original calcium signal. Calcium-binding protein (CaBP) and calmodulin (CaM) may facilitate the ability of calcium to regulate a variety of cellular processes. 1,25-(OH)$_2$D may regulate calcium entry into the cell and modulate the actions of calcium through induction of CaBP synthesis and redistribution of CaM from cytosol to membrane. 1,25-(OH)$_2$D levels are self regulated in that 1,25-(OH)$_2$D inhibits its own production and stimulates that of 24,25-(OH)$_2$D.

addition, calmodulin[36] and a vitamin D dependent calcium-binding protein[37] are found in keratinocytes. The calmodulin levels are reported to increase when the keratinocytes are induced to differentiate with calcium.[36]

We[38] have observed that keratinocytes grown to confluence in 10^{-8} M 1,25-$(OH)_2D$ have a higher intracellular free calcium concentration (330 nM) than keratinocytes grown in vehicle (230 nM). Under similar conditions, the 1,25-$(OH)_2D$-treated cells had a 30-fold higher rate of cornified envelope formation. Therefore, it seems likely that 1,25-$(OH)_2D$ made by the cell or added exogenously could stimulate keratinocyte differentiation in part by raising intracellular free calcium levels. How this is done is not known, but in analogy to the intestinal epithelium, 1,25-$(OH)_2D$ may activate the process by which calcium flux across the membrane is controlled. Alternatively, 1,25-$(OH)_2D$ may stimulate phospholipase C,[39] leading to an increase in inositol triphosphate and increased release of calcium from intracellular stores. In either case we propose that the ability of 1,25-$(OH)_2D$ to regulate keratinocyte differentiation depends on its ability to regulate intracellular free calcium. Since the keratinocyte can produce its own 1,25-$(OH)_2D$, epidermal differentiation may be well preserved even in situations such as renal failure in which circulating 1,25-$(OH)_2D$ levels are reduced.

REFERENCES

1. GRAY, T. K., G. E. LESTER & R. S. LORENC. 1979. Evidence for extra-renal 1α-hydroxylation of 25-hydroxyvitamin D_3 in pregnancy. Science (Washington, DC) **204**(4399): 1311–1312.
2. WHITSETT, J. A., M. HO, R. C. TSANG, E. J. NORMAN & K. G. ADAMS. 1981. Synthesis of 1,25-dihydroxyvitamin D_3 by human placenta in vitro. J. Clin. Endocrinol. Metab. **53**(3): 484–488.
3. WEISMAN, Y., A. HARELL, S. EDELSTEIN, M. DAVID, Z. SPIRER & A. GOLANDER. 1979. $1\alpha,25$-dihydroxyvitamin D_3 and 24,25-dihydroxyvitamin D_3 in vitro synthesis by human decidua and placenta. Nature (London) **281**(5729): 317–319.
4. HOWARD, G. A., R. T. TURNER, D. J. SHERRARD & D. J. BAYLINK. 1981. Human bone cells in culture metabolize 25-hydroxyvitamin D_3 and 24,25-dihydroxyvitamin D_3. J. Biol. Chem. **256**(15): 7738–7740.
5. TURNER, R. T., J. E. PUZAS, M. D. FORTE, G. E. LESTER, T. K. GRAY, G. A. HOWARD & D. J. BAYLINK. 1980. In vitro synthesis of $1\alpha,25$-dihydroxycholecalciferol and 24,25-dihydroxycholecalciferol by isolated calvarial cells. Proc. Natl. Acad. Sci. USA **77**(10): 5720–5724.
6. MASON, R. S., T. FRANKEL, Y. L. CHAN, D. LISSNER & S. POSEN. 1984. Vitamin D conversion by sarcoid lymph node homogenate. Ann. Intern. Med. **100**(1): 59–61.
7. ADAMS, J. S., O. P. SHARMA, M. A. GACAD & F. R. SINGER. 1983. Metabolism of 25-hydroxyvitamin D_3 by cultured pulmonary alveolar macrophages in sarcoidosis. J. Clin. Invest. **72**(5): 1856–1860.
8. FRANKEL, T. L., R. S. MASON, P. HERSEY, E. MURRAY & S. POSEN. 1983. The synthesis of vitamin D metabolites by human melanoma cells. J. Clin. Endocrinol. Metab. **57**(3): 627–631.
9. FETCHICK, D. A., D. R. BERTOLINI, P. S. SARIN, S. T. WEINTRAUB, G. R. MUNDY & J. F. DUNN. 1986. Production of 1,25-dihydroxyvitamin D_3 by human T cell lymphotrophic virus-1-transformed lymphocytes. J. Clin. Invest. **78**(2): 592–596.
10. REICHEL, H., H. P. KOEFFLER & A. W. NORMAN. 1987. 25-hydroxyvitamin D_3 metabolism by human T-lymphotropic virus-transformed lymphocytes. J. Clin. Endocrinol. Metab. **65**(3): 519–526.
11. BIKLE, D. D., M. K. NEMANIC, J. O. WHITNEY & P. W. ELIAS. 1986. Neonatal human foreskin keratinocytes produce 1,25-dihydroxyvitamin D_3. Biochemistry **25**(7): 1545–1548.
12. BIKLE, D. D., M. K. NEMANIC, E. A. GEE & P. ELIAS. 1986. 1,25-dihydroxyvitamin D_3

production by human keratinocytes: kinetics and regulation. J. Clin. Invest. **78**(2): 557–566.

13. REEVE, L., Y. TANAKA & H. F. DELUCA. 1983. Studies on the site of 1,25-dihydroxyvitamin D₃ synthesis *in vivo*. J. Biol. Chem. **258**(6): 3615–3617.

14. SHULTZ, T. D., J. FOX, H. HEATH III & R. KUMAR. 1983. Do tissues other than the kidney produce 1,25-dihydroxyvitamin D₃ *in vivo?* A re-examination. Proc. Natl. Acad. Sci. USA **80**(6): 1746–1750.

15. LAMBERT, P. W., P. H. STERN, R. C. AVIOLI, N. C. BRACKETT, R. T. TURNER, A. GREENE, I. Y. FU & N. H. BELL. 1982. Evidence for extrarenal production of 1α-,25-dihydroxyvitamin D in man. J. Clin. Invest. **69**(3): 722–725.

16. LITTLEDIKE, E. T. & R. L. HORST. 1982. Metabolism of vitamin D₃ in nephrectomized pigs given pharmacological amounts of vitamin D₃. Endocrinology **111**(6): 2008–2013.

17. FELDMAN, D., T. CHEN, C. CONE, M. HIRST, S. SHANI, A. BENDERLI & A. HOCHBERG. 1982. Vitamin D resistent rickets with alopecia; cultured skin fibroblasts exhibit defective cytoplasmic receptors and unresponsiveness to 1,25(OH)₂D₃. J. Clin. Endocrinol. Metab. **55**(5): 1020–1022.

18. FELDMAN, D., T. CHEN, M. HIRST, K. COLSTON, M. KARASEK & C. CONE. 1980. Demonstration of 1,25-dihydroxyvitamin D₃ receptors in human skin biopsies. J. Clin. Endocrinol. Metab. **51**(6): 1463–1465.

19. HOSOMI, J., J. HOSOI, E. ABE, T. SUDA & T. KUROKI. 1983. Regulation of terminal differentiation of cultured mouse epidermal cells by 1α,25-dihydroxyvitamin D₃. Endocrinology **113**(6): 1950–1957.

20. SMITH, E. L., N. C. WALWORTH & M. F. HOLICK. 1986. Effect of 1α,25-dihydroxyvitamin D₃ on the morphologic and biochemical differentiation of cultured human epidermal keratinocytes grown in serum-free conditions. J. Invest. Derm. **86**(6): 709–714.

21. PILLAI, S., D. D. BIKLE & P. M. ELIAS. 1988. 1,25-dihydroxyvitamin D production and receptor binding in human keratinocytes varies with differentiation. J. Biol. Chem. **263**(11): 5390–5395.

22. RHEINWALD, J. & H. GREEN. 1975. Serial cultivation of human epidermal keratinocytes; the formation of keratinizing colonies from single cells. Cell **6**(3): 331–343.

23. PILLAI, S., D. D. BIKLE & P. M. ELIAS. 1988. Biochemical and morphological characterization of growth and differentiation of normal human neonatal keratinocytes in a serum free medium. J. Cell. Physiol. **134**: 229–237.

24. NAPOLI, J. L., B. C. PRAMANIK, P. M. ROYAL, T. A. REINHARDT & R. L. HORST. 1983. Intestinal synthesis of 24-keto-1,25-dihydroxyvitamin D₃: a metabolite formed *in vivo* with high affinity for the vitamin D cytosolic receptor. J. Biol. Chem. **258**(15): 9100–9107.

25. MAYER, E., J. E. BISHOP, R. A. S. CHANDRARATNA, W. H. OKAMURA, J. R. KRUSE, G. POPJAK, N. OHNUMA & A. W. NORMAN. 1983. Isolation and identification of 1,25-dihydroxy-24-oxo-vitamin D₃ and 1,23,25,-trihydroxy-24-oxo-vitamin D₃. J. Biol. Chem. **258**(22): 13458–13465.

26. BIKLE, D. D., E. GEE, B. HALLORAN & J. G. HADDAD. 1984. Free 1,25-dihydroxyvitamin D levels in serum from normal subjects, pregnant subjects, and subjects with liver disease. J. Clin. Invest. **74**(6): 1966–1974.

27. RHEINWALD, J. G., E. GERMAIN & M. A. BECKETT. 1983. Expression of keratins and envelope proteins in normal and malignant human keratinocytes and mesothelial cells. *In* Human Carcinogenesis. C. Harris & H. Autrup, Eds. 85–96. Academic Press. New York, NY.

28. RHEINWALD, J. G. & M. A. BECKETT. 1981. Tumorigenic keratinocyte lines requiring anchorage and fibroblast support cultured from human squamous cell carcinomas. Cancer Res. **41**(5): 1657–1663.

29. RUBIN, A. L. & R. H. RICE. 1986. Differential regulation by retinoic acid and calcium of transglutaminases in cultured neoplastic and normal human keratinocytes. Cancer Res. **46**(5): 2356–2361.

30. PONEC, M., L. HAVEKES, J. KEMPENAAR, S. LAVRIJSEN & B. J. VERMEER. 1984. Defective low-density lipoprotein metabolism in cultured, normal, transformed, and malignant keratinocytes. J. Invest. Derm. **83**(6): 436–440.

31. HENNINGS, H., D. MICHAEL, C. CHENG, P. STEINERT, K. HOLBROOK & S. H. YUSPA. 1980.

Calcium regulation of growth and differentiation of mouse epidermal cells in culture. Cell **19**(1): 245–254.

32. HENNINGS, H., K. A. HOLBROOK & S. H. YUSPA. 1983. Factors influencing calcium-induced terminal differentiation in cultured mouse epidermal cells. J. Cell. Physiol. **116**(3): 265–281.

33. THACHER, S. M. & R. H. RICE. 1985. Keratinocyte-specific transglutaminase of cultured human epidermal cells: relation to cross-linked envelope formation and terminal differentiation. Cell **40**(3): 685–695.

34. YUSPA, S. H., T. BEN, H. HENNINGS & U. LICHTI. 1980. Phorbol ester tumor promoters induce epidermal transglutaminase activity. Biochem. Biophys. Res. Commun. **97**(2): 700–708.

35. PARKINSON, E. K., M. F. PERA, A. EMMERSON & P. A. GORMAN. 1984. Differential effects of complete and second-stage tumour promoters in normal but not transformed human and mouse keratinocytes. Carcinogenesis **5**(8): 1071–1077.

36. FAIRLEY, J. A., C. L. MARCELO, V. A. HOGAN & J. J. VOORHEES. 1985. Increased calmodulin levels in psoriasis and low Ca^{++} regulated mouse epidermal keratinocyte cultures. J. Invest. Derm. **84**(3): 195–198.

37. RIZK, M., J. H. PAVLOVITCH, L. DIDIERJEAN, J. H. SAURAT & S. BALSAN. 1984. Skin calcium-binding protein; effect of vitamin D deficiency and vitamin D treatment. Biochem. Biophys. Res. Commun. **123**(1): 230–237.

38. PILLAI, S., D. D. BIKLE, V. PATEL, M. HINCENBERGS & P. M. ELIAS. 1988. Calcium mediates the effect of 1,25 dihydroxyvitamin D_3 on keratinocyte differentiation. J. Bone Min. Res. **3**(Suppl. 1): 217.

39. TANG, W., V. A. ZIBOH, R. R. ISSEROFF & D. MARTINEZ. 1987. Novel regulatory actions of $1\alpha,25$-dihydroxyvitamin D_3 on the metabolism of polyphosphoinositides in murine epidermal keratinocytes. J. Cell. Physiol. **132**(1): 131–136.

Actions of 1α,25-Dihydroxyvitamin D₃ on Normal, Psoriatic, and Promoted Epidermal Keratinocytes[a]

TOSHIO KUROKI, SHIGETO MORIMOTO,[b]
AND TATSUO SUDA[c]

Department of Cancer Cell Research
Institute of Medical Science
University of Tokyo
Shirokanedai, Minato-ku
Tokyo 108, Japan

[b]*Department of Medicine and Geriatrics*
Osaka University Medical School
Fukushima, Fukushima-ku
Osaka 553, Japan

[c]*Department of Biochemistry*
School of Dentistry
Showa University
Hatanodai, Shinagawa-ku
Tokyo 142, Japan

INTRODUCTION

The skin is a key tissue in the metabolism of vitamin D_3 together with the liver and kidney.[1] In the epidermis, provitamin D_3 (7-dehydrocholesterol) obtained from food is coverted to vitamin D_3 (cholecalciferol) on exposure to sunlight. The resulting vitamin D_3 is hydroxylated in the liver at the 25-position and then in the kidney at the 1α-position under hormonal control to form 1α,25-dihydroxyvitamin D_3 (1α,25-$(OH)_2D_3$). Recently, epidermal keratinocytes were shown to contain 1α-hydroxylase,[2] and thus the skin, like the kidney, can be a producing organ of 1α,25-$(OH)_2D_3$.

The plasma level of 1α,25-$(OH)_2D_3$ is controlled at a concentration of about 100 pM (40 pg/ml) by hormones such as parathyroid hormone and calcitonin. A specific receptor for 1α,25-$(OH)_2D_3$ is present in the cytosol/nuclear fraction. Therefore, 1α,25-$(OH)_2D_3$ is now considered to be a hormone, with the unique character that its precursor is supplied in food like vitamins.

1α,25-$(OH)_2D_3$ has long been known to regulate the blood calcium level. However, the wide distribution of its specific receptor in various tissues including skin (for review, see REF. 3) indicates that the maintenance of calcium homeostasis is not its only physiological role. Indeed, 1α,25-$(OH)_2D_3$ has been shown to have pleiotropic actions on cell differentiation (TABLE 1) and tumor promotion (TABLE 2). In our laboratories (TK and TS), we have been working on the control of cell differentiation, tumor promotion and gene expression by 1α,25-$(OH)_2D_3$ (for review, see REF. 4). We

[a]This work was supported in part by grants for special project research, Cancer Bioscience, and Cancer Research from the Ministry of Education, Science, and Culture of Japan.

45

TABLE 1. Effects of $1\alpha,25\text{-}(OH)_2D_3$ Related to Cell Differentiation

1. Blood cells
Differentiation of leukemic cells to macrophages
Regulation of interleukin production
Cell fusion of alveolar macrophages
2. Epidermal keratinocytes
Stimulation of epidermal differentiation
Therapeutic use for psoriasis
3. Melanoma cells
4. Osteoblasts

summarize here our observations on the actions of $1\alpha,25\text{-}(OH)_2D_3$ on normal, psoriatic and promoter-treated epidermal keratinocytes.

ACTIONS OF $1\alpha,25\text{-}(OH)_2D_3$ ON NORMAL EPIDERMAL KERATINOCYTES

In the epidermis *in vivo*, epidermal keratinocytes are programmed to differentiate from proliferating basal cells into dead cornified cells through the development of high molecular weight keratin filaments, the formation of cornified envelopes, and the destruction and enucleation of cell nuclei. This process also occurs in epidermal keratinocytes in culture. Morphologically, differentiated epidermal keratinocytes form focal stratifications in which the cells become squamous and have picnotic nuclei or are enucleated (FIGS. 1B and C). The formation of a cornified envelope (FIG. 1D), a

TABLE 2. Effects of $1\alpha,25\text{-}(OH)_2D_3$ Related to Cell Transformation and Carcinogenesis

A. Cell Culture Systems:
1. Absence of induction of transformation
BALB/3T3 cells
SHE cells
2. Enhancement of chemically induced transformation
BALB/3T3 cells
SHE cells
3. Induction of anchorage-independent growth
JB6 cells
BALB/3T3 cells
NIH/3T3 cells
B. Animal Experiments:
1. Absence of carcinogenicity or tumor-promoting activity in mouse skin
2. Inhibition of ODC-induction by tumor promoters
Skin—TPA, mezerein
Stomach—NaCl
Colon—deoxycholic acid
Liver—phenobarbital
3. Inhibition of tumor promotion in mouse skin
4. Stimulation of carcinogenesis in mouse skin

chemically resistant structure formed beneath the plasma membrane by transglutaminase, is also used as a marker of epidermal differentiation.

We have reported that $1\alpha,25\text{-}(OH)_2D_3$ stimulates this process.[5] FIGURE 2 shows the time-course of events in a primary culture of epidermal keratinocytes of C57BL mice treated with 12 nM $1\alpha,25\text{-}(OH)_2D_3$. The fraction of growing basal cells decreased sharply in the presence of $1\alpha,25\text{-}(OH)_2D_3$ (FIG. 2D) and in parallel the number of squamous cells increased, first in the attached cell population (FIG. 2C) and then in the population of sloughed-off, floating cells (FIG. 2B). The number of cornified envelopes increased with time of cultivation: on day 10, about 65% of the treated cells but only 20% or less of the cells in the solvent control culture had a cornified envelope (FIG. 2A).

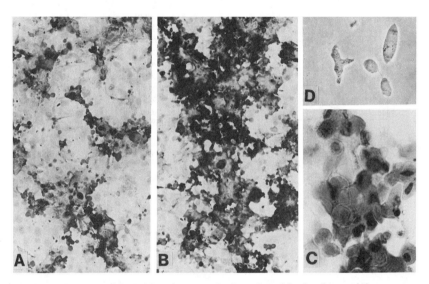

FIGURE 1. Primary epidermal keratinocytes of mice cultured in the absence (**A**) or presence (**B, C**) of 12 nM $1\alpha,25\text{-}(OH)_2D_3$ for 7 days.[5] Darkly stained focal stratification is seen in both cultures but is more pronounced in the treated culture. Enucleated or pycnotic cells are seen in the stratified foci (**C**). Differentiated cells form cornified envelopes which are resistant to a solution containing sodium dodecyl sulfate and 2-mercaptoethanol and are seen under a phase-contrast microscope (**D**). (From Hosomi *et al.*[5] Reprinted by permission from *Endocrinology.*)

We also observed an increase of transglutaminase activity in the treated cells. During differentiation, the cells become larger and less dense.

Epidermal differentiation is stimulated specifically by $1\alpha,25\text{-}(OH)_2D_3$: under the same conditions, other metabolites of vitamin D_3, *e.g.*, 24,25-dihydroxyvitamin D_3, 25-hydroxyvitamin D_3, 1α-hydroxyvitamin D_3 ($1\alpha\text{-}(OH)D_3$) and vitamin D_3, induced only partial epidermal differentiation at about 1,000 times higher concentrations than $1\alpha,25\text{-}(OH)_2D_3$.

We found that mouse epidermal keratinocytes in primary culture contain a specific receptor for $1\alpha,25\text{-}(OH)_2D_3$ with a Kd value of 54 pmol and maximum binding of 43 fmol/mg protein. The specificity of this receptor was demonstrated by experiments on competitive bindings of metabolites of vitamin D_3: as for induction of epidermal differentiation, the concentrations of other metabolites required for 50% displacement

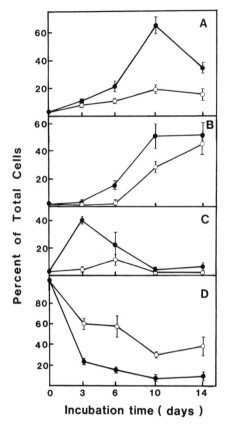

FIGURE 2. Time-dependent induction of epidermal differentiation in the presence (●) and absence (○) of $1\alpha,25\text{-}(OH)_2D_3$.[5] The vitamin at a concentration of 12 nM was added to primary cultures 24 h after plating (day 0). Differentiation was assayed morphologically by measuring the following parameters: (A) formation of a cornified envelope; (B) the number of sloughed-off squamous cells and enucleated cells; (C) the number of attached squamous cells; and (D) the number of attached basal cells. (From Hosomi et al.[5] Reprinted by permission from Endocrinology.)

of the binding were 400 to 5,000 times higher than that of $1\alpha,25\text{-}(OH)_2D_3$. These results indicate receptor-mediated induction of epidermal differentiation by $1\alpha,25\text{-}(OH)_2D_3$.

$1\alpha,25\text{-}(OH)_2D_3$ inhibits DNA synthesis dose-dependently under the conditions in which the cells differentiate. This decreased DNA synthesis may be the result of cell differentiation, rather than its cause.

Smith et al.[6] confirmed our observations using human epidermal keratinocytes grown in serum-free conditions. As with mouse keratinocytes, $1\alpha,25\text{-}(OH)_2D_3$ causes terminal differentiation of human keratinocytes, as defined by decrease of basal cells, increase of squamous cells, formation of cornified cells and increased activity of transglutaminase.

Developmental change of the receptor for $1\alpha,25\text{-}(OH)_2D_3$ in mouse skin was investigated by Horiuchi et al.[7] During the neonatal period in which the epidermal and dermal layers thicken rapidly and hair grows, they observed marked increase of the receptor number. Furthermore, Stumpf et al.[8] demonstrated by autoradiography that $1\alpha,25\text{-}(OH)_2D_3$ is concentrated in the basal cells of the epidermis, outer hair shafts and sebaceus glands, in which cells grow actively.

Recently, we found incidentally that $1\alpha,25\text{-}(OH)_2D_3$ induced mRNA of the metallothionein gene in cultured epidermal keratinocytes of mice and rats and also in skin tissue when 1α-(OH)D$_3$, a synthetic precursor of $1\alpha,25$-(OH)$_2$D$_3$, was applied *in vivo*[9] (FIG. 3). The structure and regulation of the metallothionein gene have been studied extensively, but the physiological functions of metallothionein are still under investigation. Judging from its highly efficient and specific induction, an increased level of metallothionein may be relevant to some biological actions of $1\alpha,25$-(OH)$_2$D$_3$.

Thus, all the existing data indicate that the skin is not only an effector tissue for metabolizing vitamin D$_3$, but also a target tissue.

ACTIONS OF 1α,25-(OH)₂D₃ ON PSORIATIC EPIDERMAL KERATINOCYTES

Psoriasis is a common skin disease characterized by abnormal increase in the number of epidermal keratinocytes. The skin lesions are also associated with abnormalities in the dermis including dilated tortuous capillaries and infiltration of neutrophils. The cause of psoriasis is unknown, but is suspected to be disordered regulation of differentiation of epidermal keratinocytes.

Psoriasis has been treated in many agents, such as tar or psoralen together with UV-irradiation, corticosteroids, retinoids, and methotrexate. These agents are applied topically or systemically. However, these treatments are only temporarily effective and in some cases are asssociated with adverse side effects.

During treatment of a patient with senile osteoporosis, one of the present authors (SM) found incidentally that associated psoriatic lesions were also cured by oral administration of 1α-(OH)D$_3$.[10] This finding prompted further clinical trials on the

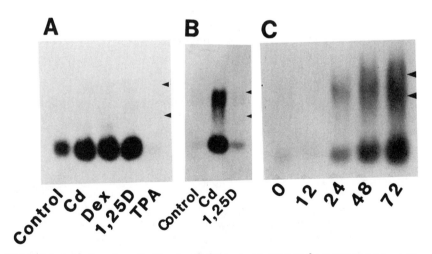

FIGURE 3. Induction of metallothionein mRNA by $1\alpha,25$-(OH)$_2$D$_3$.[9] (**A**) FRSK fetal rat skin keratinocytes; (**B**) primary epidermal keratinocytes of mice; (**C**) mouse skin *in vivo*. Cultured cells (A, B) were treated for 7 h with 1 μM CdCl$_2$(Cd), 10 μM dexamethasone (Dex), 12 nM $1\alpha,25$-(OH)$_2$D$_3$ or 100 ng/ml of TPA. In the mouse *in vivo* (C), female Sencar mice were treated orally with 5 μg of 1α-(OH)D$_3$ and the skin was removed at the indicated times. Total RNA was isolated and subjected to RNA gel blotting with a 0.4-kilobase pair *Eco*R I/*Hind* III fragment of m$_1$pEH.4 derived from mouse metallothionein-I as probe. The *upper* and *lower arrowheads* indicate the positions of 28S and 18S rRNA, respectively.

TABLE 3. Open-Design Study on Therapeutic Value of Active Forms of Vitamin D_3 for Psoriasis[11]

Group	Treatment Chemical	Dose	Period	Route	No. of Patients	No. of Cases of Effective Treatment[a]
I	1α-$(OH)D_3$	1.0 μg/day	6 mo	oral	17	13
II	$1\alpha,25$-$(OH)_2D_3$	0.5 μg/day	6 mo	oral	4	1
III	$1\alpha,25$-$(OH)_2D_3$	0.5 μg/g of base	2 mo	topical	19	16

[a]Effective treatment indicates moderate or marked improvement or complete remission of the lesions.

treatment of psoriasis with $1\alpha,25$-$(OH)_2D_3$. Morimoto *et al.* conducted an open-design study in 40 patients with psoriasis vulgaris on the effects of oral and topical treatments with $1\alpha,25$-$(OH)_2D_3$ and 1α-$(OH)D_3$.[11] As summarized in TABLE 3, improvement or complete remission of skin lesions was observed in 13 of 17 patients (76%) given 1α-$(OH)D_3$ orally (Group I), in 1 of 4 patients (25%) given $1\alpha,25$-$(OH)_2D_3$ orally (Group II), and in 16 of 19 patients (84%) treated with $1\alpha,25$-$(OH)_2D_3$ topically (Group III) (FIG. 4). No side effects were observed in any of these cases. Double-blind tests are needed for evaluating the therapeutic effects with scientific objectivity.

While this open-design study was in progress in Osaka, MacLaughlin *et al.*[12] reported that cultured fibroblasts from involved and uninvolved skin of psoriatic patients were about 100-fold more resistant than control fibroblasts to the inhibitory effect of $1\alpha,25$-$(OH)_2D_3$ on cell proliferation. The receptors for $1\alpha,25$-$(OH)_2D_3$ were the same in psoriatic fibroblasts as in those from age-matched controls. These workers suggested that there may be an inherent biochemical defect(s) in psoriatic fibroblasts and also probably in psoriatic epidermal keratinocytes. However, they were doubtful

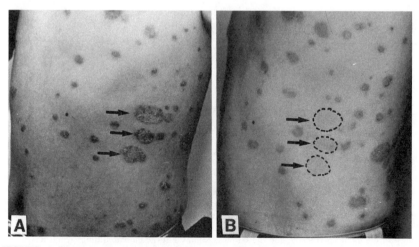

FIGURE 4. Therapy of psoriatic lesions by $1\alpha,25$-$(OH)_2D_3$. Three big psoriatic lesions (**A**) (*indicated by arrows*) disappeared after topical application of $1\alpha,25$-$(OH)_2D_3$ at 0.5 μg/g of base for 4 weeks (**B**) (*dotted areas indicated by arrows*).

about the benefit of treatment of psoriasis with $1\alpha,25\text{-}(OH)_2D_3$ because of the potent calciotropic effect of this hormone.

We are also studying the possible involvement of $1\alpha,25\text{-}(OH)_2D_3$ in psoriasis using cultured psoriatic epidermal keratinocytes. We used a new method for explant-outgrowth culture of epidermal keratinocytes from small biopsy specimens[13] and succeeded in obtaining a sheet of epidermal keratinocytes from involved and

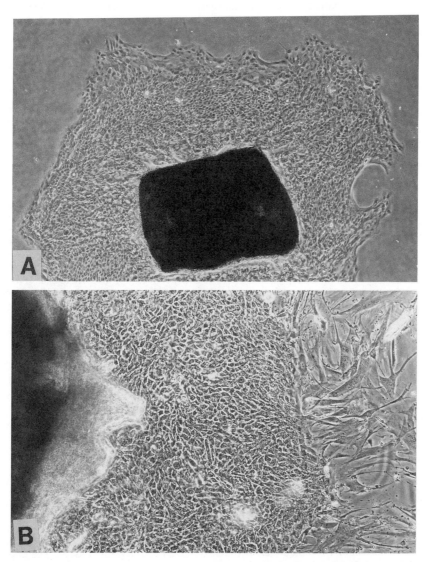

FIGURE 5. Explant-outgrowth culture of psoriatic epidermal keratinocytes from uninvolved skin without a feeder layer (**A**) and from involved skin with a feeder layer of lethally irradiated psoriatic fibroblasts (**B**).

uninvolved skin of psoriatic patients (FIG. 5). We found that a feeder layer of psoriatic fibroblasts or of BALB/3T3 cells was required for the growth of psoriatic epidermal keratinocytes from involved skin, but not for that of cells from uninvolved skin. We also found that like psoriatic fibroblasts, psoriatic epidermal keratinocytes were two orders of magnitude more resistant than control cells to the inhibitory effect of $1\alpha,25$-$(OH)_2D_3$ (unpublished data).

The therapeutic value of $1\alpha,25$-$(OH)_2D_3$ for treatment of psoriasis and the partial resistance of psoriatic epidermal keratinocytes to $1\alpha,25$-$(OH)_2D_3$ afford a new insight into the etiology and therapy of this common but enigmatic and refractory disease.

ACTIONS OF $1\alpha,25$-$(OH)_2D_3$ ON PROMOTER-TREATED EPIDERMAL KERATINOCYTES

The skin is a favorite tissue for use in studies on chemical carcinogenesis, because it is readily accessible and is highly sensitive to chemical carcinogens and tumor promoters. The idea of testing the effect of $1\alpha,25$-$(OH)_2D_3$ on tumor promotion arose from the findings in cell differentiation studies that $1\alpha,25$-$(OH)_2D_3$ has some actions in

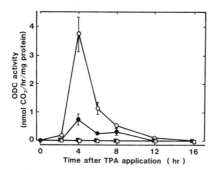

FIGURE 6. Time course of inhibition by $1\alpha,25$-$(OH)_2D_3$ of epidermal ODC activity by TPA.[18] Mice were treated first with 1 µg of $1\alpha,25$-$(OH)_2D_3$ and 30 min later with 10 µg of TPA; acetone was used as a solvent control. ●, $1\alpha,25$-$(OH)_2D_3$-TPA; ○, acetone-TPA; ■, $1\alpha,25$-$(OH)_2D_3$-acetone; □, acetone-acetone. (From Chida et al.[18] Reprinted by permission from *Cancer Research.*)

common with phorbol ester tumor promoters, *e.g.*, 12-*O*-tetradecanoylphorbol-13-acetate (TPA). First, we found that treatment of initiated BALB/3T3 cells with $1\alpha,25$-$(OH)_2D_3$ at 0.1–10 nM markedly enhanced their transformation.[14,15] Then we discovered that $1\alpha,25$-$(OH)_2D_3$ induced anchorage-independent growth of JB6 mouse epidermal cells,[16] and BALB/3T3 and NIH/3T3 cells[17] (see TABLE 2). However, the mechanism by which $1\alpha,25$-$(OH)_2D_3$ enhances cell transformation is obviously different from that of phorbol ester tumor promoters, because $1\alpha,25$-$(OH)_2D_3$ does not activate protein kinase C in a cell-free system or in intact cells.[15]

The possible promoting effect of $1\alpha,25$-$(OH)_2D_3$ was examined in mouse skin. First, we examined the induction of ornithine decarboxylase (ODC), a well established marker of tumor promotion, by topical application of $1\alpha,25$-$(OH)_2D_3$ to the dorsal skin of Sencar mice.[18] Results showed that $1\alpha,25$-$(OH)_2D_3$ did not induce ODC, but unexpectedly that it inhibited the induction of ODC by TPA: application of 1 µg of $1\alpha,25$-$(OH)_2D_3$ within 30 min before or after treatment with 10 µg of TPA resulted in 72% inhibition of ODC induction at the peak time, *i.e.*, after 4 h (FIG. 6). The dose required for 50% inhibition was 63 ng, or 150 pmol, which was about half that of retinoic acid. $1\alpha,25$-$(OH)_2D_3$ also inhibited ODC induction by teleocidin B, an indol alkaloid tumor promoter with similar effects to those of TPA, and mezerein, a

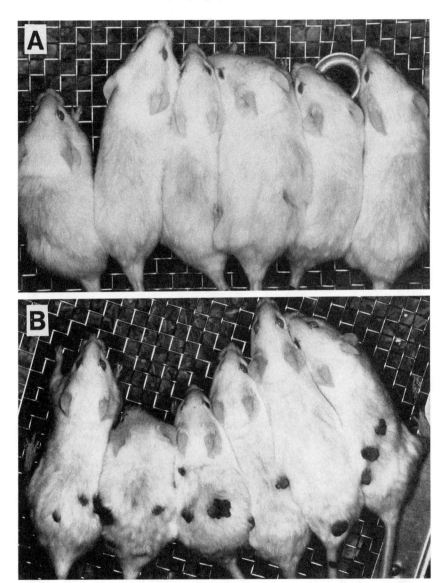

FIGURE 7. Inhibition of papilloma formation by 1α,25-(OH)₂D₃ in a two-stage promotion experiment on carcinogenesis in mouse skin.[20] Female Sencar mice were treated once with 2.5 μg of DMBA, and then once with 5 μg of TPA as a stage-I promoter, and then with 5 μg of mezerein as a stage-II promoter once a week for 19 weeks (**A**). 1α,25-(OH)₂D₃ at 1 μg was applied 30 min before treatment with the stage-I and -II promoters (**B**). (From Suda *et al.*[4] Reprinted by permission from Elsevier Science Publishers.)

reciniferol with weak tumor-promoting activity. However, $1\alpha,25\text{-}(OH)_2D_3$ did not inhibit epidermal hyperplasia induced by TPA. Similar inhibition of induction of ODC was observed in mice given $1\alpha\text{-}(OH)_2D_3$ orally.[19]

This finding suggests that $1\alpha,25\text{-}(OH)_2D_3$ is an antipromoter rather than a tumor promoter. Indeed, we found that topical application of $1\alpha,25\text{-}(OH)_2D_3$ once a week at a dose of 1 μg or less, a tolerable dose from hypercalcemia, inhibited the promotion phase of 7,12-dimethylbenz[a]anthracene-(DMBA)-induced skin carcinogenesis in mice (FIG. 7).[20] Consistent with this observation, Wood et al.[21] reported that in a conventional single-stage promotion test, topical application of $1\alpha,25\text{-}(OH)_2D_3$ suppressed formation of TPA-promotion skin tumors, although the strain of mice, the protocol of treatment, and the extent of inhibition differed from those in our study.

The apparent discrepancy between the stimulatory effect on cell transformation *in vitro* and the inhibitory effect on tumor promotion *in vivo* is hard to interpret at present, but may be explained in the near future on the basis of regulation of expression of genes that are involved in essential steps of cell transformation or tumor promotion. The recent cloning of the receptor gene for $1\alpha,25\text{-}(OH)_2D_3$[22,23] and the identification of genes regulated by $1\alpha,25\text{-}(OH)_2D_3$[9,17,24] should greatly promote studies on the biology and medicine of vitamin D.

REFERENCES

1. HOLICK, M. F. 1981. The cutaneous photosynthesis of previtamin D_3: a unique photoendocrine system. J. Invest. Dermatol. **76:** 51–58.
2. BIKLE, D. D., M. K. NEMANIC, E. GEE & P. ELIAS. 1986. 1,25-Dihydroxyvitamin D_3 production by human keratinocytes: Kinetics and regulation. J. Clin. Invest. **78:** 557–566.
3. KUROKI, T. 1985. Possible functions of $1\alpha,25$-dihydroxyvitamin D_3, an active form of vitamin D_3, in the differentiation and development of skin. J. Invest. Dermatol. **84:** 459–460.
4. SUDA, T., C. MIYAURA, E. ABE & T. KUROKI. 1986. Modulation of cell differentiation, immune responses and tumor promotion by vitamin D compounds. *In* Bone and Mineral Research, Vol. 4. W. A. PECK, Ed. Elsevier Science Publishers. Amsterdam, New York, Oxford.
5. HOSOMI, J., J. HOSOI, E. ABE, T. SUDA & T. KUROKI. 1983. Regulation of terminal differentiation of cultured mouse epidermal cells by $1\alpha,25$-dihydroxyvitamin D_3. Endocrinology **113:** 1950–1957.
6. SMITH, E. L., N. C. WALWORTH & M. F. HOLICK. 1986. Effect of $1\alpha,25$-dihydroxyvitamin D_3 on the morphologic and biochemical differentiation of cultured human epidermal keratinocytes grown in serum-free conditions. J. Invest. Dermatol. **86:** 709–714.
7. HORIUCHI, N., T. L. CLEMENS, A. L. SCHILLER & M. F. HOLICK. 1985. Detection and developmental changes of the $1,25\text{-}(OH)_2\text{-}D_3$-receptor concentration in mouse skin and intestine. J. Invest. Dermatol. **84:** 461–464.
8. STUMPF, W. E., M. SAR, F. A. REID, Y. TANAKA & H. F. DELUCA. 1979. Target cells for 1,25-dihydroxyvitamin D_3 in intestinal tract, stomach, kidney, skin, pituitary, and parathyroid. Science **206:** 1188–1190.
9. KARASAWA, M., J. HOSOI, H. HASHIBA, K. NOSE, C. TOHYAMA, E. ABE, T. SUDA & T. KUROKI. 1987. Regulation of metallothionein gene expression by $1\alpha,25$-dihydroxyvitamin D_3 in cultured cells and in mice. Proc. Natl. Acad. Sci. USA **84:** 8810–8813.
10. MORIMOTO, S. & Y. KUMAHARA. 1985. A patient with psoriasis cured by 1α-hydroxyvitamin D_3. Med. J. Osaka Univ. **35:** 51–54.
11. MORIMOTO, S., K. YOSHIKAWA, T. KOZUKA, Y. KITANO, S. IMANAKA, K. FUKUO, E. KOH & Y. KUMAHARA. 1986. An open study of vitamin D_3 treatment in psoriasis vulgaris. Br. J. Dermatol. **115:** 421–429.
12. MACLAUGHLIN, J. A., W. GANGE, D. TAYLOR, E. SMITH & M. F. HOLICK. 1985. Cultured

psoriatic fibroblasts from involved and uninvolved sites have a partial but not absolute resistance to the proliferation-inhibition activity of 1,25-hydroxyvitamin D_3. Proc. Natl. Acad. Sci. USA **82:** 5409–5412.

13. KONDO, S., Y. SATOH & T. KUROKI. 1987. Defect in UV-induced unscheduled DNA synthesis in cultured epidermal keratinocytes from xeroderma pigmentosum. Mutation Res. **183:** 95–101.

14. KUROKI, T., K. SASAKI, K. CHIDA, E. ABE & T. SUDA. 1983. 1α,25-Dihydroxyvitamin D_3 markedly enhances chemically-induced transformation in BALB 3T3 cells. Gann **74:** 611–614.

15. SASAKI, K., K. CHIDA, H. HASHIBA, N. KAMATA, E. ABE, T. SUDA & T. KUROKI. 1986. Enhancement by 1α,25-dihydroxyvitamin D_3 of chemically induced transformation of BALB 3T3 cells without induction of ornithine decarboxylase or activation of protein kinase C. Cancer Res. **46:** 604–610.

16. HOSOI, J., K. KATO & T. KUROKI. 1987. Induction of anchorage-independent growth of mouse JB6 cells by cholera toxin. Carcinogenesis **8:** 377–380.

17. HUH, N., M. SATOH, K. NOSE, E. ABE, T. SUDA, M. RAJEWSKY & T. KUROKI. 1987. 1α,25-Dihydroxyvitamin D_3 induces anchorage-independent growth and c-Ki-*ras* expression of BALB/3T3 and NIH/3T3 cells. Jpn. J. Cancer Res. (Gann) **78:** 99–102.

18. CHIDA, K., H. HASHIBA, T. SUDA & T. KUROKI. 1984. Inhibition by 1α,25-dihydroxyvitamin D_3 of induction of epidermal ornithine decarboxylase caused by 12-*O*-tetradecanoylphorbol-13-acetate and teleocidin B. Cancer Res. **44:** 1387–1391.

19. HASHIBA, H., M. FUKUSHIMA, K. CHIDA & T. KUROKI. 1987. Systemic inhibition of tumor promoter-induced ornithine decarboxylase in 1α-hydroxyvitamin D_3-treated animals. Cancer Res. **47:** 5031–5035.

20. CHIDA, K., H. HASHIBA, M. FUKUSHIMA, T. SUDA & T. KUROKI. 1985. Inhibition of tumor promotion in mouse skin by 1α,25-dihydroxyvitamin D_3. Cancer Res. **45:** 5426–5430.

21. WOOD, A. W., R. L. CHANG, M. T. HUANG, M. USKOKOVIC & A. H. CONNEY. 1983. 1α,25-Dihydroxyvitamin D_3 inhibits phorbol ester-dependent chemical carcinogenesis in mouse skin. Biochem. Biophys. Res. Commun. **116:** 605–611.

22. MCDONNELL, D. P., D. J. MANGELSDORF, J. W. PIKE, M. R. HAUSSLER & B. W. O'MALLEY. 1987. Molecular cloning of complementary DNA encoding the avian receptor for vitamin D. Science **235:** 1214–1217.

23. BURMESTER, J. K., N. MAEDA & H. F. DELUCA. 1988. Isolation and expression of rat 1,25-dihydroxyvitamin D_3 receptor cDNA. Proc. Natl. Acad. Sci. USA **85:** 1005–1009.

24. KESSLER, M. A., L. LAMM, K. JARNAGIN & H. F. DELUCA. 1986. 1,25-Dihydroxyvitamin D_3-stimulated mRNAs in rat small intestine. Arch. Biochem. Biophys. **251:** 403–412.

Keratinocytes Convert Thyroxine to Triiodothyronine[a]

MICHAEL M. KAPLAN,[b,d] PHILIP R. GORDON,[c]
CHANGYU PAN,[b] JENN-KUEN LEE,[b] AND
BARBARA A. GILCHREST[c]

[b]Endocrinology Division
Department of Medicine
New England Medical Center Hospital
and
Tufts University School of Medicine
Boston, Massachusetts 02111
and
[c]United States Department of Agriculture
Human Nutrition Research Center on Aging
Tufts University
Boston, Massachusetts 02111

Abundant evidence indicates that 3,5,3'-triiodo-L-thyronine (T3) initiates most thyroid hormone effects.[1] Inasmuch as 70–80% of the body's T3 is produced extrathyroidally by 5'-deiodination of L-thyroxine (T4),[1,2] deiodination reactions are crucial steps in the expression of thyroid hormone action. Several enzymatic iodothyronine deiodination (ID) pathways, designated ID-I, ID-Ia, ID-II, and ID-III, are common to humans and rats (TABLE 1). Each ID pathway is capable of either producing or degrading T3. These pathways are sulfhydryl-requiring reduction reactions, defined operationally by their biochemical characteristics.[3,4] The number of different enzymes involved in these pathways is not known, because no one has yet reported the purification of any deiodinase in catalytically active form, despite more than a decade of work in many laboratories. The main properties of the deiodinating pathways, reviewed in detail in several recent publications,[1,3–11] are compared in TABLE 1.

ID-I is thought to contribute an important amount of the daily extrathyroidal T3 production, as shown by the rapid, substantial fall in serum T3 caused by treatment with 6-n-propyl-2-thiouracil (PTU) to rats and humans whose only source of T4 and T3 is exogenous T4.[12,13] The relationship of ID-Ia[5] to ID-I is not clear: they may be catalyzed by different forms of the same enzyme. In vitro ID-I activities are highest in rat liver and kidney of all extrathyroidal organs, from which it has been inferred that these organs are the main sites of extrathyroidal T3 production. However, in vivo rat and human studies suggest that 40–60% of extrathyroidal T4 to T3 conversion may occur in sites other than liver and kidney.[2] Moreover, since PTU treatment results in only partial inhibition of extrathyroidal T3 production,[12,13] the PTU inhibitable pathways ID-I and ID-Ia are probably not the only important sources of extrathyroidal T3 production.

[a]Supported in part by a grant from the Tufts University Charlton Fund and by the United States Department of Agriculture Agricultural Research Service.
[d]Address for correspondence: Farmbrook Medical Two, Suite 303, 29877 Telegraph Rd., Southfield, MI 48034.

TABLE 1. Pathways of Iodothyronine Deiodination in Rat and Man

	Deiodination Pathway			
	I	Ia	II	III
Defining characteristics				
Substrate preference[a]	rT3 >T4 >T3	T4 = rT3	T4 >rT3	T3 >T4
Deiodination locus (ring)	phenolic, tyrosyl	phenolic	phenolic only	tyrosyl only
Km for T4[b]	~1 μM	~1 nM	<10 nM	50 nM
Ki for PTU	~1 μM	~100 μM	>1 mM	>1 mM
Reaction kinetics	ping-pong	sequential	sequential	sequential
Other characteristics				
Response to hypothyroidism	↓ except thyroid[c]	→	↑	↓
Subcellular localization	particulate	particulate	particulate	particulate
Thiol dependent	yes	yes	yes	yes
Inhibition by iopanoate	yes	yes	yes	yes
Rat tissue sites	liver, kidney, thyroid, many others	kidney, other?	brain, pituitary, brown fat placenta	brain, eye, placenta, skin
Human tissue sites	liver, kidney, thyroid	?	placenta, keratinocytes	placenta

[a]Determined by the ratio of the apparent maximal reaction velocity (Vmax) to the apparent Michaelis constant (Km).
[b]Determined in microsomal preparations, at 1–5 mmol/L dithiothreitol for ID-I and ID-Ia, and at 10–50 mM dithiothreitol for ID-II and ID-III.
[c]Regulated by thyroid-stimulating hormone in the thyroid gland, by thyroid hormone elsewhere.

ID-II has thus far been demonstrated only in rat central nervous system, anterior pituitary, brown adipose tissue, rat and human placenta, and human keratinocytes.[6,7,14] When ID-II is assayed using physiological T4 concentrations, its rates exceed ID-I rates in intact cells, homogenates or microsomes from these organs. In euthyroid rats, local T3 production by ID-II appears to be a major source of intracellular T3 in the central nervous sytem, anterior pituitary, and brown fat;[15,16] moreover, nearly all the T3 in the hypothyroid rat brain appears to be produced via this pathway.[15] Stimulation of ID-II by hypothyroidism thus appears to be an important mechanism by which intracellular T3 content and thyroid hormone-dependent processes in several rat tissues are at least partially defended in mild-to-moderate T4 deficiency. The applicability of these findings to man are discussed below.

ID-III has been identified in the central nervous system, placenta, and, probably, rat skin.[7-10] As a route of T3 disposal, it may act in a coordinated fashion with ID-II in the hypothyroid brain, to slow T3 degradation while T3 production is increased,[6,17] and thereby to defend intracellular T3. ID-III also converts T4 to rT3. The very high brain ID-III activity in the fetal and newborn rat brain may protect the central nervous system against too much T3 before the appropriate time in brain development.[6] Placental ID-III may be an important, or the sole, reason that the placenta is largely impermeable to T4 and T3 in late pregnancy.[8]

Generalizations about thyroid hormone metabolism from rat studies to man are made on the assumption that there are no important species differences. This assumption is risky, and probably incorrect for ID-I responses to fasting and glucocorticoids, as reviewed.[8] Hoping to develop human model systems, we screened a variety of cell types, and found conversion of T4 to T3 by human epidermal keratinocytes. The following studies were then performed to characterize the reaction pathway, with the goals of verifying the extent to which deiodination studies in the rat apply to human physiology and clinical medicine and to determine potentially important sites of T3 autoregulation by ID-II in man.

METHODS

Epidermal keratinocyte cultures were established by a modification of the method of Rheinwald and Green.[18] Foreskins from normal infants were obtained within 4 hours of elective circumcision, and punch biopsies were taken from the arm skin of adult volunteers. After removing the subcutaneous fat and deep dermis, the remaining epidermis and dermis were cut into small pieces and incubated overnight at 4°C in phophosaline containing 0.25% trypsin and 0.5 mmol/L EDTA. The epidermis was then mechanically removed from the dermis, and epidermal fragments were vortexed, pelleted by centrifugation and resuspended in Dulbecco's Modified Eagle's Medium supplemented with 20% fetal calf serum and 1.4 μmol/L cortisol. 10^5 cells were plated in 60 mm plastic dishes over a lethally irradiated feeder layer of mouse 3T3 teratoma cells (neither viable nor irradiated 3T3 cells homogenates deiodinated T4). Cultures were maintained at 37°C in 8% CO_2 and 92% air and the media were changed twice weekly. After approximately 3 weeks the keratinocytes had formed confluent stratified cultures, and were used at this time for deiodination studies.

Iodothyronine depleted fetal calf serum was prepared by treating the serum with Bio-Rad AG 1×8 anion exchange resin. By radioimmunassay analysis, culture media containing untreated and iodothyronine-depleted fetal calf serum had respective T4 concentrations of 28.4 and 1.5 nmol/L and T3 concentrations of 2.7 and 0.15 nmol/L. The 5'-deiodination assay for cell homogenates employed 3 hour incubations at 37°C, with 0.1 nmol/L [^{125}I]T4 in a buffer solution of potassium phosphate, pH 7.0—EDTA

1 mmol/L—dithiothreitol 15 mmol/L. Preliminary experiments using paper chromatographic product separation showed that the only labeled reaction products were T3 and iodide in equimolar amounts. Therefore, reaction rates were subsequently quantitated using the more sensitive and precise ion exchange method which measures radioactive iodide release from the labeled T4,[14] corrected to account for the unmeasured T3.

RESULTS

Incubation of confluent layers of intact keratinocytes with [^{125}I]T4 resulted in measurable 5'-deiodination to T3 in all cell lines studied (FIG. 1) lines 1–5 being from neonatal donors and line 6 from an adult donor. The conversion rate was inversely dependent on the albumin concentration in the medium. The albumin concentration was also inversely related to the cell/medium T4 ratio,[14] and the effect of albumin on

FIGURE 1. Conversion of T4 to T3 by intact keratinocytes. Confluent stratified cultures of keratinocytes were incubated for 48 h in medium containing 20% fetal calf serum depleted of iodothyronines by resin treatment, then for 2 days (*line 1*) or 3 days (*other lines*) in serum free medium containing 0.1 nmol/L [^{125}I]T4. *Line 1* was incubated in the absence of albumin while the medium for the *other lines* contained 0.5 or 1% bovine serum albumin. At the end of the incubation the cell pellets and the media were analyzed for [^{125}I]T3 by affinity chromatography using anti-T3 linked to agarose and by direct paper chromatography. (Adapted from REF. 14.)

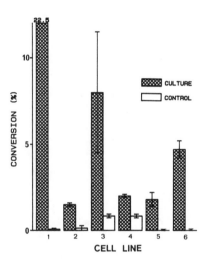

deiodination thus appears to be due to inhibition of T4 entry into the cells. In homogenates from 24 neonatal and 5 adult cell lines, the T4 5'-deiodination rates varied widely, but the means were similar in both groups, 44 and 51% T4 converted to T3/hr·mg protein. In both intact cells and homogenates, the only products of T4 metabolism were iodide and T3.

Enzyme kinetics were investigated with T4 and rT3 as substrates. Saturation curves are shown in FIGURE 2. When data from these and similar experiments were transformed into standard kinetic plots, the apparent Km values derived for T4 and rT3 were 11–14 nmol/L in keratinocyte homogenates and that for T4 was 8 nmol/L in a keratinocyte microsomal preparation. This microsomal preparation had an apparent Vmax of 0.4 pmol/hr·mg protein, 44% of the value for type II T4 5'-deiodination in microsomes from a human placental membrane assayed simultaneously. In one homogenate preparation in which both T4 and rT3 were tested as substrates, the ratio of apparent Vmax to apparent Km for T4 was 0.32, and that for rT3 was 0.19. T4 and rT3 each inhibited 5'-deiodination of the other, but neither propylthiouracil at 1

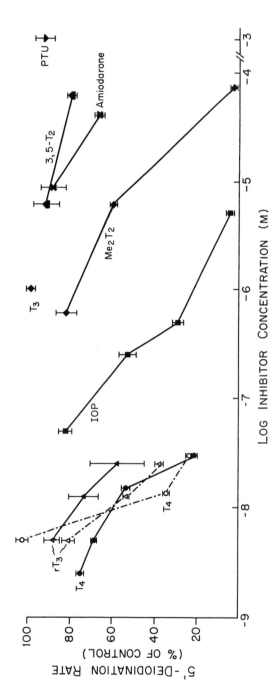

FIGURE 2. Effects of T4 analogs, PTU, iopanoic acid (IOP), and amiodarone on fractional 5'-deiodination of 0.1 nmol/L [^{125}I]T4 (*solid symbols and lines*) and 0.1 nmol/L [^{125}I]rT3 (*open symbols and broken lines*) by homogenates of human keratinocytes. Points denote the mean ±SD of triplicate determinations, expressed as percent of the control rate. (From Kaplan *et al.*[14] Reprinted by permission from the Endocrine Society).

mmol/L nor T3 at 1μmol/L had any important inhibitory effect on T4 5'-deiodination (FIG. 2). The above observations taken together establish that T4 5'-deiodination by human keratinocytes has the properties of ID-II, as previously established in rat tissues and human placental cells (TABLE 1).

Several other compounds inhibited T4 5'-deiodination in keratinocyte homogenates (FIG. 2): 3,5-diiodo-3',5'-dimethyl-L-thyronine (Me2T2), and two iodinated drugs in common clinical use, the oral cholecystographic agent iopanoic acid (Telepaque®) and the antiarrhythmic agent amiodarone (Cordarone®). However, the latter showed only partial inhibition at relatively high concentrations, being 200–300 times less potent an inhibitor than iopanoic acid.

Since ID-II is inversely related to ambient thyroid hormone concentrations *in vivo* and *in vitro* in several cell types, the effects of thyroid hormone depletion and repletion on keratinocyte deiodination were next determined (FIG. 3). Cells were cultured in the specified media for 2–4 days, then harvested, and deiodination measured in homogenates. Culturing the cells in medium with 20% thyroid hormone-depleted fetal calf serum for 3–4 days caused, on the average, a near doubling of type II T4 5'-deiodination rates. No change was seen after 2 days in the depleted medium. Adding back enough T4 to just restore that removed by the resin prevented any increase in T4 5'-deiodination. In contrast, adding an equimolar amount of T3, *i.e.*, using a greatly supraphysiological free T3 concentration, did not prevent the increase in T4 5'-deiodination.

Thus far we have not found measurable T4 5'-deiodination in fresh human skin samples, including homogenates of two trunk skin samples taken at autopsy from premature stillborn infants and one sample from a reduction mammoplasty, carried through the keratinocyte preparation to the point of plating, then homogenized instead, and direct homogenates of 4 foreskin samples containing both dermis and epidermis. Some skin from the mammoplasty sample also had the most superficial layer shaved off and homogenized without trypsinization; this was also devoid of detectable deiodinating activity.

DISCUSSION

ID-II has many properties pointing to an important role for it in T4 and T3 metabolism. The Km for T4 is in the physiological range of free T4, a typical property of regulatory enzymes. So far, wherever type II activity has been found, it increases in hypothyroidism, unlike all of the other deiodinating pathways. Regulation of type II activity by iodothyronines appears to be a nonnuclear process in man as it is in the rat, based on relative potencies of T4, T3 (*e.g.*, FIG. 3) and rT3.[7] Additional evidence for an extragenomic effect is found in *in vivo* rat studies of Leonard *et al.*[19] using cycloheximide and actinomycin-D. Furthermore, examination of the concentrations of T4, T3 and rT3 needed to modulate type II deiodination in human keratinocytes as well as in cultured human placental cells[7] reveals that only T4 is effective at physiological free hormone concentrations. Therefore, we propose that T4 is the primary effector molecule for type II deiodination. There are few, if any, other thyroid hormone actions for which T4 is the principal regulator, and it will be of considerable interest to determine the molecular mechanisms involved.

Type II deiodination in keratinocytes has potential physiological significance for epidermal tissue in particular. Skin is well known to be thyroid hormone-responsive, becoming myxedematous in hypothyroidism, as manifested by increased production and deposition of glycosaminoglycans, principally hyaluronic acid,[20] and by other, less

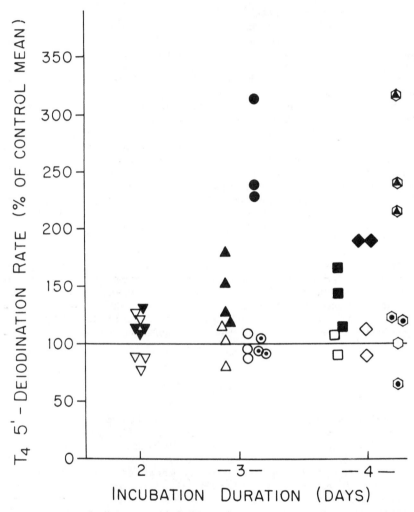

FIGURE 3. Effects of thyroid hormone depletion on T4 5'-deiodination rates in keratinocyte homogenates. Cells were cultured in medium with 20% fetal calf serum (control wells, *open symbols*), medium with 20% fetal calf serum depleted of iodothyronines by resin treatment (*closed symbols*), medium with iodothyronine depleted serum to which a normal T4 concentration (27 nmol/L of medium) was restored (*open symbols with central dot*), and medium with iodothyronine depleted serum to which an equimolar, thus supraphysiological, T3 concentration was added (*open symbols with central triangle*). The points represent rates in homogenates from individual wells, and each symbol shape indicates cells from a different donor. (From Kaplan *et al.*[14] Reprinted by permission from the Endocrine Society.)

well characterized changes. Human keratinocytes cultured in low thyroid hormone medium have reduced rates of acetate incorporation into sterols, sterol esters, free fatty acids, and triglycerides.[21] Studies in the rat indicate that organs vary in the relative contribution of plasma T3 and locally produced T3 to total intracellular T3, and that the tissues in which type II deiodination has been identified depend most heavily on local T3 production.[15] Thus, human keratinocyte type II T4 5'-deiodination may be an important source of T3 in the skin, and if so, there could be pathological conditions which selectively affect skin T4 5'-deiodination, creating a localized hypothyroid state.

Type II deiodination in epidermal cells may also have an important role in systemic thyroid hormone economy in man. Blood pool T4 and T3 kinetics in man are well described by a three pool model, consisting of the circulating pool, a pool thought to consist of liver and kidney that equilibrates rapidly with the circulating pool (fast pool), and a pool that equilibrates slowly with the circulating pool (slow pool).[2] A further analysis of blood pool kinetics indicates that a large fraction of extrathyroidal T3 production probably occurs in the slow tissue pool,[2] and skin is the second largest organ in this pool after muscle.

Evidence in man suggests that normal serum T3 concentrations, like rat brain T3, are defended ("autoregulated") in hypo- and hyperthyroxinemic states in which thyroid secretion is negligible.[22-24] Also, hypothyroidism in the rat results in reduced type I deiodination in homogenates of most extrathyroidal tissues,[8] whereas whole-body fractional conversion of T4 to T3 in hypothyroid humans and rats is either unchanged or increased.[25,26] Though ID-II is a likely mechanism for this autoregulation, the anatomic sites where this occurs in man are obscure: the existence of functional brown fat in adult humans is controversial, the central nervous system and pituitary probably cannot produce enough T3 to change circulating levels, and pregnancy is not required for this adaptation. Given the large mass of the epidermis, the present results suggest that skin might be a potentially important source of circulating T3. Moreover, the reported properties of iodothyronine deiodination in other human tissues besides placenta, including liver, kidney, fibroblasts, and peripheral blood granulocytes, are not those of ID-II, as reviewed.[14]

Although we have not yet shown type II deiodination in fresh skin, this may have been prevented by technical problems. Homogenizing whole skin could dilute type II activity with many other dermal and interstitial proteins, some of which may be inhibitory. Trypsinization could also reduce activity. Thus, at this time caution must be applied in generalizing our findings to the *in vivo* situation. Even if skin proves not to be a major source of circulating T3 in man, cultured keratinocytes are at minimum a useful model system for studying human type II iodothyronine deiodination. For example, the great difference in inhibitory potency of iopanoic acid and amiodarone, despite their similar effects (at similar oral doses) on decreasing the serum T3/T4 ratio suggests that amiodarone may act by mechanisms in addition to inhibiting deiodination, perhaps altering cellular uptake as well.[27] Future studies should illuminate other aspects of the metabolism of thyroid hormones and T3 homeostasis in the skin.

REFERENCES

1. DeGroot, L. J., P. R. Larsen, S. Refetoff & J. B. Stanbury. 1984. The Thyroid and Its Diseases. 5th edit. 36–118. John Wiley & Sons. New York, NY.
2. DiStefano, J. J. III. 1986. Modeling approaches and models of the distribution and disposal of thyroid hormones. *In* Thyroid Hormone Metabolism. G. Hennemann, Ed. 39–76. Marcel Dekker. New York, NY.

3. LEONARD, J. L. & T. J. VISSER. 1986. Biochemistry of deiodination. *In* Thyroid Hormone Metabolism. G. Hennenman, Ed. 189–229. Marcel Dekker. New York, NY.

4. HESCH, R.-D. & J. KOEHRLE. 1986. Intracellular pathways of iodothyronine metabolism. *In* Werner's The Thyroid. A Fundamental and Clinical Text. 5th edit. S. H. Ingbar & L. E. Braverman, Eds. 154–200. J. B. Lippincott. Philadelphia, PA.

5. GOSWAMI, A. & I. N. ROSENBERG. 1984. Iodothyronine 5'-deiodinase in rat kidney microsomes: kinetic behavior at low substrate concentrations. J. Clin. Invest. **74:** 2097–2106.

6. KAPLAN, M. M. 1984. The role of thyroid hormone deiodination in the regulation of hypothalamo-pituitary-thyroid function. Neuroendocrinology **38:** 254–260.

7. HIDAL, J. T. & M. M. KAPLAN. 1985. Characteristics of thyroxine 5'-deiodination in cultured human placental cells. Regulation by iodothyronines. J. Clin. Invest. **76:** 947–955.

8. KAPLAN, M. M. 1986. Regulatory influences on iodothyronine deiodination in animal tissues. *In* Thyroid Hormone Metabolism. G. Hennemann, Ed. 231–253. Marcel Dekker. New York, NY.

9. FAY, M., E. ROTI, S. L. FANG, G. WRIGHT, L. E. BRAVERMAN & C. H. EMERSON. 1984. The effects of propylthiouracil, iodothyronines, and other agents on thyroid hormone metabolism in human placenta. J. Clin. Endocrinol. Metab. **58:** 280–286.

10. KAPLAN, M. M. & E. A. SHAW. 1984. Type I iodothyronine deiodination by human and rat placenta *in vitro*. J. Clin. Endocrinol. Metabl. **59:** 253–257.

11. HUANG, T., I. J. CHOPRA, A. BEREDO, D. H. SOLOMON & G. N. CHUA TECO. 1985. Skin is an active site for the inner ring monodeiodination of thyroxine to 3,3',5'-triiodothyronine. Endocrinology **117:** 2106–2113.

12. GEFFNER, D. D., M. AZUKIZAWA & J. H. HERSHMAN. 1975. Propylthiouracil blocks extrathyroidal conversion of thyroxine to triiodothyronine and augments thyrotropin secretion in man. J. Clin. Invest. **55:** 224–229.

13. SABERI, M., F. H. STERLING & R. D. UTIGER. 1975. Reduction in extrathyroidal triiodothyronine production by propylthiouracil in man. J. Clin Invest. **55:** 218–223.

14. KAPLAN, M. M., C. PAN, P. R. GORDON, J.-K. LEE & B. A. GILCHREST. 1988. Human epidermal keratinocytes in culture convert thyroxine to 3,5,3'-triiodothyronine by type II iodothyronine deiodination. A novel endocrine function of the skin. J. Clin. Endocrinol. Metab. **66:** 815–822.

15. SILVA, J. E. & P. R. LARSEN. 1986. Regulation of thyroid hormone expression at the prereceptor and receptor level. *In* Thyroid Hormone Metabolism. G. Hennemann, Ed. 441–500. Marcel Dekker. New York, NY.

16. SILVA, J. E. & P. R. LARSEN. 1985. Potential of brown adipose tissue type II thyroxine 5'-deiodinase as a local and systemic source of triiodothyronine in rats. J. Clin. Invest. **76:** 2296–2305.

17. SILVA, J. E. & P. S. MATTHEWS. 1984. Production rates and turnover of triiodothyronine in rat-developing cerebral cortex and cerebellum. Responses to hypothyroidism. J. Clin. Invest. **74:** 1035–1049.

18. RHEINWALD, J. G. & H. GREEN. 1975. Serial cultivation of strains of human epidermal keratinocytes: the formation of keratinizing colonies from single cells. Cell **6:** 331–344.

19. LEONARD, J. L., J. E. SILVA, M. M. KAPLAN, S. A. MELLEN, T. J. VISSER & P. R. LARSEN. 1984. Acute post-transcriptional regulation of cerebrocortical and pituitary 5'-deiodinase by thyroid hormones. Endocrinology **114:** 998–1004.

20. SMITH, T. J., Y. MURATA, A. L. HORWITZ, L. PHILIPSON & S. REFETOFF. 1982. Regulation of glycosaminoglycan synthesis by thyroid hormone *in vitro*. J. Clin. Invest. **70:** 1066–1073.

21. ROSENBERG, R. M., R. R. ISSEROFF, V. A. ZIBOH & A. C. HUNTLEY. 1986. Abnormal lipogenesis in thyroid hormone-deficient epidermis. J. Invest. Dermatol. **86:** 244–248.

22. LUM, S. M. C., J. T. NICOLOFF, C. A. SPENCER & E. M. KAPTEIN. 1984. Peripheral tissue mechanism for maintenance of serum triiodothyronine values in a thyroxine-deficient state in man. J. Clin. Invest. **73:** 570–575.

23. FISCHER, H. R. A., W. H. L. HACKENG, W. S. SCHOPMAN & J. SILBERBUSCH. 1982. Analysis of factors in hyperthyroidism, which determine the duration of suppressive

treatment before recovery of thyroid stimulating hormone secretion. Clin. Endocrinol. (Oxford) **16:**575–585.

24. BRAVERMAN, L. E., A. VAGENAKIS, P. DOWNS, A. E. FOSTER, K. STERLING & S. H. INGBAR. 1973. Effect of replacement doses of sodium L-thyroxine on peripheral metabolism of thyroxine and triiodothyronine in man. J. Clin. Invest. **53:** 1010–1017.

25. SILVA, J. E., M. B. GORDON, F. R. CRANTZ, J. L. LEONARD & P. R. LARSEN. 1984. Qualitative and quantitative differences in the pathways of extra throidal triiodothyronine generation between euthyroid and hypothyroid rats. J. Clin. Invest. **73:** 898–907.

26. LoPRESTI, J. S., D. W. WARREN, E. M. KAPTEIN, M. S. CROXSON & J. T. NICOLOFF. 1982. Urinary immunoprecipitation method for estimation thyroxine to triiodothyronine conversion in altered thyroid states. J. Clin. Endocrinol. Metab. **55:** 666–670.

27. KRENNING, E. P., R. DOCTER, B. BERNARD, T. VISSER & G. HENNEMANN. 1982. Decreased transport of thyroxine (T4), 3,3',5-triiodothyronine (T3) and 3,3',5'-triiodothyronine (rT3) into rat hepatocytes in primary culture due to a decrease of cellular ATP content and various drugs. FEBS Lett. **410:** 229–233.

Steroid Metabolism by Epidermal Keratinocytes[a]

LEON MILEWICH[b], CYNTHIA B. SHAW, AND RICHARD D. SONTHEIMER

Cecil H. and Ida Green Center for Reproductive Biology Sciences
and
Departments of Obstetrics-Gynecology and Dermatology
University of Texas Southwestern Medical Center
Dallas, Texas 75235-9051

INTRODUCTION

Precursor stem cells in epidermis are the progenitors of keratinocytes; after differentiation into terminal cells, proliferation ceases, cornified envelopes develop, nuclei are lost, and the cells are shed from the uppermost epidermal layer, *i.e.*, the stratum corneum. Epidermal keratinocytes can be maintained in cell culture for relatively long times; in culture, these cells mature and undergo terminal differentiation in a manner that resembles the *in vivo* situation. We found previously that epidermal keratinocytes, in culture, metabolize C_{19}- and C_{21}-Δ^4-3-oxo steroids primarily to the corresponding 5α-reduced metabolites.[1] Thus, testosterone, androstenedione, and progesterone are metabolized by keratinocytes to 5α-dihydrotestosterone (DHT), 5α-androstanedione, and 5α-dihydroprogesterone, respectively. Thus, 5α-reductase is the most active Δ^4-3-oxo-steroid-metabolizing enzyme present in keratinocytes. The primary metabolites, in turn, undergo further metabolism in these cells.[1] In addition to 5α-reductase, the other steroid-metabolizing enzyme activities identified in keratinocytes by product formation were 17β-hydroxysteroid oxidoreductase (17β-HSOR) and 3β-hydroxysteroid oxidoreductase (3β-HSOR). 17β-Hydroxysteroid oxidoreductase in keratinocytes catalyze the interconversion of testosterone and androstenedione, but at rates that are much lower than those leading to the formation of the 5α-reduced metabolites.[1] 3β-Hydroxysteroid oxidoreductase catalyze the conversion of the 5α-reduced metabolites of testosterone and progesterone to isoandrosterone and 3β-hydroxy-5α-pregnan-20-one, respectively.[1] Also, in the conduct of these studies we found that the specific activity of 5α-reductase increased as a function of keratinocyte time in culture, up to 3 weeks. Specifically, after 1 week in cell culture, keratinocytes converted testosterone to 5α-reduced metabolites at a rate of 0.4 nmol/1×10^6 cells $\cdot$ h; after 2 weeks in culture, the rate was 4.6 nmol/1×10^6 cells $\cdot$ h; and after 3 weeks the rate was increased to 9.0 nmol/1×10^6 cells $\cdot$ h.[1] Similar results were obtained with keratinocyte 17β-HSOR by use of testosterone as the substrate, where the specific activity of the enzyme increased with keratinocyte time in culture from 0.2 nmol/1×10^6 cells $\cdot$ h after 1 week in cell culture to 1.1 nmol/1×10^6 cells $\cdot$ h after 2 weeks, and 2.1 nmol/1×10^6 cells $\cdot$ h after 3 weeks.[1] A similar increase in the

[a]This investigation was supported in part by United States Public Health Service Grants 5-P50-HD-11149 and AM-01237.

[b]Address for correspondence: University of Texas Southwestern Medical Center, Department of Obstetrics and Gynecology, 5323 Harry Hines Boulevard, Dallas, TX 75235-9051.

specific activity of keratinocyte 5α-reductase with cell time in culture was demonstrated by use of either radiolabeled androstenedione or progesterone as substrates and, also, of 17β-HSOR by use of androstenedione as the substrate.[1] It is known that the specific activities of transglutaminase and plasminogen activator increase with terminal differentiation of human epidermal keratinocytes in culture.[2-4] Thus, it is possible that the increases in the specific activities of 5α-reductase and 17β-HSOR that are associated with increased keratinocyte time in culture are related directly to the process of terminal differentiation of these cells. At the present time, however, a functional role for the observed increases in specific activities of these enzymes in keratinocytes has not been established.

Testosterone serves as a prohormone for the *in situ* synthesis of DHT in skin; and, DHT appears to be the biologically active skin androgen.[5,6] Thus, the metabolism of testosterone by keratinocytes may result in a local increase in the concentration of DHT and, also, by diffusion from these cells, may lead to an increase in the concentration of this androgen in neighboring dermal mesenchymal cells, *e.g.,* fibroblasts. Dermal fibroblasts also convert testosterone to DHT[7,8] and, in addition, to estradiol-17β.[9] Thus, it is possible that DHT, which may be synthesized *in situ* in skin cells, plays a role in processes involved in normal proliferation and differentiation of keratinocytes.

The purpose of the present study was to extend the scope of our findings in regard to the identification of steroid-metabolizing enzymes present in human keratinocytes. We conducted *in vitro* studies with cultured keratinocytes by use of various radiolabeled steroid substrates to identify the metabolites formed by these cells. With this knowledge we were able to identify putative enzymes that catalyze steroid interconversions in these cells. The tritium-labeled steroid precursors used in this study were dehydroepiandrosterone sulfate (DS), estrone sulfate (E1S), dehydroepiandrosterone (DHEA), estrone (E1), estradiol-17β (E2), pregnenolone, DHT, and deoxycorticosterone (DOC). For purposes of comparison we also studied the metabolism of radiolabeled androstenedione by human dermal fibroblasts.

MATERIALS AND METHODS

Materials

Culture media, trypsin, and antibiotics were purchased from Grand Island Biological Co. (Grand Island, NY). Cholera toxin was purchased from Sigma Chemical Co. (St. Louis, MO), mouse epidermal growth factor was obtained from Collaborative Research (Waltham, MA), and bovine dermal collagen was purchased from Collagen Corp. (Palo Alto, CA). Silica gel precoated thin-layer chromatography (TLC) sheets (Polygram Sil G-HY) were purchased from Brinkmann Instruments, Inc. (Houston, TX).

Radiolabeled steroids, [1,2,6,7-^{3}H]androstenedione (SA, 90 Ci/mmol), [1,2-^{3}H]DHT (SA, 44 Ci/mmol), [7-(N)-^{3}H]pregnenolone (SA, 22 Ci/mmol), [1,2-(N)-^{3}H]DHEA, (SA, 51 Ci/mmol), [6,7-^{3}H]E1 (SA, 49 Ci/mmol), [6,7-^{3}H]E2 (SA, 45 Ci/mmol),[6,7-^{3}H]E1S ammonium salt (SA, 59 Ci/mmol),[7-^{3}H]DS ammonium salt (SA, 22.1 Ci/mmol), and [4-^{14}C]DHT, [4-^{14}C]pregnenolone, [4-^{14}C]DHEA, [4-^{14}C]E1, [4-^{14}C]E2, [4-^{14}C]androstenedione, and [4-^{14}C]testosterone (SA, 50 mCi/mmol), were purchased from New England Nuclear Corp., (Boston, MA). The substrates, except for DS and E1S, were purified by column partition chromatography on celite columns [isooctane-*tert*-butanol-methanol-water (10:4:4:2)] before incubation. [4-^{14}C]5α-Androstanedione, [4-^{14}C]isoandrosterone, [4^{14}C]androsterone, [4-

^{14}C]5α-androstane-3α,17β-diol,[4-^{14}C]5α-androstane-3β,17β-diol, and [4-^{14}C]5-an-
drostene-3β,17β-diol (SA, 50 mCi/mmol) were synthesized.[10,11] [1,2,6,7-^{3}H]DOC was
synthesized enzymatically by use of [1,2,6,7-^{3}H]progesterone (SA, 97 Ci/mmol) as
the substrate, NADPH as the cofactor, human fetal adrenal microsomes as the source
of 21-hydroxylase, and SU-10603 (Ciba-Geigy, Summit, NJ) to inhibit 17α-hydroxy-
lase activity in the microsomes.

DS sodium salt was purchased from Searle Co. (Arlington Heights, IL). E1S
potassium salt was purchased from Sigma Chemical Co. (St. Louis, MO), and the
other nonradiolabeled steroids were purchased from Steraloids, Inc. (Wilton, NH).

Epidermal Keratinocytes

Keratinocytes were obtained from breast skin of women undergoing reduction
mammoplasty, from foreskin of newborn infants, and from leg skin of a burned male
patient. Skin was kept in Hank's balanced salt solution containing antibiotics and
transported to the laboratory packed in ice. The outer layer of the skin (0.3 mm) was
shaved off with a Castroviejo keratome and incubated with 0.3% trypsin at 37°C for 1
h. The epidermis and dermis were separated, and the epidermis was subjected to
trypsin treatment for 5 min and swirled to loosen the keratinocytes. The cells were
washed three times with Hank's balanced salt solution to remove the trypsin and
thereafter were assessed for viability using a dye exclusion technique of either trypan
blue or a combination of ethidium bromide and fluorescein diacetate. Thereafter, the
cells were plated on plastic wells (6, 16, or 35 mm in diameter) that were coated with
collagen by the technique of Liu et al.,[12] using growth medium composed of McCoy's
5A, heat-inactivated fetal calf serum (10%), penicillin (400 U/ml), streptomycin (500
μg/ml), fungizone (1 μg/ml), cholera toxin (10^{-9} M), and mouse epidermal growth
factor (0.01 μg/ml). The number of viable cells added to the wells were as follows: 1 $\times$
10^5 cells to 6-mm wells, 3 $\times$ 10^5 cells to 16-mm wells, and 1 $\times$ 10^6 cells to 35-mm wells.
The medium was changed twice a week. Cell protein was determined by the method of
Lowry et al.,[13] using bovine serum albumin as the standard.

Skin Fibroblasts

The fibroblast strains used in these studies were established from explants of
forearm skin from two healthy subjects: a 24-year-old white woman and a 35-year-old
white male. These cells were the generous gift of Dr. James E. Griffin (University of
Texas Southwestern Medical Center).

Incubations

The media from wells containing keratinocytes were removed, and the cells were
washed twice with PBS solution, pH 7.4, and incubated with either [^{3}H]E1S, [^{3}H]DS,
[^{3}H]DHEA, [^{3}H]E1, [^{3}H]E2, [^{3}H]DHT, [^{3}H]pregnenolone or [^{3}H]DOC at 37°C in an
atmosphere of humidified air-CO_2 (95:5) either in Tris buffer, pH 7.4, supplemented
with glucose (1 mg/ml), Krebs-Ringer phosphate solution, pH 7.4, supplemented with
glucose (1 mg/ml) in serum-free RPMI-1640 medium supplemented with glucose (1
mg/ml). Steroid metabolism was evaluated as a function of incubation time and, in
some cases, with time that keratinocytes were maintained in culture, up to 4 weeks.
Single point incubations were conducted in these studies. Control incubations were as

described above, but were conducted in the absence of cells. At the end of each incubation period, the media was transferred to chilled Teflon-capped glass culture tubes. The remaining cells were covered with PBS solution, detached from the dishes by scraping with a rubber policeman, and transferred to the corresponding culture tubes. The dishes were rinsed with additional PBS solution, and the rinses were pooled with the corresponding cells and media. Appropriate authentic carbon-14-labeled steroids (internal recovery standards) were added to selected samples prior to extraction with organic solvent. The harvested cells and media were extracted first with 10 vol of mixture of chloroform-methanol (2:1, vol/vol) and then twice with chloroform, except for the samples used to determine sulfatase activity by an abbreviated technique. The pooled chloroform-methanol extracts were washed twice with water (2 ml) and used for identification and quantification of metabolites. The media from the control incubations were processed in a similar manner.

Identification and Quantification of Metabolites

The metabolites of [³H]DHT, [³H]pregnenolone, [³H]E1, [³H]E2, [³H]DHEA, and [³H]DOC that were present in the organic solvent extracts were separated initially by gradient elution chromatography using Celite-ethylene glycol columns.[14] Thereafter, authentic nonradiolabeled steroids (100 μg each) were added to the corresponding peaks of radioactivity, and the metabolites were purified further by TLC, as previously described.[15] The carrier steroids were visualized on the TLC plates by spraying with a water mist. The areas of the TLCs containing the steroids were removed by scraping and transferred to pasteur pipettes containing glass wool plugs, and the steroids were eluted with ethyl acetate (5 ml). Small aliquots of these eluates (0.3 ml/5) were assayed for radioactivity to monitor the ³H to ¹⁴C ratios. After additional purification by TLC as the free steroids, the metabolites were acetylated by use of a mixture of pyridine and acetic anhydride (1:1, vol/vol; 0.4 ml) at 24°C for 15 h and subjected to further purification by TLC.[15] 5α-Dihydrodeoxycorticosterone acetate was purified by TLC using methylene chloride-ethyl acetate (99:1, vol/vol) as the developing solvent mixture. Once a constant ³H to ¹⁴C ratio was obtained, the metabolites were mixed with corresponding nonradiolabeled steroids (50 mg each) and crystallized five times, as described.[15] 5α-Dihydrodeoxycorticosterone acetate was crystallized from acetone-petroleum ether (20–40°C). The constancy of the ³H to ¹⁴C ratios (or of the specific activities) was verified throughout the various purification steps to confirm the identity of the metabolites. The specific activities of the substrates used for the incubations together with the ³H to ¹⁴C ratios for each purified metabolite and the amounts of radioactivity in the [¹⁴C]steroids added as internal recovery standards were used to compute the corresponding rates of metabolite formation.

The metabolites obtained in some studies were quantified by an abbreviated technique that was described previously.[15] In brief, aliquots of the organic solvent extracts that contained the mixtures of substrate and metabolites (~250,000 dpm) were mixed with authentic carrier steroids (10 μg each) and applied to TLC plates. The solvent system used for development of the metabolites of [³H]DHT, [³H]DHEA, [³H]pregnenolone, [³H]androstenedione, [³H]E1 and [³H]E2 was a mixture of methylene chloride-ethyl acetate-methanol (85:15:1, vol/vol/vol).[15] The metabolite of [³H]DOC was separated using a mixture of isopropyl ether-ethyl acetate (6:4, vol/vol; two ascents). The carrier steroids were located by spraying with a mist of an acid mixture [acetic acid (100 ml), sulfuric acid (2 ml), and anisaldehyde (1 ml)] and heating at 105°C for 20 min. The amounts of radioactivity that migrated with each carrier steroid were expressed as a fraction of the total radioactivity recovered in each chromatographic lane. Radioactivities from the corresponding areas of chromatograms of control incubations were expressed as a fraction of the total radioactivity in each lane and substracted from corresponding areas of experimental samples. These fractional conversions and the specific activities of the tritium-labeled substrates used for incubation were used to compute the rates of metabolite formation.

The formation by keratinocytes of E1 and E2 by metabolism of [1,2,6,7-³H]androstenedione was investigated as described previously.[16,17]

Radioactivity was determined in a Packard Tri-Carb liquid scintillation spectrometer (model 3330, Downers Grove, IL) in scintillation fluid (12 ml) prepared with toluene (3.8 liters), Omnifluor (New England Nuclear, Boston, MA; 15.0 g), and methanol (76 ml). The counting efficiency was 40% for tritium and 68% for carbon-14.

Sulfatase Activity

For determination of steroid sulfatase activity, epidermal keratinocytes in culture were incubated in a mixture of Tris buffer (0.1 M; pH 7.4; 1.0 ml), glucose (5.6 mM), and either [³H]E1S (3.37 μCi; 15 μM) or [³H]DS (3.37 μCi; 15 μM) at 37°C with air-CO_2 (95:5) as the gas phase. Single point incubations were conducted in each study.

Identification of Metabolites of Estrone Sulfate and Dehydroepiandrosterone Sulfate

For purposes of metabolite characterization, appropriate ¹⁴C-labeled steroids, viz. [4-¹⁴C]E1, with [³H]E1S as the substrate, and [4-¹⁴C]DHEA, with [³H]DS as the substrate, were added as internal recovery standards to samples obtained in 4-h incubations. After chloroform extraction, the metabolites were separated and isolated by gradient elution chromatography on columns of Celite-ethylene glycol, as described previously.[14] Thereafter, the metabolites were mixed with appropriate nonradiolabeled steroid carriers (100 μg each) and purified further by TLC on silica gel-coated plastic plates.[18] After each purification step, the steroid markers on the TLC plates were detected by spraying with a water mist. The steroids were eluted from the adsorbent with ethyl acetate (5 ml), and aliquots (0.3 ml) were assayed for radioactivity to determine the ³H to ¹⁴C ratios. The developing solvent system used for the purification of the products of sulfatase action, E1 and DHEA, was made of a mixture of methylene chloride, ethyl acetate, and methanol (85:15:1, vol/vol/vol). The R_f values for E1 and DHEA were 0.78 and 0.52, respectively. These metabolites were eluted and acetylated by reaction at 24°C for 15 h with a mixture of pyridine and acetic anhydride (1:1, vol/vol; 0.4 ml). After evaporation of the acetylating mixture under a stream of nitrogen at 40°C, the acetylated derivatives were purified by TLC on silica gel-coated plates with a developing system of isooctane-ethyl acetate (6:4, vol/vol). The R_f values for estrone acetate and dehydroepiandrosterone acetate were 0.45 and 0.54, respectively. Thereafter, the acetylated metabolites were mixed with authentic steroid carriers (50 mg each) and crystalized five times from a mixture of ethyl ether-petroleum ether (boiling point, 20–40°C). The mother liquors and final crystals obtained were assayed for radioactivity, and the ³H to ¹⁴C ratios were computed. To calculate the rates of hydrolysis of [³H]E1S and [³H]DS in the samples analyzed by this technique, we took into consideration the ³H to ¹⁴C ratios, the [¹⁴C] radioactivity added with the internal recovery standard, the specific activity of the tritium-labeled substrate, and the number of cells used for incubation.

Quantification of Steroid Sulfatase Activity by an Abbreviated Technique

An abbreviated technique[18] was used to determine steroid sulfatase activity in keratinocytes. In brief, after incubation, the samples were treated with chloroform (three times with 5 ml) to extract the unconjugated steroids formed; thereafter, the pooled organic solvent extracts were washed repeatedly with water (five times with 5 ml) to remove the contaminating water-soluble substrate. Aliquots of the chloroform extracts (1 ml) were transferred to scintillation vials, the solvent was evaporated to dryness, and the residues were dissolved in scintillation fluid and, thereafter, assayed for radioactivity. After correcting for the total volume of chloroform used for extraction (15 ml), the tritium radioactivity in extracts of control incubations (without cells) was subtracted from the radioactivity in extracts obtained in incubations with cells. The radioactivities associated with the unconjugated steroid metabolites and the specific activity of the substrates used for incubation were used to compute steroid sulfatase activity.

Characterization of [³H]Androstenedione Metabolites Formed in Incubations with Fibroblasts

Fibroblasts at confluence were rinsed twice with PBS solution and, thereafter, incubated with [1,2,6,7-³H]androstenedione in RPMI-1640 serumless medium supplemented with glucose (1 mg/ml) at 37°C with air-CO_2 (95:5) as the gas phase. Control incubations were conducted in the absence of fibroblasts. After incubation, the media and cells were transferred to glass culture tubes, appropriate ¹⁴C-labeled steroids were added as internal recovery standards, and the steroids were extracted from the medium and cells by use of chloroform-methanol (2:1, vol/vol) and chloroform. The metabolites were separated initially by gradient elution column chromatography on Celite-ethylene glycol,[14] and were purified further by TLC, before and after acetylation, as described.[15] Finally, the metabolites were mixed with corresponding nonradiolabeled steroids (50 mg) and crystallized 5 times[15] to constant ³H to ¹⁴C ratios or constant specific activity to ascertain radiochemical homogeneity.

RESULTS

[³H]Estrone and [³H]Estradiol-17β Metabolism by Keratinocytes

The patterns of metabolism of [³H]E1 and [³H]E2 by cultured keratinocytes are illustrated in FIGURE 1A. The metabolism of E1 yielded E2 (FIG. 1A), and that of E2 resulted in the formation of E1 (FIG. 1C). The radiochemical homogeneities of these products were determined by successive purification steps that involved TLC, before and after acetylation, and crystallation (TABLE 1). The rates of metabolite formation were linear with incubation time up to 4 h, and the rate of E1 formation from E2 was approximately 10-fold greater than the rate of formation of E2 from E1 (FIG. 1E).

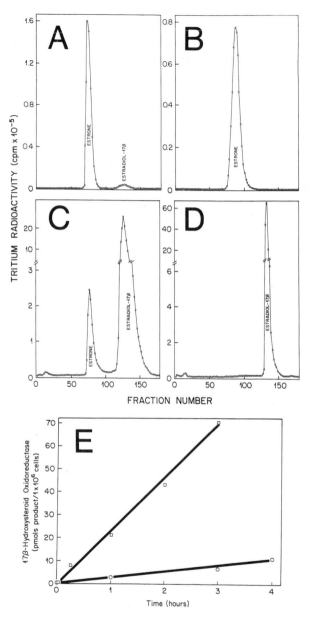

FIGURE 1. Chromatographic separation by gradient elution chromatography on Celite-ethylene glycol columns of metabolites of estrone and estradiol-17β. (**A**) The product, radiolabeled estradiol-17β, was produced by incubation of [6,7-^{3}H]estrone (1 μM; 1.35 μCi) with cultured epidermal keratinocytes for 3 h at 37°C. (**B**) Unmetabolized [6,7-^{3}H]estrone was recovered in a control incubation conducted as described above, but in the absence of cells. (**C**) Radiolabeled estrone was produced by incubation of [6,7-^{3}H]estradiol-17β (1 μM; 1.35 μCi) with cultured keratinocytes for 3 h at 37°C. (**D**) Unmetabolized [6,7-^{3}H]estradiol-17β was recovered in an incubation as described above, but conducted in the absence of keratinocytes. (**E**) Rates of [^{3}H]estrone formation from [6,7-^{3}H]estradiol-17β (1 μM; 1.35 μCi) (□——□) and [^{3}H]estradiol-17β formation from [6,7-^{3}H]estrone (1 μM; 1.35 μCi) (○——○) by cultured epidermal keratinocytes as a function of incubation time. The incubation medium was RPMI-1640 (1 ml) supplemented with glucose (1 mg).

TABLE 1. Steroid Metabolism by Human Epidermal Keratinocytes in Culture: Evidence for Radiochemical Homogeneity of Tritium-labeled Metabolites[a]

Substrate	Metabolite	[¹⁴C]Steroid Added Corresponding to the Metabolite (cpm)	$^3H{:}^{14}C$ Ratios[f]				
			Last TLC (after Acetylation)	Crystallization			
				ML3	ML4	ML5	Final Crystals
[³H]Estrone[b]	Estradiol-17β	3,130	9	9	9	11	13
[³H]Estradiol-17β[b]	Estrone	2,030	83	89	87	89	90
[³H]Dehydroepiandrosterone[c]	5-Androstene-3β,17β-diol	2,960	14	17	16	16	16
[³H]Dehydroepiandrosterone[c]	Androstenedione	1,500	1.7	1.8	1.7	1.8	1.6
[³H]Estrone sulfate[d]	Estrone	14,500	38	40	36	40	38
[³H]Dehydroepiandrosterone sulfate[d]	Dehydroepiandrosterone	9,100	11	12	12	11	11
[³H]5α-Dihydrotestosterone[e]	5α-Androstanedione	6,340	4.6	4.2	3.2	3.4	3.2
	Isoandrosterone	7,000	1.7	1.7	1.0	0.8	0.9
	5α-Androstane-3α,17β-diol	6,350	1.9	1.2	1.3	1.5	1.8
	5α-Androstane-3β,17β-diol	7,930	6.4	6.6	7.4	6.4	7.2

[a]Appropriate authentic carbon-14-labeled steroids were added to the samples after incubation. After extraction with organic solvents the products were separated by gradient elution chromatography on Celite-ethylene glycol columns and, thereafter, purified by TLC, before and after acetylation. Derivatized and underivatized metabolites were mixed with corresponding nonradiolabeled steroids (50 mg) and crystallized 5 times to constant $^3H{:}^{14}C$ ratios.

[b]Keratinocytes at confluence were incubated with either [6,7-³H]E1 (1 μM; 1.35 μCi) or [6,7-³H]E2 (1 μM; 1.35 μCi) in RPMI-1640 medium (1 ml) supplemented with glucose (1 mg) at 37°C for 3 h, with air-CO₂ (95:5) as the gas phase. The samples were processed as described in the text.

[c]Keratinocytes were incubated with [1,2-(N)-³H]DHEA (30 μM; 1.13 μCi) in 0.1 M Tris buffer (1 ml), pH 7.4, supplemented with glucose (1 mg) at 37°C for 18 h, with air-CO₂ (95:5) as the gas phase.

[d]Keratinocytes at confluence were incubated with either [6,7-³H]E1S (20 μM; 5.63 μCi) or [7-³H]DS (20 μM; 5.63 μCi) in 0.1 M Tris buffer (1 ml), pH 7.4, containing glucose (1 mg) at 37°C for 2 h, with air-CO₂ (95:5) as the gas phase. The products were processed as explained in the text.

[e]Keratinocytes at confluence were incubated with [1,2-³H]DHT (30 μM; 1.13 μCi) in RPMI-1640 medium (1 ml) containing glucose (1 mg) at 37°C for 24 h, with air-CO₂ (95:5) as the gas phase; the products were processed as described in the text.

[f]The ³H to ¹⁴C ratios obtained after the last TLC and those of the last three mother liquors (ML) and final crystals obtained by crystallization are given in the table.

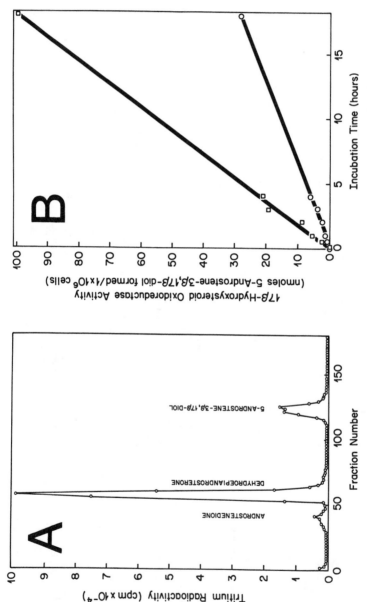

FIGURE 2. (A) Chromatographic pattern obtained by Celite-ethylene glycol column chromatography of the products of metabolism of [1,2-(N)-³H]dehydroepiandrosterone (30 μM; 1.13 μCi) produced by cultured epidermal keratinocytes. The incubation was conducted at 37°C for 18 h in 0.1 M Tris buffer (1 ml), pH 7.4, supplemented with glucose (1 mg). (B) Rate of formation of radiolabeled 5-androstene-3β,17β-diol by keratinocyte metabolism of [1,2-(N)-³H]dehydroepiandrosterone (30 μM; 1.13 μCi) as a function of incubation time, and of keratinocyte time in culture for 1 week (O——O) and 4 weeks (□——□).

[³H]Dehydroepiandrosterone Metabolism by Keratinocytes

The pattern of metabolite formation obtained by gradient elution chromatography on a Celite-ethylene glycol column of an organic solvent extract from cells and medium obtained after a 24-h incubation of cultured keratinocytes and [³H]DHEA is illustrated in FIGURE 2A. The major metabolite formed was the product of 17β-hydroxysteroid oxidoreductase activity, *viz.*, 5-androstene-3β,17β-diol, and a minor metabolite was androstenedione. Radiochemical homogeneity was achieved after purification by TLC, prior and after acetylation, and crystallization, as illustrated in TABLE 1. The rate of formation of 5-androstene-3β,17β-diol was linear with time of

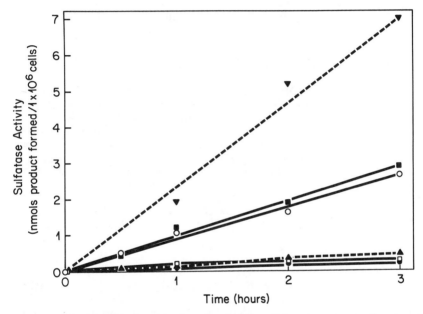

FIGURE 3. Steroid sulfatase activity in cultured epidermal keratinocytes as a function of incubation time. The substrates used were [7-³H]dehydroepiandrosterone sulfate (20 μM; 5.63 Ci) (□——□, foreskin keratinocytes; ●——●, breast skin keratinocytes; ▲---▲, leg skin keratinocytes) and [6,7-³H]estrone sulfate (20 μM; 5.63 μCi) (O——O, foreskin keratinocytes; ■——■, breast skin keratinocytes; ▼---▼, leg skin keratinocytes). The incubations were conducted in 0.1 M Tris buffer (1 ml), pH 7.4, supplemented with glucose (1 mg).

incubation up to 18 h, and the specific activity of 17β-hydroxysteroid oxidoreductase was greater in keratinocytes maintained in culture for 4 weeks compared with keratinocytes kept in culture for 1 week (100 nmols 5-androstene-3β,17β-diol formed/$1 \times 10^6 \cdot$ 18 h *vs* 27 nmols/$1 \times 10^6 \cdot$ 18 h) (FIG. 2B).

Steroid Sulfatase Activity in Cultured Keratinocytes

The products of the enzymatic hydrolysis of [³H]E1S and [³H]DS by epidermal keratinocytes in culture were [³H]E1 and [³H]DHEA. Product formation was not detected in control incubations without keratinocytes (data not shown). The metabolites were characterized by their isopolarity and isomorphism with authentic steroid standards; the constancy of the ^{3}H to ^{14}C ratios throughout the various chromatographic steps, either as the free steroids or as the acetates, and after five crystallizations served to establish their identity, as illustrated in TABLE 1.

The specific activity of keratinocyte steroid sulfatase, with E1S as the substrate, was approximately 5- to 14-fold greater compared with DS (FIG. 3). The rates of hydrolysis of DS by cultured keratinocytes obtained from either breast skin, foreskin or leg skin were approximately the same (FIG. 3), and those of foreskin and breast skin keratinocytes, determined by use of E1S as the substrate, also were comparable, while that of the enzyme in leg skin keratinocytes was two-fold higher than in cells obtained from foreskin or breast skin (FIG. 3). Sulfatase activity in these cells, with either E1S or DS as the substrate, was linear with incubation time up to 3 h (FIG. 3).

[^{3}H]5α-Dihydrotestosterone Metabolism by Keratinocytes

The pattern of [^{3}H]DHT metabolism by cultured epidermal keratinocytes is illustrated in FIGURE 4A, and that of nonmetabolized [^{3}H]DHT recovered from a control incubation is presented in FIGURE 4B. The metabolites, 5α-androstane-3,17-dione, androsterone, isoandrosterone, 5α-androstane-3α,17β-diol, and 5α-androstane-3β,17β-diol were characterized either by the constancy of the ^{3}H to ^{14}C ratios obtained in successive purification steps or by constant specific activity (TABLES 1 and 2). The rates of formation of 5α-androstane-3α,17β-diol and 5α-androstane-3β,17β-diol, the primary metabolites of DHT metabolism, increased linearly with incubation time up to 24 h, while the levels of DHT fell linearly with incubation time (FIG. 4C). The specific activities of 3α-hydroxysteroid oxidoreductase (3α-HSOR) and 3β-HSOR did not appear to change with keratinocyte time in culture up to 3 weeks (FIG. 4C).

[^{3}H]Deoxycorticosterone Metabolism by Keratinocytes

The pattern of metabolism of [^{3}H]DOC by cultured epidermal keratinocytes is presented in FIGURE 5A, and that of nonmetabolized [^{3}H]DOC recovered from a control incubation is illustrated in FIGURE 5B: the only metabolite identified in this study was 5α-dihydrodeoxycorticosterone (TABLE 2). The rate of formation of 5α-dihydrodeoxycorticosterone was linear with time of incubation up to 4 h (FIG. 5C).

Estrogen Formation in Keratinocytes

Estrone and E2 were sought in the organic solvent extracts of media and cells obtained from epidermal keratinocytes incubated with [1,2,6,7-^{3}H]androstenedione. This substrate can be converted to estrogens that contain the same number of tritium atoms, since very little tritium loss occurs in the process of aromatization of [1,2,6,7-^{3}H]androstenedione (~1.83%),[16] but even with this sensitive methodology E1 and E2 could not be detected in incubations with keratinocytes (data not shown).

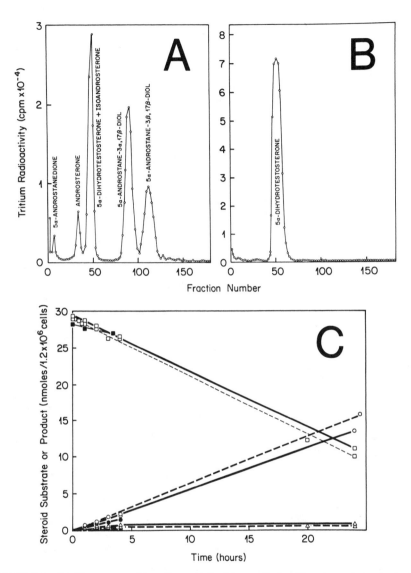

FIGURE 4. (A) Chromatographic pattern obtained by gradient elution chromatography on a Celite-ethylene glycol column of the products of metabolism of [1,2-³H]dihydrotestosterone (30 μM; 1.13μCi) produced by cultured epidermal keratinocytes. The medium used was RPMI-1640 (1 ml) supplemented with glucose (1 mg), and the incubation was conducted at 37°C for 24 h. (B) An incubation conducted as described above, but in the absence of keratinocytes, resulted in the recovery of unmetabolized [1,2-³H]dihydrotestosterone. (C) Rates of formation of 5α-androstane-3β,17β-diol plus 5α-androstane-3α,17β-diol, and of 5α-androstane-3,17-dione obtained in incubations of [1,2-³H]dihydrotestosterone (30 μM; 1.13 μCi) with cultured keratinocytes as a function of incubation time up to 24 h, and of keratinocyte time in culture up to 3 weeks: 5α-androstane-3β,17β-diol plus 5α-androstane-3α,17β-diol (●——●, 1 week; ○——○, 2 weeks; ○---○, 3 weeks); 5α-androstane-3,17-dione (▲——▲, 1 week; △——△, 2 weeks; △---△, 3 weeks). Recovered 5α-dihydrotestosterone (■——■, 1 week; □——□, 2 weeks; □---□, 3 weeks).

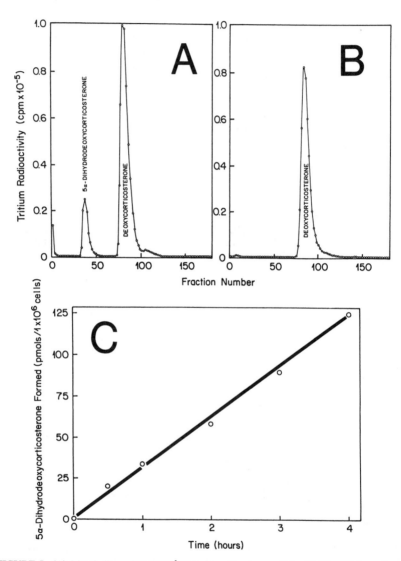

FIGURE 5. (A) Metabolism of [1,2,6,7-³H]deoxycorticosterone (0.2 μM; 19 μCi) by cultured epidermal keratinocytes in RPMI-1640 medium (2 ml) supplemented with glucose (1 mg) obtained in a 4 h incubation at 37°C. Gradient elution on a Celite-ethylene glycol column was used to obtain the chromatographic pattern shown in the figure. (**B**) An incubation as above but conducted in the absence of keratinocytes resulted in the recovery of unmetabolized [1,2,6,7-³H]deoxycorticosterone. (**C**) Rate of formation of radiolabeled 5α-dihydrodeoxycorticosterone by metabolism of [1,2,6,7-³H]deoxycorticosterone (0.2 μM; 19μCi) by cultured epidermal keratinocytes as a function of incubation time.

TABLE 2. Steroid Metabolism by Human Epidermal Keratinocytes in Culture: Evidence for Radiochemical Homogeneity of Radiolabeled Metabolites[a]

Substrate	Product	Specific Activity (cpm/mg)					
		ML1	ML2	ML3	ML4	ML5	Final Crystals
[³H]5α-Dihydrotestosterone[b]	Androsterone	123	181	210	180	150	130
[³H]Deoxycorticosterone[c]	5α-Dihydrodeoxycorticosterone	2,010	2,590	1,930	2,350	2,510	2,250

[a]The radiolabeled products were purified by gradient elution chromatography on Celite-ethylene glycol columns and subsequently by TLC, before and after acetylation. The acetylated metabolites were mixed with the corresponding authentic nonradiolabeled steroids (50 mg) and crystallized 5 times to constant specific activity. The specific activities of the products in mother liquors (ML) and final crystals are given in the table.
[b]Keratinocytes were incubated with [1,2-³H]DHT (30 μM; 1.13 μCi) in RPMI-1640 medium (1 ml) containing glucose (1 mg) at 37°C for 24 h, with air-CO$_2$ (95:5) as the gas phase.
[c]Keratinocytes were incubated with [1,2,6,7-³H]DOC (0.2 μM; 19 μCi) in RPMI-1640 medium (2 ml) containing glucose (1 mg) at 37°C for 4 h, with air-CO$_2$ (95:5) as the gas phase.

[³H]Pregnenolone Metabolism by Keratinocytes

Pregnenolone was recovered unchanged in time course incubations up to 24 h with cultured epidermal keratinocytes (data not shown).

[³H]Androstenedione Metabolism by Dermal Fibroblasts

The pattern of [³H]androstenedione metabolism by forearm skin fibroblasts from a 35-year-old man, obtained by gradient elution chromatography on Celite-ethylene glycol of an organic solvent extract obtained from a sample incubated for 4 h, is illustrated in FIGURE 6A. The chromatographic pattern obtained from a control incubation is presented in FIGURE 6B. The products formed were 5α-androstanedione, androsterone, isoandrosterone, DHT, 5α-androstane-3α,17β-diol, 5α-androstane-3β,17β-diol, and testosterone. The same metabolites were formed during incubation of [³H]androstenedione with forearm skin fibroblasts from a 24-year-old woman. Radiochemical homogeneity of the isolated metabolites was established after successive TLC purifications, before and after acetylation, and by crystallization to constant ³H to ¹⁴C ratios (TABLE 3). The rates of metabolite formation (FIG. 6C) and the specific activities of the androstenedione-metabolizing enzymes (FIG. 7A) were higher with the fibroblasts of the 24-year-old woman than from those of the 35-year-old man. The enzymatic activities, however, when expressed as a percent of total androstenedione-metabolizing enzyme activities, were the same in both strains of fibroblasts, with 5α-reductase activity constituting ~58% of the total, 3α-HSOR ~30%, 17β-HSOR ~9%, and 3β-HSOR ~3% (FIG. 7B). By comparison, 5α-reductase determined in keratinocytes with either androstenedione, testosterone or progesterone as the substrates, constituted between 80 and 90% of the total steroid-metabolizing enzymatic activities in these cells[1] (FIG. 8).

DISCUSSION

In an attempt to gain some insight into the metabolic fate in cultured human keratinocytes of steroids known to be present in circulation and, probably, skin,[19] *viz.*, E1S, DS, E1, E2, DHEA, DHT pregnenolone, androstenedione, and DOC, we conducted *in vitro* studies to identify and quantify the metabolites of these steroids formed by these cells. Also, for purposes of comparison, we investigated the metabolism of radiolabeled androstenedione in human dermal fibroblasts.

The interconversion of E1 and E2 has been studied in various organs, including mouse and human skin.[20,21] In the present study, we demonstrated that E1 is converted to E2 by cultured keratinocytes at a rate that is approximately 10-fold lower than that of the conversion of E2 to E1.

The major metabolite of DHEA produced by keratinocytes was 5-androstene-3β,17β-diol. As demonstrated previously with testosterone and androstenedione as substrates,[1] the specific activity of 17β-HSOR with DHEA as the substrate was greatly increased with keratinocyte time in culture from 1 to 4 weeks. It is possible that 5-androstene-3β,17β-diol serves as a substrate for keratinocyte 3β-hydroxysteroid oxidoreductase-Δ⁵⁻⁴-isomerase to give testosterone; however, this reaction has not been studied.

Pregnenolone is not metabolized by keratinocytes. This is noteworthy in view of the fact that DHEA, also a 3β-hydroxy-Δ⁵-steroid, serves as substrate for keratinocyte 3β-hydroxysteroid oxidoreductase-Δ⁵⁻⁴-isomerase to produce androstenedione. It

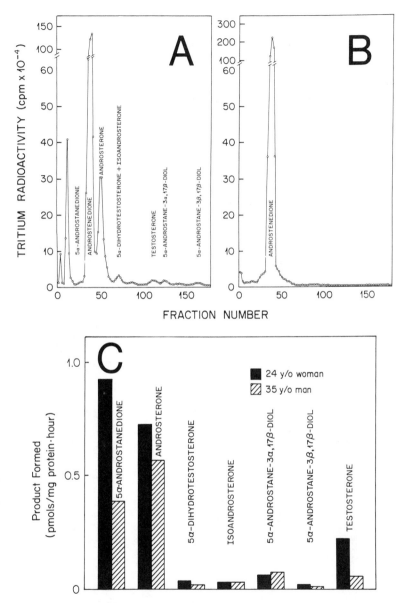

FIGURE 6. (A) Chromatographic separation by gradient elution chromatography on a Celite-ethylene glycol column of metabolites obtained by incubation of [1,2,6,7-³H]androstenedione (0.13 μM; 21 μCi) with forearm skin fibroblasts of a 35-year-old man. The incubation medium was RPMI-1640 (2 ml) supplemented with glucose (2 mg), and the incubation was conducted at 37°C for 4 h. (B) Chromatographic pattern obtained from an incubation as described above, but conducted in the absence of fibroblasts. (C) Rates of product formation obtained in 4 h incubations at 37°C of [1,2,6,7-³H]androstenedione (0.13 μM; 21 μCi) with forearm skin fibroblasts of a 35-year-old man and a 24-year-old woman.

TABLE 3. Androstenedione Metabolism by Forearm Skin Fibroblasts Obtained from a 35-Year-Old Man and a 24-Year-Old Woman: Evidence for Radiochemical Homogeneity of the Tritium-labeled Metabolites[a]

Fibroblasts Obtained from	Metabolite	[¹⁴C]Steroid Added (cpm)	³H:¹⁴C ratios				
			Last TLC (after Acetylation)	Crystallization			
				ML3	ML4	ML5	Final Crystals
male	5α-Androstanedione	13,000	13	10	9	9	10
female			24	24	22	21	21
male	Androsterone	9,000	20	20	20	19	19
female			30	31	29	29	26
male	5α-Dihydrotestosterone	2,600	3.9	2.3	2.1	2.1	2.1
female			4.1	3.7	4.4	3.9	3.5
male	Isoandrosterone	2,400	4.7	4.2	3.9	3.4	3.8
female			5.6	3.9	4.4	3.2	3.5
male	Testosterone	3,140	7.2	5.1	5.1	5.3	4.9
female			32	29	28	28	26
male	5α-Androstane-3α,17β-diol	2,580	15	10	11	12	11
female			9.5	9.5	9.0	8.4	8.4
male	5α-Androstane-3β,17β-diol	2,700	0.7	0.7	0.9	0.9	0.9
female			3.2	2.2	2.0	2.3	2.3

[a]Fibroblasts at confluence were incubated with [1,2,6,7-³H]androstenedione (0.13 μM; 21 μCi) in RPMI-1640 serumless medium enriched with glucose (5.6 mM) in a total volume of 2 ml. The incubations were conducted at 37°C for 4 h with air-CO₂ (95:5) as the gas phase. After incubation, appropriate authentic carbon-14-labeled steroids were added to the media, and the steroids were extracted with organic solvent, separated by column chromatography on celite-ethylene glycol columns, and purified by TLC, before and after acetylation, and thereafter, crystallized five times. The ³H to ¹⁴C ratios obtained after the last TLC and those of the last three mother liquors (ML) and final crystals are given.

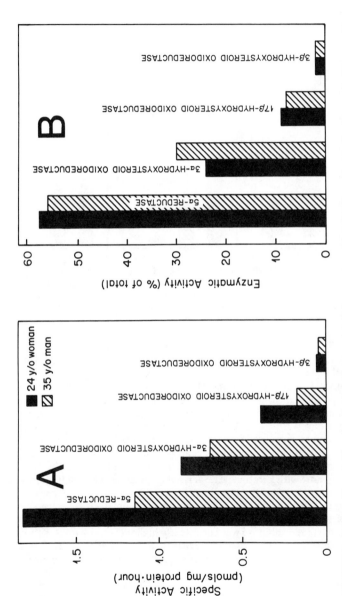

FIGURE 7. (A) Specific activities of androstenedione-metabolizing enzymes present in forearm skin fibroblasts of a 35-year-old man and a 24-year-old woman. Incubations with [1,2,6,7,-³H]androstenedione were conducted as described in FIGURE 6A. (B) The enzymatic activities in forearm skin fibroblasts of a 35-year-old man and a 24-year-old woman are expressed as a percent of total androstenedione-metabolizing enzymes in these cells.

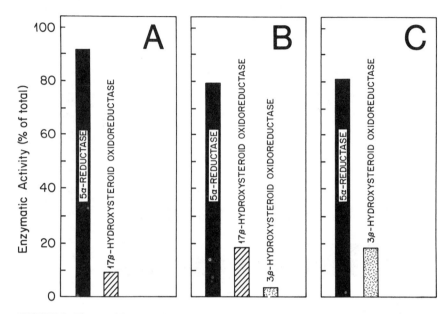

FIGURE 8. The steroid-metabolizing enzyme activities in cultured keratinocytes, which were identified by use of various substrates, are expressed as a percent of total enzymatic activities in these cells. (A) Substrate: [1,2,6,7-³H]androstenedione (2.0 µM; 21 µCi). (B) Substrate: [1,2,6,7-³H]testosterone (0.2 µM; 10 µCi). (C) Substrate: [1,2,6,7-³H]progesterone (0.5 µM; 8.4 µCi). (Based on data from REF. 1.)

would appear that the C-17 acetyl group of pregnenolone interferes with the metabolism of this steroid to produce progesterone in keratinocytes.

The steroid sulfoconjugates E1S and DS are found in relatively high concentrations in human peripheral circulation.[22,23] These conjugated steroids are weakly bound to plasma proteins and undergo enzymatic hydrolysis at peripheral tissues to give E1 and DHEA, respectively.[22,24–26] For example, the enzymatic hydrolysis of E1S and DS in a very large number of mouse tissues, including skin, has been demonstrated.[26] We found that steroid sulfatase activity also is expressed in cultured human keratinocytes derived from breast skin, foreskin, and leg skin. The specific activity of steroid sulfatase was the same with keratinocytes, derived from either breast skin, foreskin, or leg skin with DS as the substrate; however, the specific activity of the enzyme with E1S as the substrate was twice as great in keratinocytes obtained from leg skin of a man compared with that of breast skin or foreskin keratinocytes. In addition, the specific activity of steroid sulfatase was consistently higher (5- to 14-fold) with E1S as the substrate than with DS. The possibility exists that the rate differences obtained by use of these two substrates are a reflection of substrate preference by the enzyme or, otherwise, that there are two steroid sulfatases in keratinocytes that have differing substrate specificities, viz., one that catalyzes the hydrolysis of steroid alkyl sulfates, such as DS, and the other that catalyzes the hydrolysis of steroid aryl sulfates, such as E1S, as demonstrated in sheep brain tissue.[27]

The metabolism of DHT by human skin has been reported to result in the formation of 5α-androstane-3α,17β-diol glucuronide;[28] a direct correlation between serum concentrations of this steroid conjugate and hirsutism in women also has been demonstrated.[29] Because DHT is the major product formed by metabolism of testosterone in cultured keratinocytes,[1] we investigated the metabolic fate of DHT in these cells to find out whether 5α-androstane-3α,17β-diol was formed as a product. We were able to demonstrate that, in addition to 5α-reductase, 17β-HSOR and 3β-HSOR keratinocytes also express 3α-HSOR activity, as evidenced by the formation of 5α-androstane-3α,17β-diol and androsterone from DHT. Thus, we found that the primary pathways of DHT metabolism in keratinocytes involve the reduction of the 3-oxo group by the actions of 3α-HSOR and 3β-HSOR to produce 5α-androstane-3α,17β-diol and 5α-androstane-3β,17β-diol, respectively, and, also, the oxidation of the 17β-hydroxy group of DHT by 17β-HSOR action to give 5α-androstane-3,17-dione. The specific activities of 3α-HSOR and 3β-HSOR did not appear to change with keratinocyte time in culture (FIG. 4), contrary to what was found previously with 5α-reductase[1] and 17β-HSOR[1] (FIG. 9) .

The extraadrenal formation of DOC from plasma progesterone has been demonstrated in humans. During the third trimester of pregnancy the concentrations of DOC in maternal plasma are 4- to 50-fold greater than those found in nonpregnant women and men,[30] and reflect the increased levels of plasma progesterone in pregnancy. Also, during the luteal phase of the ovarian cycle in the human the plasma levels of DOC are markedly increased compared with those found in the follicular phase.[31] We studied the metabolic fate of DOC in cultured epidermal keratinocytes and found that the primary metabolite was 5α-dihydrodeoxycorticosterone: this finding served to ascertain the presence of 5α-reductase in these cells.

The extraglandular aromatization of androstenedione to E1 and the enzymatic hydrolysis of blood-borne E1S to E1, with subsequent metabolism of E1 to E2 *in situ* in various tissues[26,32,33] contribute to the maintenance of total plasma and tissue estrogen levels.[32] We attempted to demonstrate the presence of aromatase activity in cultured keratinocytes; however, this enzymatic activity was not expressed in these cells. This is in contrast with the presence of the enzyme in fibroblasts derived from either skin,[9]

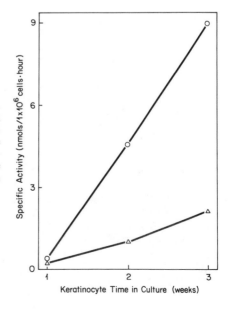

FIGURE 9. The specific activities of cultured epidermal keratinocyte 5α-reductase (O——O) and 17β-hydroxysteroid oxidoreductase ($\triangle$——$\triangle$), obtained by use of [1,2,6,7-^{3}H]testosterone (20 μM; 10 μCi) as the substrate, are expressed as a function of keratinocyte time in culture (Based on data from REF. 1.)

endometrial stroma,[33] adipose tissue stroma,[34] or prostate stroma.[8] Human lung fibroblasts, however, do not appear to have aromatase activity[35] and, thus, at the present time it is not possible to ascertain that keratinocytes from sites other than breast, leg, or foreskin do not express aromatase activity.

For comparison purposes we studied the metabolism of radiolabeled androstenedione by fibroblasts obtained from forearm skin of a 24-year-old woman and a 35-year-old man. The metabolites formed were 5α-androstanedione, androsterone, isoandrosterone, DHT, 5α-androstane-3α,17β-diol, 5α-androstane-3β,17β-diol, and testosterone. The rates of metabolite formation and the specific activities of the enzymes identified were higher with fibroblasts from the woman than from the man.

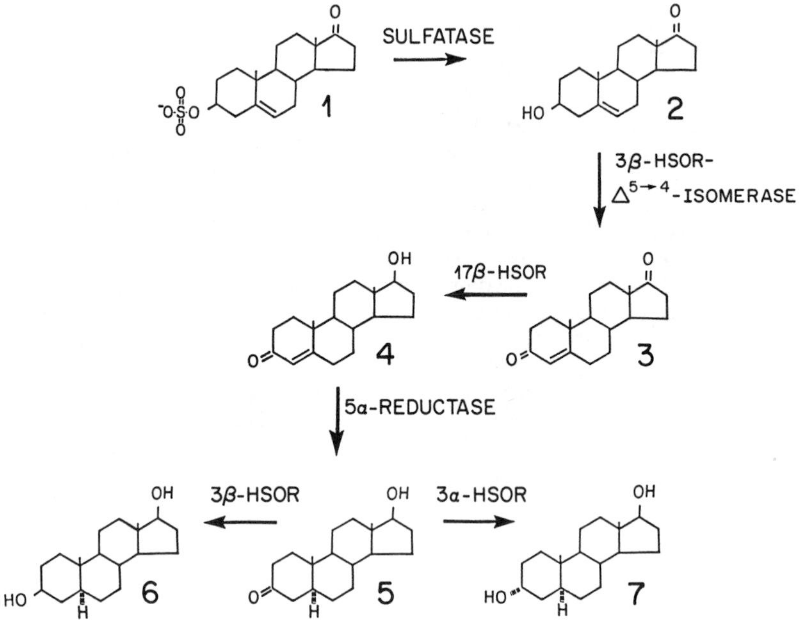

FIGURE 10. The metabolic pathway serves to illustrate the steroid-metabolizing enzymes identified in cultured epidermal keratinocytes and some of the metabolites that were isolated and characterized in these studies: (**1**) Dehydroepiandrosterone sulfate; (**2**) dehydroepiandrosterone; (**3**) androstenedione; (**4**) testosterone; (**5**) 5α-dihydrotestosterone; (**6**) 5α-androstane-3β,17β-diol; (**7**) 5α-androstane-3α,17β-diol.

The enzymatic activities, however, when expressed as a percent of total androstenedione-metabolizing enzyme activities, were the same in both fibroblast strains, with 5α-reductase activity constituting approximately 58%. By comparison, in keratinocytes, with either androstenedione, testosterone, or progesterone as the substrates, 5α-reductase constituted between 80 and 90% of total enzymatic activity.[1]

The present studies serve to ascertain that epidermal keratinocytes in culture express 5α-reductase, 17β-HSOR, 3β-HSOR, 3α-HSOR, 3β-hydroxysteroid oxidoreductase-Δ$^{5 \to 4}$-isomerase, and sulfatase activities, but no aromatase activity; the general pathways of steroid metabolism in cultured keratinocytes demonstrated in these studies are illustrated in FIGURE 10. The 17β-HSOR-catalyzed reaction, demonstrated by the interconversions of E1 and E2, androstenedione and testosterone, and DHEA and 5-androstene-3β,17β-diol, and those of 3β-HSOR (progesterone → 3β-hydroxy-5α-pregnan-20-one; DHT → 5α-androstane-3β,17β-diol) and 3α-HSOR (5α-DHT → 5α-androstane-3α,17β-diol) require the obligatory participation of reduced and oxidized cofactors, *viz.*, NAD(P)H and NAD(P). In addition, the 5α-reductase-catalyzed reaction (androstenedione → 5α-androstanedione; testosterone → DHT; progesterone → 5α-dihydroprogesterone; DOC → 5α-dihydrodeoxycorticosterone) requires NADPH as cofactor. These cofactors are generated intracellularly in intact cells, including keratinocytes and fibroblasts. Thus, the relative amounts of products reported in these studies are representative of both the intracellular levels of steroid-metabolizing enzymes found in the cultured epidermal keratinocytes as well as of the concentrations of reduced and oxidized cofactors.

SUMMARY

The metabolism of various radiolabeled steroids by cultured human epidermal keratinocytes was studied in an attempt to identify the steroid-metabolizing enzymes present in these cells. Sulfatase activity was demonstrated in keratinocytes with either E1S or DS as substrates. The products of sulfatase action were E1 and DHEA, respectively. The specific activity of the enzyme was approximately 5- to 14-fold greater with E1S as the substrate compared with DS, and the rates of hydrolysis were linear with incubation time up to 3 h. The metabolism of DHEA by the keratinocyte 17β-HSOR-catalyzed reaction resulted in the predominant formation of 5-androstene-3β,17β-diol. The rate of formation of 5-androstene-3β,17β-diol was linear with time of incubation up to 18 h, and the specific activity of 17β-HSOR, with DHEA as the substrate, was greater in keratinocytes maintained in culture for 4 weeks compared with keratinocytes kept in culture for 1 week. Androstenedione was a minor product of DHEA metabolism. The metabolism of DHT by epidermal keratinocytes resulted in the formation of 5α-androstanedione, 5α-androstane-3α,17β-diol, 5α-androstane-3β,17β-diol, androsterone, and isoandrosterone: the rates of formation of 5α-androstane-3α,17β-diol and 5α-androstane-3β,17β-diol were linear with incubation time up to 24 h, and the specific activities of 3α-HSOR and 3β-HSOR did not appear to change with keratinocyte time in culture up to 3 weeks. The metabolism of DOC by epidermal keratinocytes resulted in 5α-dihydrodeoxycorticosterone production: the rate of formation of this metabolite was linear with incubation time up to 4 h. The metabolism of E1 by epidermal keratinocytes yielded E2, and that of E2 resulted in the formation of E1. The rate of E1 formation from E2 was approximately 10-fold greater than the rate of formation of E2 from E1; these rates were linear with incubation time up to 4 h. Epidermal keratinocytes maintained in culture did not metabolize androstenedione to either E1 or E2, and pregnenolone was not metabolized by these cells. This study serves to ascertain that epidermal keratinocytes express steroid 5α-reductase, 17β-HSOR, 3β-HSOR, 3α-HSOR, 3β-hydroxysteroid oxidoreductase-$\Delta^{5\rightarrow4}$-isomerase, and sulfatase activities.

REFERENCES

1. MILEWICH, L., V. KAIMAL, C. B. SHAW & R. D. SONTHEIMER. 1986. Epidermal keratinocytes: a source of 5α-dihydrotestosterone production in human skin. J. Clin. Endocrinol. Metab. **62:** 739–746.
2. RICE, R. H. & H. GREEN. 1978. Relation of protein synthesis and transglutaminase activity

to formation of the cross-linked envelope during terminal differentiation of the cultured human epidermal keratinocyte. J. Cell. Biol. **76**: 705–711.

3. GOLDSMITH, L. A. 1983. Human epidermal transglutaminase. J. Invest. Dermatol. **80**: 39s–41s.

4. ISEROFF, R. R., N. E. FUSENIG & D. B. RIFKIN. 1982. Plasminogen activator in differentiating mouse keratinocytes. J. Invest. Dermatol. **80**: 217–222.

5. EBLING, F. J., E. EBLING, V. McCAFFERY & J. SKINNER. 1971. The response of the sebaceous glands of the hypophysectomized-castrated male rat to 5α-dihydrotestosterone, androstenedione, dehydroepiandrosterone and androsterone. J. Endocrinol. **51**: 181–190.

6. VOIGT, W. & S. L. HSIA. 1973. The antiandrogenic action of 4-androsten-3-one-17α-carboxylic acid and its methyl ester on hamster flank organ. Endocrinology **92**: 1216–1222.

7. MOORE, R. J. & J. D. WILSON. 1976. Steroid 5α-reductase in cultured human fibroblasts. J. Biol. Chem. **251**: 5895–5900.

8. SCHWEIKERT, H. U., H. J. HEIN, J. C. ROMIJN & F. H. SCHRODER. 1982. Testosterone metabolism of fibroblasts grown from prostatic carcinoma, benign prostatic hyperplasia and skin fibroblasts. J Urol. **127**: 361–365.

9. SCHWEICKERT, H. U., L. MILEWICH & J. D. WILSON. 1976. Aromatization of androstenedione by cultured human fibroblasts. J. Clin Endocrinol. Metab. **43**: 785–795.

10. MILEWICH, L. & H. U. SCHWEIKERT. 1977. Synthesis of carbon-14 labelled C_{19}-steroids. J. Labelled Compd. Radiopharm. **14**: 427–434.

11. MILEWICH, L., A. J. WINTERS, P. STEPHENS & P. C. MACDONALD. 1977. Metabolism of dehydroisoandrosterone and androstenedione by the human lung *in vitro*. J. Steroid Biochem. **8**: 277–284.

12. LIU, S. C., M. J. EATON & M. A. KARASEK. 1979. Growth characteristics of human epidermal keratinocytes from newborn foreskin in primary and serial cultures. In Vitro **15**: 813–822.

13. LOWRY, O. H., N. J. ROSEBROUGH, L. A. FARR & R. J. RANDALL. 1951. Protein measurement with the Folin phenol reagent. J. Biol. Chem. **193**: 265–275.

14. SIITERI, P. K. 1975. A universal chromatographic system for the separation of steroid hormones and their metabolites. Methods Enzymol. **36**: 485–489.

15. MILEWICH, L. & M. G. WHISENANT. 1982. Metabolism of androstenedione by human platelets: a source of potent androgens. J. Clin. Endocrinol. Metab. **54**: 969–974.

16. SCHWEIKERT, H. U., L. MILEWICH & J. D. WILSON. 1975. Aromatization of androstenedione by isolated human hairs. J. Clin. Endocrinol. Metab. **40**: 413–417.

17. MILEWICH, L., F. G. GEORGE & J. D. WILSON. 1977. Estrogen formation by the ovary of the rabbit embryo. Endocrinology **100**: 187–196.

18. MILEWICH, L. & J. C. PORTER. 1987. *In situ* sulfatase activity in human epithelial carcinoma cells of vaginal, ovarian, and endometrial origin. J. Clin. Endocrinol. Metab. **65**: 164–169.

19. TOTH, I. & I. FAREDIN. 1985. Concentrations of androgens and C_{19}-steroid sulfates in abdominal skin of healthy women and men. Acta Med. Hungar. **42**: 13–20.

20. MILEWICH, L., R. L. GARCIA & L. W. GERRITY. 1985. 17β-Hydroxysteroid oxidoreductase: a ubiquitous enzyme. Interconversion of estrone and estradiol-17β in BALB/c mouse tissues. Metabolism **34**: 938–944.

21. WEINSTEIN, G. D., P. FROST & S. L. HSIA. 1968. *In vitro* interconversion of estrone and 17β-estradiol in human skin and vaginal mucosa. J. Invest. Dermatol. **51**: 4–10.

22. RUDDER, H. J., L. LORIAUX & M. B. LIPSETT. 1972. Estrone sulfate: production rate and metabolism in men. J. Clin. Invest. **51**: 1020–1033.

23. MILEWICH, L., C. GOMEZ-SANCHEZ, J. D. MADDEN, D. J. BRADFIELD, P. M. PARKER, S. L. SMITH, B. R. CARR, C. D. EDMAN & P. C. MACDONALD. 1978. Dehydroisoandrosterone sulfate in peripheral blood of premenopausal, pregnant, and postmenopausal women and men. J. Steroid Biochem. **9**: 1159–1164.

24. LONGCOPE, C. 1972. The metabolism of estrone sulfate in normal males. J. Clin. Endocrinol. Metab. **34**: 113–122.

25. SANDBERG, E., E. GURPIDE & S. LIEBERMAN. 1964. Quantitative studies on the metabolism of dehydroisoandrosterone sulfate. Biochemistry **3**: 1256–1267.

26. MILEWICH, L., R. L. GARCIA & L. W. GERRITY. 1984. Steroid sulfatase and 17β-hydroxysteroid oxidoreductase activities in mouse tissues. J. Steroid Biochem. **21:** 529–538.

27. LAKSHMI, S. & A. S. BALASUBRAMANIAN. 1981. The distribution of estrone sulphatase, dehydroepiandrosterone sulphatase and arylsulphatase C in the primate (Macaca radiata) brain and pituitary. J. Neurochem. **37:** 358–362.

28. LOBO, R. A., W. L. PAUL, E. GENTZSCHEIN, P. C. SERAFINI, J. A. CATALINO, R. J. PAULSON & R. HORTON. 1987. Production of 3α-androstanediol glucuronide in human genital skin. J. Clin. Endocrinol. Metab. **65:** 711–714.

29. HORTON, R., D. HAWKS & R. LOBO. 1981. 3α,17β-Androstanediol glucuronide in plasma: a marker of androgen action in idiopathic hirsutism. J. Clin. Invest. **69:** 1203–1206.

30. WINKEL, C. A., L. MILEWICH, C. R. PARKER, JR., N. F. GANT, E. R. SIMPSON & P. C. MACDONALD. 1980. Conversion of plasma progesterone to deoxycorticosterone in men, nonpregnant and pregnant women, and adrenalectomized subjects. J. Clin. Invest. **66:** 803–812.

31. PARKER, C. R., JR., C. A. WINKEL, A. J. RUSH, J. C. PORTER & P. C. MACDONALD. 1981. Plasma concentrations of 11-deoxycorticosterone in women during the menstrual cycle. Obstet. Gynecol. **58:** 26–30.

32. SIITERI, P. K. & P. C. MACDONALD. 1973. Role of extraglandular estrogen in human endocrinology. *In* Handbook of Physiology. R. O. Greep & E. B. Astwood, Eds. Section 7, Endocrinology. Vol 2: 615–629. American Physiological Society. Washington, DC.

33. TSENG, L., J. MAZELLA & B. SUN. 1986. Modulation of aromatase activity in human endometrial stromal cells by steroids, tamoxifen and RU486. Endocrinology **118:** 1312–1318.

34. MENDELSON, C. R., M. E. SMITH, W. H. CLELAND & E. R. SIMPSON. 1984. Regulation of aromatase activity of cultured adipose stromal cells by catecholamines and adrenocorticotropin. Mol. Cell. Endocrinol. **37:** 61–72.

35. MILEWICH, L., V. KAIMAL, C. B. SHAW & A. R. JOHNSON. 1986. Androstenedione metabolism in human lung fibroblasts. J. Steroid Biochem. **24:** 893–897.

Estimating the Contribution by Skin to Systemic Metabolism

J. KAO[a]

Department of Drug Metabolism
Smith Kline & French Laboratories
King of Prussia, Pennsylvania

INTRODUCTION

The skin is the largest organ in the body. As the primary interface between the body and its external environment it serves as a living protective envelope surrounding the body. Although it is an effective barrier, it is becoming increasingly apparent that this multilayered organ is not a complete barrier, and that it is an important portal for entry of chemicals into the systemic circulation. Indeed, the increasing awareness that adverse health effects can result from dermal exposure has highlighted our inadequate understanding of skin function as it relates to percutaneous absorption and the cutaneous toxicity of chemicals.

Not long ago studies in skin toxicology were primarily concerned with the development of methods for producing and evaluating skin irritation and allergic reactions in both animal and human skin. However, the significant advances made in tissue culture techniques, cellular biology and toxicokinetic principles, have enormously expanded our horizons in studies concerning skin function and toxicity, and we are just beginning to appreciate some of the more novel, but important physiological functions and metabolic capabilities of this organ.

Studies with tissue slices, isolated cell preparations and subcellular fractions from the skin have shown this organ to be capable of performing a variety of metabolic functions including those involved in the metabolism of drugs, carcinogens and hormones. A full complement of drug metabolizing enzyme activities has been shown to be present in the skin,[1-7] and the skin is now recognized as an important site for the extrahepatic metabolism of xenobiotics.[8-13] Since skin contains the enzymes capable of metabolizing xenobiotics and hormones, questions are raised concerning the importance of skin metabolism relative to systemic metabolism. What, if any, is the functional significance of skin metabolism in the fate of both endogenous and exogenous chemicals? From a dermatopharmaceutic standpoint, will the ability of the skin to metabolize drugs prove to be a useful tool in the development of novel cutaneous therapy? Can cutaneous metabolism influence skin absorption, and be an important determinant in the development of local and systemic toxicity? What are the modulating factors that may affect the contribution of skin in the physiological disposition of chemicals?

In order to ascertain the role of skin metabolism in the systemic disposition of chemicals it will be necessary to have some knowledge of the magnitude of the metabolic process which can act on the chemical during its translocation through the

[a]Address for correspondence: Smith Kline & French Laboratories, P.O. Box 1539, King of Prussia, PA 19406-0939.

skin. Furthermore, some understanding of the processes that influence the passage of chemicals into and across the skin will also be essential. Although the diffusional bases governing skin absorption have been defined[14,15] and the metabolic capabilities of the skin well appreciated, experimental evaluation of skin metabolism as it pertains to skin absorption remains to be systematically investigated.

What then are the experimental approaches that can be employed for assessing the role of skin in systemic disposition? Unfortunately, few if any investigations have been conducted in which studies were designed to specifically address this question. However, the recent developments, and the success of the use of the transdermal route as a means for the controlled delivery of potent drugs for systemic therapy have resulted in a renewed interst in skin permeability and the significance of skin metabolism. Various approaches have been described for determining percutaneous absorption of topically applied chemicals, and some of these include an assessment of the influence of cutaneous metabolism. The purpose of this paper is to review and identify the potentials and problems associated with these approaches, and to assess their utility for estimating the contribution by the skin, focusing on the role of skin metabolism in the percutaneous fate of topically applied chemicals.

PHARMACOKINETIC APPROACHES

Skin absorption is the translocation of a topically applied chemical into and through the various layers of the epidermis and dermis to a location where the compound can enter the systemic circulation via the dermal microvasculature and lymphatics, or remain in the deeper layers of the skin. It is the net result of skin penetration, cutaneous metabolism, binding and permeation of the topically applied agent. It is a complex phenomenon, involving a myriad of diffusional and metabolic processes that are proceeding either concurrently or sequentially; consequently, theoretical models describing the overall process of percutaneous absorption will be approximations, and will be based on our current knowledge concerning the most relevant events that are important in the dermal absorption processes.

Traditionally, diffusion is recognized as the major determinant in skin absorption.[16-18] The important steps involved have been identified as the partitioning of the compound from the delivery vehicle to the stratum corneum, transport through the stratum corneum, partitioning from the lipophilic stratum corneum into the more aqueous viable epidermis, transport across the epidermis, and uptake by the cutaneous microvasculature with subsequent systemic distribution.[19] Mathematical models, of varying complexity, describing skin absorption in terms of diffusional parameters have been developed by various investigators.[16-18,20-22] From these models mathematical expressions have been derived that relate the degree of percutaneous absorption to physiochemical parameters of the penetrants and the biophysical properties of the skin, and have provided the theoretical bases for assessing dermal absorption. Indeed they have been invaluable as tools in the design and development of transdermal drug systems. Appreciation of the skin as a drug metabolizing organ is a relatively new phenomenon, and interest in the role of cutaneous metabolism during skin absorption has only recently led to the development of models which describe simultaneous diffusion and metabolism in biological membranes.[23,24] Unfortunately, complete evaluation of these models often requires a degree of mathematical sophistication and analytical measurements that cannot be matched by current experimental techniques in skin absorption.

An alternative approach, which permits experimental evaluation, involves the application of linear kinetic models.[25-28] By and large these models are based on sorption and permeation characteristics of the penetrant in the skin *in vitro,* and the elimination kinetics of the penetrant from pharmacokinetics studies. In these models, the various events involved in the overall process of percutaneous absorption are characterized by first order rate constants. The rate constants associated with events in the skin are assumed to be proportional to important diffusional parameters and may be estimated by *in vitro* experiments. Other rate constants associated with the systemic distribution and elimination are estimated from pharmacokinetic assessment following intravenous administration of the penetrant. It is recognized that this methodology lacks theoretical purity, but as first approximations they may provide acceptable predictions on the percutaneous absorption of chemicals. The utility of this approach was demonstrated in recent studies where the theoretical predictions of the plasma concentration-time profile for transdermally delivered nitroglycerin, clonidine and estradiol compared favorably with the experimental observations.[29-31] In consideration of the influence of skin metabolism in percutaneous absorption extensions describing rate constants for cutaneous metabolism and elimination of metabolites have been added to these kinetics models.[32-34] Using the extended model, computational simulations of the consequence of varying degrees of skin metabolism, either by the epidermal enzymes or cutaneous microflora have been described.[35] These simulations, however, are only theoretical, and experimental confirmation requires a better understanding of the enzymes involved in cutaneous metabolism, and this will necessitate more work on skin as a drug metabolizing organ.

There have been attempts to quantify the degree of cutaneous metabolism during percutaneous absorption by pharmacokinetic means, and this was illustrated during an investigation of the pharmacokinetics of topical and intravenous ^{14}C-nitroglycerin in the monkey.[36] Nitroglycerine is a drug that exhibits unusually rapid metabolism by a nonspecific enzymatic process.[37] In the monkey, its apparent bioavailability, based on the excretion of radioactivity, was estimated to be 72.7% following topical application. This was similar to the value (77.2%) estimated from the ratio of the area under the plasma radioactivity-time profile (AUC) derived from topical and intravenous administration. Thus, absorption estimates from plasma and urinary radioactivity measurements were in good agreement. However, absolute bioavailability estimated from AUC of plasma nitroglycerin (*i.e.,* unchanged parent compound) versus time plot was 56.6%. The difference between the values of the radioactivity measurements and the specific determination of nitroglycerin was claimed by the investigators to represent cutaneous first pass metabolism during its absorption and permeation through the skin. This then represents a somewhat oversimplified approach in which an estimate of about 20% was attributed to the skin for the overall metabolism of nitroglycerin.[35]

In a recent study, a physiologically based perfusion-limited pharmacokinetic model was constructed and it was used to simulate the transdermal bioavailability of nitroglycerine and its cutaneous first pass metabolism in man.[37] The results of the simulations produced an estimate of bioavailability and cutaneous metabolism that is similar to the results observed in the monkey. This suggested that a more sophisticated approach involving the development of physiologically based pharmacokinetic models for skin may provide another experimentally feasible alternative for estimating the contribution of skin to systemic metabolism.

IN VITRO APPROACHES

A review of the percutaneous absorption literature shows that *in vitro* methods, with excised skin mounted in diffusion chambers, are the most frequently employed techniques used in the assessment of dermal absorption. Generically, these diffusion chambers consist of a donor and a receptor compartment. Skin absorption is then determined based on the assumption that recovery of the compound of interest in the receptor compartment following application in the donor compartment will provide an accurate measure of penetration and permeation. The success and popularity of the *in vitro* approach stem from the fact that the methodology used in the *in vitro* studies is relatively simple to follow. The skin preparations are viewed as an inert physical barrier to diffusion, and studies are performed in accordance with the principles of diffusion. At steady state conditions skin absorption of a given compound is characterized in terms of diffusional characteristics and mass transfer coefficients.[14-16] In these experiments the investigator is provided with the ability to monitor the rate and extent of dermal absorption in skin removed from the rest of the body. Experimental conditions can be readily manipulated and controlled, and compared to *in vivo* studies, *in vitro* results can be obtained relatively quickly. Furthermore, it is recognized that *in vitro* methodology has contributed significantly in defining the important physiochemical parameters underlying percutaneous penetration and is responsible for much of our current understanding on the mechanism of dermal absorption.[14-16]

Fundamental to the validity of *in vitro* results are the assumptions that excised skin retains its barrier functions, diffusion across the nonviable stratum corneum is the rate determining step, and metabolic activities in the skin contribute little to the dermal absorption process. The justifications for this opinion are derived from studies demonstrating the importance of diffusion laws in percutaneous absorption, and from favorable *in vivo* and *in vitro* comparisons of dermal absorption. Using *in vitro* conditions in which the viability of the excised skin may be seriously compromised, reasonable correlations between *in vivo* and *in vitro* permeation results have been reported for a wide range of compounds.[39-42] Consequently, the emphasis of *in vitro* skin absorption studies has traditionally been concerned with the physiochemical parameters that influence the diffusion of chemicals across the nonviable skin barrier. Until recently, the possible influence of cutaneous metabolic transformations on dermal absorption have been largely neglected.

In a recent study with mouse skin in organ culture it was demonstrated that skin permeation of benzo(a)pyrene was determined largely by epidermal viability.[43] Subsequently, *in vitro* investigations with metabolically viable and structurally intact skin preparations from various animal species showed that dermal absorption of benzo(a)pyrene in cultured skin preparations was accompanied by extensive cutaneous first pass metabolism; it was influenced by the metabolic status of the skin, and the results suggested that skin absorption of benzo(a)pyrene was dependent on skin metabolism.[44-46] Furthermore dermal absorption of selected steroids in skin maintained as short term organ cultures was also shown to be associated with varying degrees of cutaneous metabolism, and there were extensive species differences.[46,47] These observations demonstrated that skin metabolism can play a significant role in the percutaneous fate of topically applied agents, and have led to new emphasis being placed on the metabolic activities of the skin as important determinants in skin absorption.

The techniques described for maintaining viability of the excised skin used in these studies were relatively simple. Basically, the skin preparations are maintained under appropriate conditions as short-term organ culture. They are supported over the culture medium, so that their epidermal surfaces are not covered. Material of interest can be applied topically in a manner similar to exposure *in vivo*; the material reaches the epidermal cells by diffusion where they may be metabolized, and recovery of both metabolites and parent compounds in the culture fluid then provides a measure of skin permeation and the extent of cutaneous first pass metabolism. Two systems have been described. In the "static" system discs of freshly excised skin are maintained epidermal side up on filter paper on a stainless steel ring support within individual culture dishes containing a suitable culture fluid.[45] In the "flow through" system the skin discs, supported on a stainless grid, form the upper seal of tissue wells of a compact, water jacketed multisample skin penetration chamber. Fresh, oxygenated culture medium is continuously perfused through the tissue wells and the well effluents may be collected at timed intervals.[44] The dermal absorption and metabolism studies described above utilizing this methodology demonstrated that, by maintaining the metabolic viability of the excised skin under appropriate culture conditions, the *in vitro* approach used provided the means whereby the potential influence of skin metabolism may be evaluated in conjunction with the diffusional aspect of percutaneous absorption. This methodology therefore offers a possible way to estimate the contribution by skin to the percutaneous fate of topically applied chemicals.

However, the simplicity of the *in vitro* methodology belies many unresolved issues related to dermal absorption *in vitro*. When absorption is accompanied with an evaluation of cutaneous metabolism there are many more questions that need to be answered, and we are only just beginning to appreciate some of the issues involved. A number of the important issues relating to the relevance of *in vitro* methodology such as the appropriate receptor fluid, thickness of the skin preparation, and the significance of tissue viability and integrity have been discussed in recent reviews on dermal absorption.[41,48]

With respect to the role of the skin in the systemic disposition of topically applied chemicals, one of the more important issues concerns the potential influence of the dermal vasculature in the assessment of cutaneous metabolism and skin function. In this regard the skin organ culture system described above is severely limited. This is particularly evident in experiments where the effect of blood flow may be important, or there is a need to determine differences in venous and arterial concentrations of chemicals and metabolites. However, promising techniques involving the use of isolated perfused porcine skin flaps,[49] and in situ rat/human skin flap[50] with defined and accessible vasculature are being developed. Once validated these techniques should provide additional approaches to further our understanding of the mechanisms of skin absorption and disposition, and provide more sophisticated tools for assessing the contribution of the skin to systemic metabolism.

CONCLUSIONS

It is clear that the skin is not just a protective envelope surrounding the body. It is an organ active in many metabolic processes, and the significance of these cutaneous reactions in skin function are the subject of an increasing number of investigations. From the perspective of dermal absorption the skin, in addition to being a drug metabolizing organ, is also a target organ for local toxicity and a portal of entry for a variety of topically applied chemicals. Thus, knowledge of the processes involved in the translocation of chemicals through the skin into the systemic circulation, coupled with the response of the skin to such chemicals, and the effects of such response on the physiological disposition and availability of topically delivered chemicals are important aspects of cutaneous pharmacology and toxicology. Moreover, knowledge of the metabolic capability and capacity of the skin are also important factors for consideration in examining the functional significance of skin relative to systemic metabolism.

In this paper some of the theoretical models and *in vitro* methodology employed in dermal absorption studies have been described, albeit briefly, and it is suggested that a combination of these techniques may provide the basis for the development of experimental approaches which can be used as probes for estimating the contribution of skin to systemic metabolism. It should be emphasized that research in this area is in its infancy. Experimental techniques are evolving and the rationale by which *in vivo* and *in vitro* models are selected and developed is continually being revised. Further development in this area, however, will necessitate improvements in analytical techniques and a better understanding of the interplay between skin penetration, permeation and metabolism, and the role of modulating factors that may influence skin metabolism and functions.

REFERENCES

1. COOMES, M. W., R. W. SPARKS & J. R. FOUTS. 1984. J. Invest. Dermatol. **82:** 598–601.
2. HAPER, K. & G. CALCUTT. 1960. Nature (London) **186:** 80–81.
3. MOLONEY, S. J., J. M. FROMSON & J. W. BRIDGES. 1982. Biochem. Pharmacol. **31:** 4011–4018.
4. MOLONEY, S. J., J. W. BRIDGES & J. M. FROMSON. 1982. Xenobiotica **12:** 481–487.
5. MOLONEY, S. J., J. M. FROMSON & J. W. BRIDGES. 1982. Biochem. Pharmacol. **31:** 4005–4009.
6. MUKHTAR, H. & P. R. BICKERS. 1981. Drug Metab. Dispos. **9:** 311–314.
7. VIZETHUM, W., T. RUZICAK & G. GOERZ. 1980. Chem.-Biol. Interact. **31:** 215–219.
8. PANNATIER, A., P. JENNER, B. TESTA & J. C. ELTER. 1978. Drug Metab. Rev. **8:** 319–343.
9. BICKERS, D. R. 1980. *In* Current Concepts in Cutaneous Toxicity. V.A. Dill & P. Lazar, Eds.: 95–126. Academic Press. New York, NY.
10. BICKERS, D. R. 1983. *In* Biochemistry and Physiology of Skin. L.A. Goldsmith, Ed. Vol. **2:** 1169–1186. Oxford University Press. New York, NY.
11. NOONAN, P. K. & R. C. WESTER. 1983. *In* Dermatotoxicology. 2nd edit. F. N. Marzulli & H. I. Maibach, Eds.: 71–90. Hemisphre. New York, NY.
12. WESTER, R. C. & H. I. MAIBACH. 1986. *In* Progress in Drug Metabolism. J. W. Bridges & L. F. Chasseaud, Eds. Vol. **9:** 95–109. Taylor & Francis. London.
13. MARTIN, R. J., S. P. DENYER & J. HADGRAFT. 1987. Int. J. Pharmaceut. **39:** 23–32.
14. BARRY, B. W. 1983. Dermatological Formulation: Percutaneous Absorption. Dekker. New York, NY.
15. SCHAEFER, H., A. ZESCH & G. STUTTGEN. 1982. Skin Permeability. Springer-Verlag. New York, NY.
16. SCHEUPLEIN, R. J. 1977. *In* Handbook of Physiology—Reaction to Environmental Agents.: 299–322. Amer. Physiol. Soc. Washington, DC.
17. MICHAELS, A. S., S. K. CHANDRUSEKARAN & J. E. SHAW. 1975. AIChE J. **21:** 285–996.
18. DUGARD, P. H. 1981. *In* Dermatotoxicology. 2nd edit. F. N. Marzulli & H. I. Maibach, Eds.: 91–129. Hemisphere. New York, NY.
19 WESTER, R. C. & H. I. MAIBACH. 1983. Drug Metab. Rev. **14:** 169–205.
20. SCHEUPLEIN, R. J. & R. L. BRONAUGH. 1983. *In* Biochemistry and Physiology of Skin. L. A. Goldsmith, Ed. Vol. **2:** 1255–1295.
21. FLYNN, G. L. 1985. *In* Percutaneous Absorption. R. L. Bronaugh & H. I. Maibach, Eds.: 17–42. Dekker. New York, NY.
22. ZATZ, J. L. 1985. *In* Percutaneous Absorption. R. L. Bronaugh & H. I. Maibach, Eds.: 165–181. Dekker. New York, NY.
23. ANDO, Y. W., N. F. H. HO & W. I. HIGUCHI. 1977. J. Pharm. Sci. **66:** 1525–1528.
24. FOX, J. L. C. D. YU., W. I. HIGUCHI & N. F. H. HO. 1979. Int. J. Pharmaceut. **2:** 41–57.
25. GUY, R. H., J. HADGRAFT & H. I. MAIBACH. 1982. Int. J. Pharmaceut. **11:** 119–129.
26. GUY, R. H. & J. HADGRAFT. 1983. J. Pharmacokin Biopharm. **11:** 189–203.
27. KUBOTA, K. & T. ISHIZAKI. 1985. J. Pharmacokin Biopharm. **13:** 55–72.
28. GUY, R. H. & J. HADGRAFT. 1985. Int. J. Pharmaceut. **24:** 267–274.
29. GUY, R. H. & J. HADGRAFT. 1985. Pharmaceut. Res. **2:** 206–211.
30. GUY, R. H. & J. HADGRAFT. 1985. J. Pharm. Sci. **74:** 1016–1018.
31. GUY, R. H. & J. HADGRAFT. 1986. Int. J. Pharmaceut. **32:** 159–163.
32. HADGRAFT, J. 1980. Int. J. Pharmaceut. **4:** 229–239.
33. GUY, R. H. & J. HADGRAFT. 1984. Int. J. Pharmaceut. **20:** 43–51.

34. GUY, R. H., J. HADGRAFT & D. A. W. BUCKS. 1987. Xenobiotica 17: 325–343.
35. DENYER, S. P., R. H. GUY, J. HADGRAFT & W. B. HUGO. 1985. Int. J. Pharmaceut. 26: 89–97.
36. WESTER, R. C., P. K. NOONAN, S. SMEACH & L. KOSOBUD. 1983. J. Pharm. Sci. 72: 745–748.
37. NEEDLEMAN, P. 1976. Annu. Rev. Pharmacol. Toxicol. 16: 81–93.
38. NAKASHIMA, E., P. K. NOONAN & L. Z. BENET. 1987. J. Pharmacokin. Biopharm. 15: 423–437.
39. FRANZ, T. J. 1975. J. Invest. Dermatol. 54: 190–195.
40. BRONAUGH, R. L. & H. I. MAIBACH. 1983. In Dermatotoxicology. 2nd edit. F. N. Marzulli & H. I. Maibach, Eds.: 117–129. Hemisphere. New York, NY.
41. BRONAUGH, R. L. 1985. In Percutaneous Absorption. R. L. Bronaugh & H. I. Maibach, Eds.: 267–279. Dekker. New York, NY.
42. BRONAUGH, R. L. & T. J. FRANZ. 1986. Br. J. Dermatol. 115: 1–8.
43. SMITH, L. H. & J. M. HOLLAND. 1981. Toxicology 21: 47–57.
44. HOLLAND, J. M., J. KAO & M. J. WHITAKER. 1984. Toxicol. Appl. Pharmacol. 68: 206–217.
45. KAO, J., J. HALL, L. R. SHUGART & M. J. HOLLAND. 1984. Toxicol. Appl. Pharmacol. 75: 289–298.
46. KAO, J., F. K. PATTERSON & J. HALL. 1985. Toxicol. Appl. Pharmacol. 81: 502–516.
47. KAO, J. & J. HALL. 1987. J. Pharmacol. Exp. Ther. 241: 482–487.
48. KAO, J. 1988. In Fundamentals and Methods of Dermal and Ocular Toxicology. D. W. Hobson, Ed. Telford Press. Caldwell, NJ. In press.
49. RIVIERE, J. E., K. F. BOWMAN, N. A. MONTEIRO-RIVIERE, L. P. DIX & M. P. CARVER. 1986. Fund. Appl. Toxicol. 7: 444–453.
50. KRUEGER, G. G., Z. J. WOJCIECHOWSKI, S. A. BURTON, A. GILHAR, S. E. HUETHER, L. G. LEONARD, U. D. ROHR, T. J. PETELENZ, W. I. HIGUCHI & L. K. PERSHING. 1985. Fund. Appl. Toxicol. 5: S112–S121.

Sulfation Reactions of the Epidermis

ERVIN H. EPSTEIN, JR., ALEX W. LANGSTON,
AND JESSICA LEUNG

Department of Dermatology
San Francisco General Hospital
and
University of California, San Francisco
1001 Potrero Street
San Francisco, California 94110

In recent years, we have had two reasons to study epidermal sulfation of various substrates. The first is our interest in the role of cholesterol sulfate in stratum corneum cohesion and desquamation, and the second our work on the metabolism of exogenous compounds absorbed through the skin.

Interest in cholesterol sulfate as an important molecule in the epidermis can be traced to the mid 1970s, when investigators in Amsterdam and Los Angeles discovered independently that patients with recessive X-linked ichthyosis (RXLI) lack activity of the enzyme steroid sulfatase.[1,2] Steroid sulfatase is a microsomal enzyme that normally is active in stratum corneum and is widely distributed in other organs as well. In RXLI, the lack of enzyme activity is due to the lack of enzyme protein,[3] and in 90% of patients this in turn is due to large DNA deletions that include the gene that codes for the enzyme.[4-8] In the absence of steroid sulfatase, patients accumulate high levels of the steroid sulfatase substrate cholesterol sulfate in both blood and stratum corneum.[9-11] In fact, there are several lines of evidence that desulfation of cholesterol sulfate is necessary for normal stratum corneum desquamation. First is the correlation between its accumulation in the stratum corneum of patients with RXLI and the thickening of stratum corneum that is characteristic of that disease. Second, application of cholesterol sulfate to the skin of hairless mice causes thickening of their stratum corneum.[12] Third, the lipids of shed corneocytes are relatively poor in cholesterol sulfate by comparison with the lipids of corneocytes still adherent to the skin.[13] Thus, it appears that cholesterol sulfate desulfation is a necessary step in normal desquamation.

This important role of cholesterol sulfate in stratum corneum shedding prompted us to examine its derivation. The first evidence that cholesterol sulfate is synthesized locally in the epidermis came from studies showing that lipids of the upper epidermis contain more cholesterol sulfate than do those of the lower epidermis, thus suggesting a local site of synthesis rather than mere ingress from blood.[14] We next conducted studies of mouse epidermis to assay directly for cholesterol sulfotransferase (CST) activity. This enzyme can be assayed by mixing tissue with an appropriate acceptor molecule and commercially-available ^{35}S-phospho-adenosine phosphosulfate (PAPS) and noting increased hydrophilicity of the acceptor or increasing hydrophobicity of the ^{35}S. Thus, ^{35}S-cholesterol sulfate is readily separated from ^{35}S-sulfate and ^{35}S-PAPS by partitioning between a lower chloroform-methanol phase and an upper aqueous potassium chloride phase and can be quantitated by liquid scintillation chromatography.[15]

Using this assay, we readily demonstrated cholesterol sulfotransferase activity in epidermis of newborn mice separated from the underlying dermis after incubation with medium containing dithiothreitol.[16] This enzyme is present in the cytosolic fraction of the epidermis and is present in undiluted cytosol at a specific activity greater than that of adult or newborn mouse liver. Dilution of cytosol or partial purification of enzyme

97

by chromatography on AffiGel Blue® increased the specific activity markedly, presumably by removing endogenous inhibitors, and the enzyme specific activity is equivalent in such partially purified preparations from newborn mouse skin and liver. Staphylococcal epidermolytic toxin was used to fractionate the epidermis, and we found CST enzyme activity in the basal and spinous cell fraction but none in the stratum granulosum or stratum spinosum. Cultured human keratinocytes also express cholesterol sulfotransferase activity with a specific activity similar to that of newborn mouse epidermis. Not only do cultured keratinocytes express cholesterol sulfotransferase activity but also Mary Williams has shown in a careful series of experiments that they do synthesize cholesterol sulfate.[17,18] Thus the belief that epidermis synthesizes its own cholesterol sulfate *in vivo* rests on the higher levels of cholesterol sulfate in upper vs lower epidermis, the demonstration of cholesterol sulfotransferase activity in the basal-spinous cell fraction of newborn mouse epidermis and in cultured human keratinocytes, and the production of cholesterol sulfate by cultured keratinocytes.

We have studied the range of substrates sulfated by our newborn mouse epidermal cytosol preparations and found that sulfate is added to the 3β position of dehydroepiandrosterone and androsterone but β-estradiol and corticosterone are not sulfated. In fact, sulfation of cholesterol is inhibited by β-estradiol, corticosterone, dexamethosone, and desoximetasone at high levels, a fact of yet uncertain pharmacologic relevance since it is not clear whether these steroids could reach the 300–600 μM range necessary for inhibition of cytosolic enzyme.[16]

Although the focus of this volume is on the epidermal keratinocyte, it is appropriate to consider recent studies of the regulation of differentiation of pulmonary epithelial cells. Jetten and co-workers at the National Institute of Environmental Health Sciences in North Carolina have been interested in the squamous metaplasia that the normal mucociliary epithelial lining of the trachea and bronchi undergoes following injury or when depleted of vitamin A.[19-21] Endothelial cells cultured from rabbit trachea or human bronchi on collagen and fibronectin proliferate rapidly. When the cells reach confluence, they markedly reduce their rate of cell division; they express greatly increased levels of transglutaminase; they form cross-linked envelopes; and they change the keratins they synthesize. In addition, they express cholesterol sulfotransferase activity that is 20–30-fold higher than in proliferating cells, and they incorporate 50–100-fold more sulfate into cholesterol sulfate than do proliferating cells. This group also has demonstrated that the development of these markers of squamous differentiation all are inhibited by addition of retinoids to the culture medium. Specifically, retinoic acid blocks the expression of cholesterol sulfotransferase activity with a Ki of 5×10^{-10} M. However, although retinoids inhibit the expression of CST as part of their inhibition of squamous differentiation, they are not direct inhibitors of this enzyme.[21]

Thus cholesterol sulfate can be formed by epithelial cells not only of the skin but also of the lung. Furthermore, normal rat gastric mucosa also contains cholesterol sulfate and expresses cholesterol sulfotransferase activity.[15] Thus it appears that at least the potential for cholesterol sulfate synthesis is a general property of epithelial cells.

In contrast to the studies previously discussed, sulfation reactions traditionally have been studied primarily in their role of detoxifying noxious chemicals. The metabolic transformation of many lipophilic molecules can be divided into two phases. The second phase makes the molecule more hydrophilic, and the first phase adds a "hook" to which the hydrophilic group can be attached. Like glucuronidation, sulfation renders lipophilic molecules more hydrophilic. The increased hydrophilicity reduces the toxicity both by increasing the excretion of the compound in urine and bile and also by reducing the molecule's ingress into cells through the lipophilic plasma membrane. These reactions have been most studied in liver, but some biotransformations have been described in cells lining the gut and lung. Because we have found active sulfation of endogenous sterols, we wished to determine whether the skin also can sulfate exogenous compounds as well. As a model of such an exogenous compound, we studied phenols. We were particularly interested in phenol as a substrate for conjugation reactions that potentially detoxify compounds because phenol is used topically to reduce itching and because it rapidly penetrates stratum corneum. Therefore, we assayed epidermal cytosolic sulfotransferase activity using phenols as substrates and indeed found sulfation of para-nitrophenol, of meta-nitrophenol, and of phenol itself.[16] Next, in order to examine in situ sulfoconjugation, we turned to the use of percutaneous absorption chambers. Fresh newborn mouse skin was clamped into such chambers, and the lower well was filled with culture medium. ^{14}C-phenol was applied in acetone to the dry upper surface of the epidermis, and four hours later we assayed the culture medium for radiolabelled compounds that had passed completely through the skin. Thin layer chromatography readily separates the resultant radiolabelled phenyl glucuronide and phenyl sulfate. We estimate that only a few percent of the tracer dose applied to our mouse skin preparation actually is conjugated under these conditions,[22] but we do not know whether this limited capacity for conjugation reflects the *in vivo* situation or whether it is an artifact of the *in vitro* health of the preparation. The answer to that question might come from *in vivo* studies in preparations such as those of the rat skin flaps devised by Gerald Krueger.[23]

We were interested in assaying phenol conjugation across intact skin also because it could afford us a system in which we could study the modification of conjugation by topical agents. We reasoned that if indeed cholesterol sulfate is an important cohesive agent in the stratum corneum and if it is produced locally, then inhibition of cholesterol sulfation might enhance stratum corneum desquamation. We chose first to study salicylic acid, an agent already well known to enhance stratum corneum shedding.[24] In cytosol preparations salicylic acid indeed does inhibit cholesterol sulfotransferase activity but is effective only at high concentrations—the Ki is approximately 15 mM.[16] It inhibits phenol sulfotransferase at the same high concentration. Therefore, we pretreated our newborn mice by applying salicylic acid to the skin surface and then assessed conjugation of transcutaneously passed ^{14}C-phenol. Such pretreatment inhibited both glucuronidation and sulfation by 90%.[22] This indicates that, at least in newborn mouse skin, topically-applied salicylic acid can accumulate in the epidermis at levels sufficiently high to inhibit sulfoconjugation of cholesterol and argues in favor of the hypothesis that salicylic acid may promote desquamation, at least in part, by inhibiting formation of a putative "glue"—cholesterol sulfate.

It is clear that the range of substrates which the epidermis can sulfate is not limited to sterols-steroids and phenols. Several laboratories have demonstrated that cultured mouse and human normal and malignant keratinocytes can synthesize sulfated glycosaminoglycans.[25,26] These include hyaluronic acid as well as heparan sulfate and chondroitin sulfate. Synthesis is stimulated by switching from low to high calcium medium, but precise details of enzyme activity have not been published. Furthermore, it is of interest that the form of minoxidil that is active in producing antihypertensive effects is minoxidil sulfate,[27] and there is some evidence that skin can form this compound. Perhaps sulfation of topically applied minoxidil might be important in its stimulation of hair growth. Lastly, sulfation is one of the steps of metabolism of polyaromatic hydrocarbons in other tissues,[28] and these compounds represent important candidates for epidermal sulfation as well. Sulfation of polyaromatic hydrocarbons renders them less mutagenic, but the complex role of such reactions is underscored by the finding that sulfation of aromatic amides and amines may enhance their carcinogenicity.[29]

In summary, there is good evidence that the epidermis synthesizes sulfated molecules likely to be of importance in epidermal structure and integrity. Furthermore, sulfation may play a significant role as a second line of defense against noxious compounds that penetrate the first line of defense—the stratum corneum—a function analogous to that which it is postulated to serve in lung and gut.[30] Deliberate modulation of these reactions has the potential to be of real therapeutic significance.

REFERENCES

1. JOBSIS, A. C., C. Y. VAN DUUREN, G. P. DE VRIES, J. G. KOPPE, Y. RIJKEN, G. M. J. VAN KEMPEN & W. P. DE GROOT. 1976. Trophoblast sulphatase deficiency associated with X-chromosomal ichthyosis. Ned. Tijdschr. Geneeskd. **120:** 1980.
2. SHAPIRO, L. J., R. WEISS, M. M. BUXMAN, J. VIDGOFF & R. L. DIMOND. 1978. Enzymatic basis of typical X-linked ichthyosis. Lancet **2:** 756–757.
3. EPSTEIN, E. H., JR. & J. M. BONIFAS. 1985. Recessive X-linked ichthyosis: lack of immunologically detectable steroid sulfatase enzyme protein. Hum. Genet. **71:** 201–205.
4. YEN, P. H., E. ALLEN, B. MARSH, T. MOHANDAS, N. WANG, R. T. TAGGART & L. J. SHAPIRO. 1987. Cloning and expression of steroid sulfatase cDNA and the frequent occurrence of deletions in STS deficiency: implications for X-Y interchange. Cell **49:** 443–454.
5. BALLABIO, A., G. PARENTI, R. CARROZZO, G. SEBASTIO, G. ANDRIA, V. BUCKLE, N. FRASER, I. CRAIG, M. ROCCHI, G. ROMEO, A. C. JOBSIS & M. G. PERSICO. 1987. Isolation and characterization of a steroid sulfatase cDNA clone: genomic deletions in patients with X-chromosome-linked ichthyosis. Proc. Natl. Acad. Sci. USA **84:** 4519–4523.
6. CONARY, J. T., G. LORKOWSKI, B. SCHMIDT, R. POLHMANN, G. NAGEL, H. E. MEYER, C. KRENTLER, J. CULLY, A. HASILIK & K. VON FIGURA. 1987. Genetic heterogeneity of steroid sulfatase deficiency revealed with cDNA for human steroid sulfatase. Biochem. Biophys. Res. Commun. **144:** 1010–1017.
7. GILLARD, E. F., N. A. AFFARA, J. R. W. YATES, D. R. GOUDIE, J. LAMBERT, D. A. AITKEN & M. A. FERGUSON-SMITH. 1987. Deletion of a DNA sequence in eight of nine families with X-linked ichthyosis (steroid sulfatase deficiency). Nucleic Acids Res. **15:** 3977–3985.
8. BONIFAS, J. M., B. J. MORLEY, R. E. OAKEY, Y. W. KAN & E. H. EPSTEIN, JR. 1987. Cloning of a cDNA for steroid sulfatase: frequent occurrence of gene deletions in patients with recessive X chromosome-linked ichthyosis. Proc. Natl. Acad Sci. USA **84:** 9248–9251.
9. BERGNER, E. A. & L. J. SHAPIRO. 1981. Increased cholesterol sulfate in plasma and red blood cell membranes of steroid sulfatase deficient patients. J. Clin. Endocrinol. Metab. **53:** 221–223.
10. EPSTEIN, E. H., JR., R. M. KRAUSS & C. H. L. SHACKLETON. 1981. X-linked ichthyosis: increased blood cholesterol sulfate and electrophoretic mobility of low-density lipoprotein. Sciences **214:** 659–660.
11. WILLIAMS, M. L. & P. M. ELIAS. 1981. Increased cholesterol sulfate content of stratum corneum in recessive X-linked ichthyosis. J. Clin. Invest. **68:** 1404–1410.
12. MALONEY, M. E., M. L. WILLIAMS. E. H. EPSTEIN, JR., M. Y. L. LAW, P. O. FRITSCH & P. M. ELIAS. 1984. Lipids in the pathogenesis of ichthyosis: topical cholesterol sulfate-induced scaling in hairless mice. J. Invest. Dermatol. **83:** 252–256.
13. RANASINGHE, A. W., P. W. WERTZ, D. T. DOWNING & I. C. MACKENZIE. 1986. Lipid composition of cohesive and desquamated corneocytes from mouse ear skin. J. Invest. Dermatol. **86:** 187–190.
14. ELIAS, P. M., M. L. WILLIAMS, M. E. MALONEY, J. M. BONIFAS, B. E. BROWN, S. GRAYSON & E. H. EPSTEIN, JR. 1984. Stratum corneum lipids in disorders of cornification: steroid sulfatase and cholesterol sulfate in normal desquamation and the pathogenesis of recessive X-linked ichthyosis. J. Clin. Invest. **74:** 1414–1421.

15. LIN, Y. N. & M. I. HOROWITZ. 1980. Enzymatic sulfation of cholesterol by rat gastric mucosa. Steroids **36:** 697–708.
16. EPSTEIN, E. H., JR., J. M. BONIFAS, T. C. BARBER & M. HAYNES. 1984. Cholesterol sulfotransferase of newborn mouse epidermis. J. Invest. Dermatol. **83:** 332–335.
17. WILLIAMS, M. L., S. L. RUTHERFORD, M. PONEC, M. HINCENBERGS, D. R. PLACZEK & P. M. ELIAS. 1988. Density-dependent variations in the lipid content and metabolism of cultured human keratinocytes. J. Invest. Dermatol. In press.
18. WILLIAMS, M. L., B. E. BROWN, D. J. MONGER, S. GRAYSON & P. M. ELIAS. 1988. Enhanced differentiation of human keratinocyte cultures grown at the air-medium interface: lipid content and metabolism. J. Cell. Physiol. In press.
19. REARICK, J. I. & A. M. JETTEN. 1986. Accumulation of cholesterol 3-sulfate during *in vitro* squamous differentiation of rabbit tracheal epithelial cells and its regulation by retinoids. J. Biol. Chem. **261:** 13898–13904.
20. REARICK, J. I., T. W. HESTERBERG & A. M. JETTEN. 1987. Human bronchial epithelial cells synthesize cholesterol sulfate during squamous differentiation *in vitro*. J. Cell. Physiol. **133:** 573–578.
21. REARICK, J. I., P. W. ALBRO & A. M. JETTEN. 1987. Increase in cholesterol sulfotransferase activity during *in vitro* squamous differentiation of rabbit tracheal epithelial cells and its inhibition by retinoic acid. J. Biol. Chem. **262:** 13069–13074.
22. LANGSTON, A. W., J. M. BONIFAS & E. H. EPSTEIN, JR. 1986. Inhibition of epidermal sulfation is a mechanism by which topical salicylic acid exerts its keratolytic effect. J. Invest. Dermatol. **86:** 487.
23. WOJCIECHOWSKI, Z., L. K. PERSHING, S. HUETHER, L. LEONARD, S. A. BURTON, W. I. HIGUCHI & G. G. KRUEGER. 1987. An experimental skin sandwich flap on an independent vascular supply for the study of percutaneous absorption. J. Invest. Dermatol. **88:** 439–446.
24. GOLDSMITH, L. A. 1979. Salicylic acid. Int. J. Dermatol. **18:** 32–36.
25. KING, I. A. 1981. Characterizaton of epidermal glycosaminoglycans synthesized in organ culture. Biochim. Biophys. Acta **674:** 87–95.
26. LAMBERG, S. I., S. H. YUSPA & V. C. HASCALL. 1986. Synthesis of hyaluronic acid is decreased and synthesis of proteoglycans is increased when cultured mouse epidermal cells differentiate. J. Invest. Dermatol. **86:** 659–667.
27. JOHNSON, G. A., K. J. BARSUHN & J. M. MCCALL. 1982. Sulfation of minoxidil by liver sulfotransferase. Biochem. Pharmacol. **31:** 2949–2954.
28. NEMOTO, N. 1981. Glutathione, glucuronide, and sulfate transferase in polycyclic aromatic hydrocarbon metabolism. *In* Polycyclic Hydrocarbons and Cancer. Vol. 3. M. V. Gelboin & P. O. P. Tso, Eds. 213–258. Academic Press. New York, NY.
29. DEBAUN, J. R., E. C. MILLER & J. A. MILLER. 1970. N-Hydroxy-2-acetylaminofluorene sulfotransferase: its probable role in carcinogenesis and in protein-(methion-S-yl) binding in rat liver. Cancer Res. **30:** 577–595.
30. CASSIDY, M. K. & J. B. HOUSTON. 1980. Phenol conjugation by lung *in vivo*. Biochem. Pharmacol. **29:** 471–474.

Metabolic Activation of Carcinogens by Keratinocytes

DAVID R. BICKERS

Department of Dermatology
Case Western Reserve University
School of Medicine
Cleveland, Ohio 44106

Traditional consideration of the skin has questioned the possibility that biologically significant rates of metabolic activity occur in this tissue. In recent years, it has become clear that the major determinant of biologically significant metabolism is not the amount but rather the type(s) of metabolites that are generated, and it is now known that certain metabolic species even when produced in trace amounts can evoke unique biological responses which may be either beneficial or detrimental to the host.

The skin is a major body interface with the environment and as such is directly exposed on a continuing basis to an enormous array of physical and chemical agents capable of altering cells resident within the skin itself as well as cells that traffic through cutaneous tissue in the intravascular space. Furthermore, metabolites generated in the skin may diffuse into the vascular space and subsequently exert an influence at sites distant from the point of environmental exposure. Examples of this latter phenomenon include the ability of incident solar ultraviolet radiation to photocatalyze the transformation of epidermal 7-dehydrocholesterol to pre-vitamin D_3, a metabolite which then moves into the vascular space, binds to a carrier protein which transports it to the liver and the kidney where subsequent enzyme-mediated hydroxylations result in the formation of 1,25-dihydroxyvitamin D_3 (1,25-$(OH)_2D_3$) which is crucial for calcium homeostasis. The 1,25-$(OH)_2D_3$ can then return to the skin where it can bind to receptors of epidermal keratinocytes and influence proliferation/differentiation of these cells. Furthermore, human keratinocytes can hydroxylate 25-hydroxyvitamin D_3 to 1,25-$(OH)_2D_3$.[1] These insights into the endocrine aspects of the skin are discussed elsewhere in this volume.

A second example of this type of reaction is the ability of incident solar radiation to facilitate trans-cis isomerization of urocanic acid in the stratum corneum.[2] This photocatalyzed product may then circulate to the spleen where it is capable of altering the immunologic function of dendritic antigen-presenting cells such that suppressor pathways are activated. Thus, events occurring in the skin can alter distant immunologic reactions and function as a modulator of the body's immune function.

Our ability to survive in an increasingly complicated chemical environment is to a significant extent dependent upon the structural barrier function of environmental interfaces such as the skin which preclude the entry of many noxious chemicals into the body. Furthermore, the body possesses additional subcellular "barriers" that diminish the toxicity of xenobiotics, and one of these is the family of heme-proteins known as cytochrome(s) P-450.

Cytochrome(s) P-450 are enzymes located in the endoplasmic retriculum that are responsible for the oxidatitve metabolism of a broad range of endogenous and exogenous substrates including steroid hormones, fatty acids, prostaglandins, leukotrienes as well as virtually limitless numbers of drugs, chemical carcinogens and mutagens.[3] Molecular biological studies have revealed multiple P-450s and the so-called P-450 superfamily is presently known to comprise at least 10 gene families of which 8 are present in mammals.[4] Experimental studies have shown that an increase in certain P-450 proteins may correlate with environmental carcinogenesis. For example, mouse P_1450 (P450 IAI) is inducible by polyaromatic hydrocarbons (PAHs) such as 3-methylcholanthrene (3-MC), and enzymes dependent upon this particular isozyme are capable of metabolizing this class of carcinogenic chemicals to a broad range of metabolites including phenols, quinones, and epoxides some of which are highly reactive and can evoke toxic, mutagenic and carcinognic effects in cells. The P^1-450-dependent enzyme responsible for the metabolic activation of PAHs is known as aryl hydrocarbon hydroxylase (AHH). Like most P-450-dependent enzymes, AHH activity can be induced by enhanced transcription of the gene that codes for it. AHH inducton is dependent upon a cytosolic receptor which binds the inducer, and then the inducer-receptor complex is translocated to the nucleus where it associates with specific regions of the DNA. The induction response is activated transcriptionally and mRNA increases with resulting enhancement of specific P-450 isozymes in the endoplasmic reticulum. Prior studies by Nebert and others had shown that in mice, the so-called Ah (aromatic hydrocarbon) cytosolic receptor is encoded by the Ah locus and that lack of induction of AHH is a function of absent or aberrant receptor.

A major response to polyhalogenated hydrocarbons which also bind to the Ah receptor and induce P-450 isozymes is the development of chloracne in the skin. These yellow cystic lesions typically develop on the face and neck of individuals exposed to halogenated hydrocarbons as a result of occupational or recreational activities. The pathologic changes that accompany such lesions include epidermal thickening, squamous metaplasia of the sebaceous follicle, hyperkeratosis, and cyst formation.[5] It is generally agreed that chloracne is a sensitive indicator of human exposure to halogenated hydrocarbons.

Prior studies in human epidermis had shown that topical application of mixtures of polycyclic aromatic hydrocarbons (crude coal tar) to human skin resulted in the induction of AHH.[6] Furthermore, using freshly keratomed human epidermis, in an *in vitro* organ culture system, AHH activity was shown to be induced following incubation with a PAH. On the basis of these responses, it seemed reasonable to speculate that human skin cells possess an Ah receptor, but no unequivocal evidence for this has been available until very recently.[8] Using a squamous cell carcinoma line (A431), it was found that by modifying standard techniques employed for receptor isolation it became possible to verify the presence of the Ah receptor in these human epidermal cells. These modifications included reduction of charcoal to one-tenth the amount used in rodent liver cytosols and the use of buffers containing sodium molybdate which appeared to stabilize the cytosolic receptor from these cells when held for prolonged periods at 4°C to 20°C. The concentration of Ah receptor was found to be approximately 100 fmol/mg cytosolic protein. Despite its relatively high concentration in this cell line the apparent affinity of the receptor for the potent inducer of AHH, tetrachlorodibenzo[p]dioxin (TCDD), was substantially less than that previously identified in a mouse hepatocyte line and in fact, was shifted as much as one log unit to the right of the saturation curve observed in hepatocytes. This reemphasized the fact that the binding affinity of an inducer for its receptor may be a more important determinant of AHH responsiveness than the absolute amount of receptor. These results underline that at least one human cell has high levels of the cytosolic Ah receptor, which undergoes inducer-receptor transformation into a form which can bind to nuclei and is followed by high induction of AHH in these cells. They also suggest that human skin cells possess the same induction mechanism for AHH as previously shown in other cells and tissues in laboratory animals.

Evidence for the existence of multiple P-450 isozymes in the liver has led to attempts to define the pattern of P-450s in extrahepatic tissues where in general enzyme activity is considerably below that in the liver. Studies in the skin have been hampered by the apparently small amount of catalytic activity which could also be an artifact due to the drastic preparative procedures necessary to obtain satisfactory subcellular fractions for experimental study.

While it would clearly be preferable to investigate human cells, it has been necessary to employ animal models for these experiments because of ethical considerations and because of the very low catalytic activity in the skin. For the past several years, we have utilized neonatal rodents (rats and mice) for studies of skin P-450 because these animals are readily available thereby simplifying the accumulation of the large amounts of tissue needed for this kind of work. Furthermore, it is essential to obtain pure epidermis for study and clean unequivocal dermal-epidermal separation can be accomplished by a relatively simple and nontraumatic procedure in which sheets of freshly removed whole skin are incubated at 4°C in 0.1 M phosphate buffer containing 20% glycerol, 10 mM EDTA and 10 mM dithiothreitol.[9] After being shaken at 4°C for 2-3 hours the epidermis and dermis can be teased apart, and the epidermal sheets are then ready for further preparative manipulation.

In prior studies using this approach, we showed that there is spectrally detectable P-450 in neonatal rat epidermis and that this is inducible following topical application of halogenated hydrocarbons and carcinogenic PAHs.[10] These studies were then extended to Balb/C mice in an effort to directly compare P-450-dependent activity in cultured keratinocytes and intact epidermis from this strain of animal.[11] Our results indicated that the cultured murine cells retained 30–40% of the activity of intact epidermis insofar as AHH and other P-450-dependent enzymes were concerned. These cells had a pH optimum of 7.4 for catalytic activity and the Vmax for AHH was 15-fold greater in carcinogen-treated cells than in vehicle-treated controls.

Subsequently, Coomes *et al.* described a method for isolating mouse skin cells in which enzyme digestion was followed by separation on metrizamide and Percoll gradients which yielded at least 3 cell fractions each of which was enriched to varying degrees with sebaceous cells, basal cells and differentiated keratinocytes.[12] The different cell populations were shown to metabolize xenobiotics at different rates. The sebaceous cells were most active whereas the basal population was most inducible following exposure to an inducer of AHH. These studies were among the first to show that there is heterogeneity among epidermal cells insofar as P-450-dependent metabolism is concerned.

In addition, Pohl *et al.* described a technique in which epidermal cell suspensions were prepared from hairless mouse epidermis and metrizamide gradients and elutriation used to obtain subpopulations of cells.[13] Cells of greater size which were more differentiated had higher P-450-dependent metabolizing activity suggesting that the largest and most differentiated cells are those most likely to metabolize xenobiotics.

Cultured murine keratinocytes have also proved to be a useful system for the study of carcinogen metabolism and its inhibition by various compounds.[14,15] By employing isotopically labeled carcinogens to aid HPLC separation techniques, it has been possible to measure the intracellular and extracellular metabolism of these compounds and to evaluate both Phase I (P-450-dependent) and Phase II (conjugation reaction) metabolism.

The use of skin cells has permitted more precise characterization of the pattern of P-450-dependent metabolites elaborated by these cells. The major organic solvent-soluble metabolites of BP identified in the cultured cells were trans-7,8-dihydro-7,8-dihydroxybenzo(a)pyrene (BP-7,8-diol); 9-hydroxybenzo(a)pyrene (9-OH-BP); 3-hydroxybenzo(a)pyrene (3-OH-BP) as well as trace amounts of trans-4,5-dihydro-4,5-dihydroxybenzo(a)pyrene, BP-quinones and trans-9,10-dihydro-9,10-hydroxybenzo(a)pyrene. The major organic solvent-extractable metabolites of BP found in the extracellular culture medium were primarily the diols with smaller quantities of phenols and quinones. The major water-soluble metabolites of BP present both intracellularly and extracellularly were glucuronide conjugates of 3-OH-BP, 9-OH-BP, BP-3,6-quinone and to a lesser extent, sulfate conjugates of 7,8-diol. Covalent BP binding to keratinocyte DNA and protein was also shown to occur.

It is important to emphasize that there are alternate pathways for the oxidation of polyaromatic hydrocarbons by keratinocytes.[16] Labeled (+)-BP-7,8-diol was added to fresh murine keratinocytes and shown to be oxidized to the anti- and syn-stereoisomers (anti- and syndiolepoxides) in greater than a 4:1 ratio. The time course for the formation of the antidiolepoxide was closely correlated with peroxyl radical formation (as indicated by inhibition of this reaction by butylated hydroxyanisole, an inhibitor of peroxyl radical-dependent metabolism). Treatment of the cells with beta-naphthoflavone, a known inducer of P_1-450, altered the metabolism of BP-7,8-diol such that the major metabolite was the syndiolepoxide. These results indicate that peroxyl-radical-mediated metabolism is primarily responsible for the oxidation of (+)-BP-7,8-diol in control animals while the cytochrome P-450 system is primarily responsible for oxidation in animals pretreated with inducers.

The development of monoclonal antibodies and their application to studies of metabolic processes in the skin has provided valuable probes for the assessment of P-450-dependent reactions in cutaneous tissue. We have been attempting to characterize the P-450 isozyme distribution in the epidermis using highly specific monoclonal antibodies directed against purified P-450s.[17,18] SDS-PAGE of solubilized epidermal microsomes has shown clearly increased precipitation of protein in the P-450 region with preparations obtained from rats treated with a single topical application of 3-methylcholanthrene or of the polychlorinated biphenyl Aroclor 1254.[19] Western blot analysis employing monoclonal antibody 1-7-1 directed against 3-MC-inducible hepatic P-450 (P-450c) demonstrated a sharp immunoprecipitin band with microsomes obtained from the inducer-treated, but not the vehicle-treated animals. However, a barely detectable band was observed in the microsomes from vehicle-treated controls indicating that there may be small amounts of this heme-protein present constitutively in neonatal rats.

In further studies, we have used these monoclonal antibodies to reaction phenotype cutaneous tissue. Monoclonal antibody 1-7-1 is a potent inhibitor of epidermal microsomal AHH and 7-ethoxycoumarin de-ethylase, both of which are dependent upon P-450c. This further confirms that this isozyme of P-450 is present and inducible in mammalian epidermis. In addition, we have shown that this monoclonal antibody can also inhibit the metabolic activation (DNA binding) of benzo(a)pyrene.

Another type of monoclonal antibody that has proved useful in studies of skin xenobiotic metabolism is that directed against BP-diol-epoxide-I-DNA adducts (BPDE-I-dg). In collaboration with Dr. Regina Santella of Columbia University, we have shown that BPDE-I-dG adduct formation increased in the epidermis of SENCAR mice following topical application of BP and/or crude coal tar (Mukhtar *et al.* 1986).[20] Topical application of two doses of BP (20 mg) at 72-hour intervals resulted in the formation of 197 fmol of BPDE-I-dg adducts per mg DNA. It is of interest that combined treatment with BP and crude coal tar resulted in a smaller number of adducts than would have been expected, suggesting that crude coal tar may contain inhibitors of this type of metabolic activation.

Knowledge that metabolic activation by P-450-dependent enzymes is a critical step in the initiation of tumor formation by PAHs such as BP has led to the development of the concept that by inhibiting this type of catalytic activity, it may be possible to diminish the risk of carcinogenesis. Several years ago, it was shown that ellagic acid, a polyphenolic compound widely distributed in plants, could inhibit the mutagenicity of bay region diol-epoxides of several PAHs in the Ames mutagen assay and in Chinese hamster cells.[21] Studies from our laboratory then revealed that ellagic acid could inhibit BP metabolism and its enzyme-mediated binding to DNA and that topical application of this compound substantially diminished the carcinogenicity of topically applied 3-MC in Balb/C mice.[22,23]

In collaboration with Dr. Cynthia Marcelo of the University of Michigan, we assessed the effects of ellagic acid on BP metabolism in cultured Balb/C mouse keratinocytes.[14] Dose-dependent inhibition of AHH and ECD metabolism as well as the intracellular enzyme-mediated binding of BP to mouse keratinocyte DNA was observed, suggesting that naturally occurring polyphenolic compounds such as ellagic acid could prove useful in modulating the risk of cutaneous cancer resulting from environmental exposure to these agents.

Using the same cell culture system, we also showed that the antifungal imidazole, clotrimazole, was a potent inhibitor of P-450-dependent metabolism.[15]

In summary, these studies amply illustrate that the skin contains certain P-450 isozymes that can convert PAHs into highly reactive metabolites that can bind to DNA and thereby initiate a neoplastic response in the skin. Future challenges in this field of research include the further definition of the complete isozyme distribution in human skin with emphasis on defining the possible role of this system in human health. As discussed elsewhere in this volume, it may prove feasible to insert genes into skin cells, and the cloning of genes for the P-450 isozymes may permit the development of carefully programmed patterns of P-450s that could be placed in such cells in an effort to diminish the potential toxicity of environmental agents that require metabolic activation. This could lead to novel strategies in, for example, reducing the risk of environmentally-induced cancer.

REFERENCES

1. BIKLE, D. D., M. K. NEMANIC, E. GEE & P. ELIAS. 1988. 1,25-dihydroxyvitamin D_3 production by human keratinocytes. J. Clin. Invest. **78**: 557–566.
2. NOONAN, F. P., E. C. DeFABO & H. MORRISON. 1988. Cis-urocanic acid, a product formed by ultraviolet B irradiation of the skin, initiates an antigen presentation defect in splenic dendritic cells *in vivo*. J. Invest Dermatol. **99**: 92–99.
3. BOOBIS, A. & J. CALDWELL. 1985. Microsomes and Drug Oxidations. F. DEMatteis & Davies, Eds. Taylor & Francis. London
4. NEBERT, D. W., A. K. JAISWAL, U. A. MEYER & F. J. GONZALEZ. 1987. Human P-450 genes: Evolution, regulation and possible role in carcinogenesis. Biochem. Soc. Trans. **15**: 586–589.
5. DUNAGIN, W. G. 1984. Cutaneous signs of systemic toxicity due to dioxins and related chemicals. J. Ann. Acad. Dermatol. JAAD **10**: 688–700.
6. BICKERS, D. R. & A. KAPPAS. 1978. Human aryl hydrocarbon hydroxylase. Induction by coal tar. J. Clin. Invest. **62**: 1061–1068.
7. BICKERS, D. R., H. MUKHTAR, T. DUTTA-CHOUDHURY, C. L. MARCELO & J. J. VOORHEES. 1984. J. Invest. Dermatol. **83**: 51–56.
8. HARPER, P. A., C. L. COLAS & A. B. OKEY. 1988. Characterization of the Ah receptor and aryl hydrocarbon hydroxylase induction by 2,3,6,8-tetrachlorodibenzo-p-dioxin and benz(a)anthracene in the human A431 squamous cell carcinoma line. Cancer Res. **48**: 2388–2395.
9. EPSTEIN, E. H., JR., N. L. MUNDERLOH & K. FUKUYAMA. 1979. Dithiothreitol separation of newborn dermis and epidermis. J. Invest. Dermatol. **73**: 208–210.
10. BICKERS, D. R., T. DUTTA-CHOUDHURY & H. MUKHTAR. 1982. Epidermis: A preferential site of drug metabolism in neonatal rat skin. Studies on cytochrome P-450 content, the mixed function oxidase and epoxide hyrolase activity. Mol. Pharmacol. **21**: 239–247.

11. BICKERS, D. R., C. L. MARCELO, T. DUTTA-CHOUDHURY & H. MUKHTAR. 1982. Studies on microsomal cytochrome P-450, monooxygenases and epoxide hyrolase in cultured keratinocytes and intact epidermis from Balb-C mice. J. Pharmacol. Exp. Ther. 223: 163–168.
12. COOMES, M. W., A. H. NORLING, R. J. POHL, D. MULLER & J. R. FOUTS. 1983. Foreign compound metabolism by isolated skin cells from the hairless mouse. J. Pharmacol. Exp. Ther. 225: 770–777.
13. POHL, R. J., M. W. COOMES, R. W. SPARKS & J. R. FOUTS. 1984. 7-ethoxycoumarin o-deethylation activity in viable basal and differentiated keratinocytes isolated from the skin of the hairless mouse. Drug Metab. Dispos. 12: 25–34.
14. MUKHTAR, H., B. J. DELTITO, JR., C. L. MARCELO, M. DAS & D. R. BICKERS. 1984. Ellagic acid: A potent naturally occurring inhibitor of benzo(a)pyrene metabolism and its subsequent glucuronidation, sulfation and covalent binding to DNA in cultured Balb/C mouse keratinocytes. Carcinogenesis 5: 1565–1571.
15. DAS, M., H. MUKHTAR, B. J. DELTITO, JR., C. L. MARCELO & D. R. BICKERS, 1986. Clotrimazole, an inhibitor of benzo(a)pyrene metabolism and its subsequent glucuronidation, sulfation and macromacular binding in Balb/C mouse cultured keratinocytes. J. Invest. Dermatol. 87: 4–10.
16. FRIEDMAN, F. K., S. S. PARK, B. J. SONG, K. C. CHENG, T. FUJINO & H. V. GELBOIN. 1986. Monoclonal antibody directed analysis of cytochrome P-450. Adv. Exp. Med. Biol. 197: 145–154.
17. ELING, T., J. CURTIS, J. BATTISTA & L. J. MARNETT. 1986. Oxidation of (+)-7-8dihydroxy-7,8-dihydrobenzo(a)pyrene by mouse keratinocytes: Evidence for peroxyl radical-and monooxygenase dependent metabolism. Carcinogenesis 7: 1957–1963.
18. ROBINSON, R. C., K. C. CHENG, S. S. PARK, H. V. GELBOIN & F. K. FRIEDMAN. 1986. Structural comparison of monoclonal antibody immunopurified pulmonary and hepatic cytochrome P-450 from 3-methylcholanthrene-treated rats. Biochem. Pharmacol. 35: 3827–3830.
19. KHAN, W. A., S. S. PARK, H. V. GELBOIN, D. R. BICKERS & H. MUKHTAR. 1987. Characterization of cytochrome P-450 isozymes in neonatal rat epidermis using monoclonal antibodies. Clin. Res. 35: 694A.
20. MUKHTAR, H., P. ASOKAN, M. DAS, R. SANTELLA, D. R. BICKERS. 1986. Benzo(a)pyrene diol-epoxide-I-DNA adduct formation in the epidermis and lung of SENCAR mice following topical application of crude coal tar. Cancer Lett. 33:287–294.
21. WOOD, A. W., M. T. HUANG, R. L. CHANG, H. C. NEWMARK, R. E. LEHR, H. YAGI, J. M. SAYER, D. M. JERINA & A. H. CONNEY. 1983. Inhibition of the mutagenicity of bay-region diol-epoxides of polycyclic aromatic hydrocarbons by naturally occurring plant phenols: Exceptional activity of ellagic acid. Proc. Natl. Acad. Sci. USA 79: 5513–5517.
22. DELTITO, B. J., JR., H. MUKHTAR & D. R. BICKERS. 1983. Inhibition of epidermal metabolism and DNA-binding of benzo(a)pyrene by ellagic acid. Biochem. Biophys. Res. Commun. 114: 388–394.
23. MUKHTAR, H., M. DAS, B. J. DELTITO, JR. & D. R. BICKERS, 1984. Protection against 3-methylcholanthrene-induced skin tumorigenesis in Balb/C mice. Biochem. Biophys. Res. Commun. 119: 751–757.

Human Fibroblast and Keratinocyte Synthesis of Eicosanoids in Response to Interleukin 1

Evidence for Fibroblast Heterogeneity

MARC E. GOLDYNE

Department of Dermatology
University of California, San Francisco
and
San Francisco Veterans Administration Medical Center
4150 Clement Street
San Francisco, California 94121

The generation of prostaglandin E_2 (PGE_2) by skin fibroblasts[1] coupled with the ability of PGE_2 to stimulate keratinocyte proliferation[2,3] suggest that this eicosanoid may, under certain conditions, serve as a fibroblast-derived autocoid that could affect keratinocyte function. Conversely, interleukins 1α and 1β (IL-1α, IL-1β) have been shown to be released by keratinocytes[4-6] and can enhance eicosanoid synthesis by fibroblasts,[7,8] apparently by stimulating the synthesis of cyclooxygenase.[9] These observations suggest the potential for a keratinocyte-fibroblast interaction involving keratinocyte-derived IL-1 stimulation of fibroblast-derived PGE_2 with subsequent feedback by PGE_2 to influence keratinocyte function (FIG. 1).

Recent studies by Korn[10] have suggested the existence of heterogeneity in the capacity of skin fibroblasts to generate eicosanoids. Accordingly, we decided to evaluate the ability of purified human IL-1 (a mixture of IL-1α and IL-1β) to induce PGE_2 generation by human skin fibroblasts from different anatomical sites. Our results demonstrate that skin fibroblasts exhibit heterogenity in their capacity to spontaneously generate PGE_2 from exogenously introduced ^{14}C-arachidonic acid (^{14}C-AA). Whereas all fibroblast populations responded to IL-1 by increasing PGE_2 generation, the degrees of response were different. In contrast to the fibroblasts, human preconfluent foreskin keratinocytes failed to show an increase in PGE_2 synthesis in response to IL-1 at a concentration that affects the fibroblast synthesis of eicosanoids.

MATERIALS AND METHODS

Cells

Human foreskin fibroblasts isolated from circumcision specimens were grown and passaged in DMEH 21 medium containing 4.5 g glucose/L, 5% newborn bovine calf serum (NBCS) and 2.5 μg/ml fungizone, 1 μg/ml of an equal mixture of penicillin and streptomycin and 0.01% glutamine.[11]

Human skin fibroblasts from other sites were obtained from skin biopsy explants and grown and passaged as previously described.[11]

Human foreskin keratinocytes were grown and passaged in the same medium as the fibroblasts except for the substitution of 5% fetal calf serum (FCS) for NBCS plus the use of 0.4 μg/ml hydrocortisone, 20 μg/ml epidermal growth factor, and 10^{-9} M cholera toxin which are required for optimum keratinocyte growth.[12]

Incubations

Fibroblasts or keratinocytes were incubated using two different protocols:

The first protocol, used to evaluate metabolism of exogenously introduced ^{14}C-arachidonic acid (^{14}C-AA), involved incubating nonconfluent cultures of fibroblasts with 1×10^6 cpm ^{14}C-AA (Sp. Act. 56 μCi/mmol) for 30 min in DMEH-21 without any serum or other additives. Supernatant lipid extracts were then performed and the extracts subjected to thin-layer chromatography as previously described.[11] For some experiments, the fibroblasts were incubated with three half-maximal units/ml of purified human IL-1 (Collaborative Research, Inc., Beford, MA) for 24 hr prior to the incubation with ^{14}C-AA.

The second protocol involved incubation of fibroblasts or keratinocytes for 72 hr in the presence of three half-maximal units/ml of IL-1 with or without 10^{-6} indomethacin. At the conclusion of the incubation, the supernatants were collected, centrifuged to

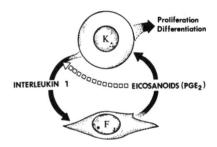

FIGURE 1. This diagram illustrates possible cell-cell interactions occurring between keratinocytes and fibroblasts that may affect keratinocyte function (see text). The *open arrow* signifies a potential negative feedback (see REF. 13).

remove any cells, diluted 1:5 and assayed for PGE$_2$ using a previously described radioimmunoassay.[11] Results are expressed as ng PGE$_2$/10^6 cells plated/72 hrs.

RESULTS

FIGURE 2 shows the lipid extract chromatograms from supernatants of 30-min incubations of neonatal foreskin fibroblasts and adult ear skin fibroblasts (all between the 5th and 7th passage) with ^{14}C-AA in the absence and presence of a 24-hr preincubation with IL-1. Two major results are evident: 1) whereas the neonatal foreskin fibroblasts spontaneously generate ^{14}C-PGE$_2$ from the ^{14}C-AA, the adult fibroblasts failed to show either phospholipid uptake (peak at origin of chromatogram) or metabolism of ^{14}C-AA; and 2) IL-1 not only increased foreskin fibroblast generation of PGE$_2$ but stimulated the adult ear skin fibroblasts to take up ^{14}C-AA into their phospholipids as well as metabolize it to PGE$_2$. Only the cells pre-exposed to the IL-1 for 24 hr prior to incubation with ^{14}C-AA showed the increase in ^{14}C-PGE$_2$. Incubation of IL-1 with the fibroblasts only during the 30-min incubation did not have an effect on PGE$_2$ generation.

FIGURE 3 graphically shows the difference in absolute levels of PGE$_2$ synthesized

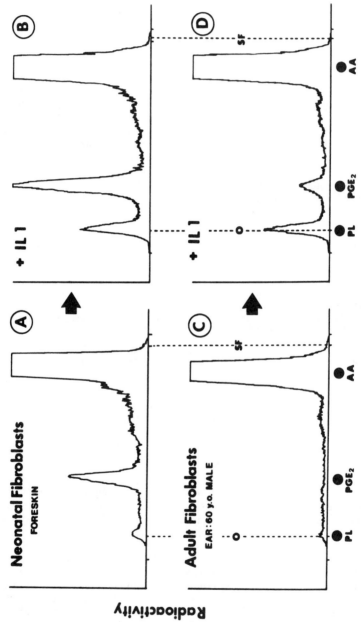

FIGURE 2. Differences in metabolism of ^{14}C-arachidonic acid by neonatal and adult fibroblasts in the absence of, and following a 24-hr preincubation with interleukin 1. Abbreviations: PL, phospholipids labeled with ^{14}C-AA; PGE$_2$, prostaglandin E$_2$; AA, arachidonic acid; O, origin of chromatogram; S.F., solvent front of chromatogram; IL-1, interleukin 1.

by neonatal foreskin (NFF) and adult back skin fibroblasts (ABF) in the absence or presence of IL-1 with or without indomethacin. Both the basal and IL-1-stimulated levels of PGE_2 generated over a 72-hr incubation were significantly greater for the neonatal foreskin than for the corresponding incubations of the adult back skin fibroblasts ($p < 0.05$, (one tailed Student t test). In both cases, however, the IL-1-stimulated cells generated significantly more PGE_2 than their unstimulated counterparts ($p < 0.05$).

To evaluate whether the above differences were purely a function of donor age in contrast to anatomic origin, we compared the results of the studies with neonatal foreskin fibroblasts (NFF) to results obtained using 3rd passage foreskin fibroblasts from two adults (AFF) who underwent elective circumcisions. FIGURE 4 shows that the unstimulated neonatal foreskin fibroblasts generated significantly higher levels of PGE_2 over 72 hr ($p < 0.05$, one tailed Student t test) than did the adult foreskin

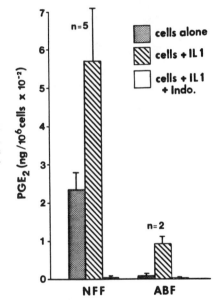

FIGURE 3. Levels of PGE_2 generated over 72 hr by neonatal foreskin fibroblasts (NFF) and adult back skin fibroblasts (ABF) in the absence and in the presence of IL-1 with or without 10^{-6} M indomethacin.

fibroblasts. However, the stimulated (IL-1) fibroblasts from both neonates and adults showed statistically equivalent increases in the generation of PGE_2.

Furthermore, we attemped to see if IL-1 also stimulated PGE_2 synthesis among preconfluent human neonatal foreskin keratinocytes (NFK). Similar incubation studies to those carried out with the fibroblasts were performed. As shown in FIGURE 5, IL-1, at a concentration of three half-maximal units/ml., had no stimulatory effect on baseline levels of PGE_2 synthesized by the keratinocytes over a 72-hr period.

DISCUSSION

The generation of PGE_2 by fibroblasts, the enhanced PGE_2 synthesis triggered by IL-1, and the concept of heterogeneity among human fibroblasts in their capacity to

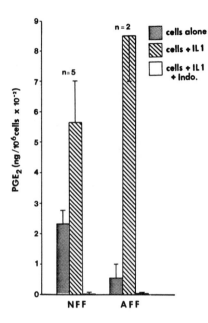

FIGURE 4. Levels of PGE$_2$ generated over 72 hr by neonatal (NFF) versus adult (AFF) foreskin fibroblasts in the absence and in the presence of IL-1 with or without 10^{-6} M indomethacin.

FIGURE 5. Levels of PGE$_2$ generated over 72 hr by human preconfluent foreskin keratinocytes in the absence and in the presence of IL-1 with or without 10^{-6} M indomethacin.

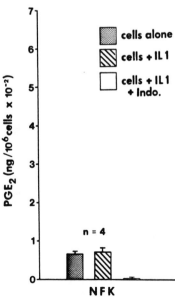

generate PGE_2 are not novel observations. The ability of IL-1 to stimulate eicosanoid generation was documented in previous studies.[7,8] In addition, Korn reported the existence of subpopulations of foreskin fibroblasts, generated using limiting dilution techniques, that synthesize differing quantities of eicosanoids.[10]

Our studies suggest that both age-related and anatomic heterogeneity may exist among human skin fibroblasts in the capacity to generate eicosanoids from exogenously introduced, as well as endogenously derived arachidonic acid. We have observed that both adult ear skin fibrblasts (FIG. 2) and back skin fibroblasts (data not shown) are, under the experimental conditions employed, unable to spontaneously convert exogenous free arachidonic acid to eicosanoids; this is not the case with neonatal foreskin fibroblasts. The adult fibroblasts, however, did respond to IL-1 with both an increased uptake of AA into phospholipids and the generation of PGE_2. Thus there is a difference between the neonatal foreskin and adult back and ear skin fibroblasts in the ability to convert exogenous AA to eicosanoids. It also appears that the response of the neonatal foreskin fibroblasts to IL-1 is greater judging by the peak heights in FIGURE 2. Since data has been published documenting the ability of IL-1 to enhance the synthesis of prostaglandin generating cyclooxygenase,[9] our studies suggest some other regulatory steps that may influence the final generation of eicosanoids in response to IL-1.

The data summarized in FIGURE 4 support the same conclusions as suggested in FIGURE 3, this time, however, in the context of metabolism of endogenously-derived AA. Whereas both the neonatal foreskin and adult back skin fibroblasts spontaneously generate PGE_2, and respond to IL-1, the levels generated by the neonatal foreskin cells under both circumstances, were significantly greater ($p > 0.05$ 1, tailed Student t test) than those generated by the adult back skin cells. As with the chromatographic studies, heterogeneity in the basal and IL-1-stimulated generation of PGE_2 among the fibroblasts studied was demonstrated.

The question of whether the above changes are the result of age-dependent or anatomically-dependent heterogeneity among human skin fibroblasts was in part answered by the studies summarized in FIGURE 4.

These studies compared neonatal (NFF) to adult foreskin fibroblasts (AFF) in regard to both the basal as well as IL-1 stimulated generation of PGE_2. Whereas the neonatal foreskin cells generated statistically greater basal levels of PGE_2 than did the adult cells, stimulation with IL-1 produced quantitatively equivalent PGE_2 levels among both neonatal and adult cells. Thus, while there is an apparent age-related difference in the spontaneous generation of PGE_2 by neonatal versus adult foreskin fibroblasts, this is not seen, as with the adult back skin fibroblasts, when IL-1-stimulated PGE_2 is evaluated.

In summary, the results of these studies are compatible with the concept that both anatomic as well as age-related heterogeneity exists in the capacity of human skin fibroblasts to generate PGE_2 and at least from preliminary studies (data not shown), 6-keto-PGF_{1a} (the stable metabolite of prostaglandin (also called PGI_2)). Because of the known stimulatory effect of PGE_2 on early keratinocyte growth,[2] as well as its inhibitory effect on IL-1 generation by macrophages,[13] it is important to consider how fibroblasts with differing capacities to generate PGE_2 may affect keratinocyte growth, or differentiation, as well as the generation of IL-1 by keratinocytes. In recalling the studies of Billingham and Silvers (14) showing that dermal tissue can influence the regional histogenesis of the epidermis, one cannot help but wonder if fibroblast-derived eicosanoids may be involved as "rheostatic" agents participating in the modulation of keratinocyte growth and differentiation.

The demonstration that IL-1, at the concentration studied, failed to affect PGE_2

synthesis by keratinocytes (FIG. 5) raises the intriguing possibility that in the context of stimulating PGE_2 synthesis, IL-1 can be a keratinocyte-derived product with a specific fibroblast effect not necessarily exerted on other keratinocytes. This finding encourages the exploration of keratinocyte-derived IL-1 and fibroblast-derived PGE_2 as interrelated messengers in keratinocyte-fibroblasts interactions.

ACKNOWLEDGMENTS

The author would like to thank Ms. Laidler Rea, M.S., M.T., for her technical assistance.

REFERENCES

1. BAENZIGER, N. L., M. J. DILLENDER & P. W. MAJERUS. 1977. Cultured human skin fibroblasts and arterial cells produce a labile platelet-inhibitory prostaglandin. Biochem. Biophys. Res. Commun. **78:** 294–301.
2. PENTLAND, A. P. & P. NEEDLEMAN. 1986. Modulation of keratinocyte proliferation *in vitro* by endogenous prostaglandin synthesis. J. Clin. Invest. **77:** 246–256.
3. FURSTENBURGER, G. & F. MARKS. 1980. Early prostaglandin E synthesis is an obligatory event in the induction of cell proliferation in mouse epidermis *in vivo* by the phorbol ester TPA. Biochem. Biophys. Res. Commun. **92:** 749–756.
4. SAUDER, D. N., C. S. CARTER, S. I. KATZ & J. J. OPPENHEIM. 1982. Epidermal cell production of thymocyte activating factor (ETAF). J. Invest. Dermatol. **79:** 34–39.
5. HAUSER, C., J. H. SAURAT, A. SCHMITT, F. JAUNIN & J.-M. DAYER. 1986. Interleukin I is present in normal epidermis. J. Immunol. **136:** 3317–3323.
6. KUPPER, T. S., D. W. BALLARD, A. O. CHUA, J. McGUIRE, P. M. FLOOD, M. C. HOROWITZ, L. LIGHTFOOT & A. AUBLER. 1986. Human keratinocytes contain an RNA indistinguishable from monocyte interleukin 1a and β mRNA. J. Exp. Med. **164:** 2095–2100.
7. ZUCALI, J. R., C. A. DINARELLO, D. J. OBLON, M. A. GROSS, L. ANDERSON & R. S. WEINER. 1986. Interleukin-I stimulates fibroblasts to produce granylocyte-macrophage colony stimulating activity and prostaglandin E_2. J. Clin. Invest. **77:** 1857–1863.
8. BALAVOINE, J.-F., B. DE ROCHEMONTEIX, K. WILLIAMSON, P. SECKINGER, A. CRUCHAND & J.-M. DAYER. 1986. Prostaglandin E_2 and collagenase production by fibroblasts and synovial cells is regulated by urine-derived interleukin I and inhibitor(s). J. Clin. Invest. **78:** 1120–1124.
9. RAZ, A., A. WYCHE, N. SIEGEL & P. NEEDLEMAN. 1988. Regulation of fibroblast cyclooxygenase synthesis by interleukin I. J. Biol. Chem. **263:** 3022–3028.
10. KORN, J. H. 1985. Substrain heterogeneity in prostaglandin E_2 synthesis of human dermal fibroblasts. Arthritis Rheum. **28:** 315–322.
11. BLACKER, K. L., M. L. WILLIAMS & M. E. GOLDYNE. 1987. Mitomycin C-treated 3T3 fibroblasts used as feeder layers for human keratinocyte culture retain the capacity to generate eicosanoids. J. Invest. Dermatol. **89:** 536–539.
12. RHEINWALD, J. G. & H. GREEN. 1975. Serial cultivation of strains of human epidermal keratinocytes: The formation of keratinizing colonies from single cells. Cell **6:** 331–334.
13. KUNKEL, S. L., S. W. CHENSUE & S. H. PHAN. 1986. Prostaglandins as endogenous mediators of interleukin I production. J. Immunol. **136:** 186–192.
14. BILLINGHAM, R. & W. K. SILVERS. 1967. Studies on the conservation of epidermal specificities of skin and certain mucosas in adult mammals. J. Exp. Med. **125:** 429–446.

Local and Systemic Implications of Thymidine Catabolism by Human Keratinocytes[a]

PAULINE M. SCHWARTZ, HAIM REUVENI, AND
LEONARD M. MILSTONE

Dermatology Service
Veterans Administration Medical Center
West Spring Street
West Haven, Connecticut 06516
and
Department of Dermatology
Yale University School of Medicine
New Haven, Connecticut 06510

INTRODUCTION

The epidermis is a continuously renewing tissue. Thymidine in the epidermis is utilized for DNA synthesis by the proliferating basal keratinocytes[1-3] and is generated by the degradation of DNA that occurs in the terminally differentiating granular keratinocytes.[4,5] Little attention has been given to the traffic of thymidine within the epidermis. The recent finding of an enzyme, thymidine phosphorylase, that actively catabolizes thymidine in human keratinocytes[6,7] prompted us to begin an analysis of thymidine traffic throughout the epidermis and to consider the role of epidermis in systemic homeostasis of thymidine.

Thymidine is potentially available to epidermal keratinocytes from two sources (see FIG. 1 for anatomic pathways): from the systemic circulation[8] and from the action of 5'-nucleotidase on deoxythymidine monophosphate[5,9] (see FIG. 2 for metabolic pathways). The amount of thymidine available from these sources is not known, but the maximum amount of available thymidine can be estimated from reported data on blood flow to the skin and the turnover of cells in the epidermis.

Thymidine could have four possible fates within the epidermis (FIG. 1). It could be utilized by proliferating keratinocytes as a precursor for DNA synthesis.[1-3] It could diffuse out of the epidermis and enter the systemic circulation.[10] It could be lost from the body, trapped in desquamating keratinocytes. It could be catabolized by the keratinocyte thymidine phosphorylase to thymine, a molecule with no known fate other than further catabolism.[6,7] Data are available to support the existence of each of these fates for thymidine, except loss through the desquamating keratinocyte; although neither thymidine nor thymine has been detected in human scale or callus,[11,12] thymidine could be removed from the epidermis via desquamation as small molecular weight thymine catabolites trapped in scale.

We recently described the rapid catabolism of thymidine in the human epidermis and attributed this activity to thymidine phosphorylase.[6,7] The physiological role of the

[a]This work was supported by the Veterans Administration Medical Center, West Haven, CT.

115

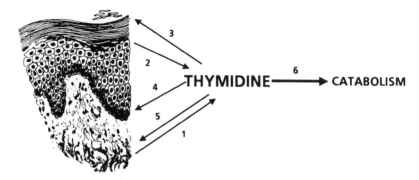

FIGURE 1. Traffic of thymidine within the epidermis. Within the epidermis, thymidine is available from the circulation.[1] Thymidine is generated from the degradation of DNA of differentiating keratinocytes in the granular layer[2] and may be removed from the epidermis by desquamating keratinocytes.[3] Thymidine is utilized by replicating keratinocytes in the basal layer of the epidermis[4] or may enter the systemic circulation.[5] Thymidine can be catabolized by epidermal cells.[6]

thymidine phosphorylase in keratinocytes has not been established. However, it could have important systemic as well as local effects. The activity of the enzyme could alter the local availability of thymidine, from either the circulation or the degradation of DNA. The enzyme also could have a systemic effect on circulating levels of thymidine either by contributing to the catabolism of circulating thymidine or by limiting the contribution of keratinocyte-derived thymidine to the systemic circulation. Studies described below were designed to estimate the capacity of human keratinocytes to catabolize thymidine. Data on the capacity of keratinocytes to catabolize thymidine can be compared to calculations of locally available and circulating concentrations of

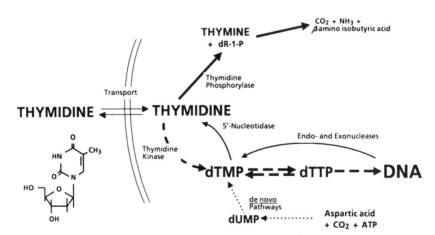

FIGURE 2. Biochemical pathways for thymidine. Thymidine can be transported in and out of the cell ($\rightleftharpoons$), salvaged and incorporated into DNA ($-\rightarrow$) and catabolized ($\rightarrow$). Thymidine can be generated from DNA degradation ($\rightarrow$). The cell can synthesize dTMP via *de novo* pathways in the absence of extracellular thymidine ($\cdots\rightarrow$).

thymidine. With this information, we can make a quantitative estimate of the effect that thymidine phosphorylase could have on the traffic of thymidine within the epidermis.

The capacity of keratinocytes to catabolize thymidine has practical implications for the use of thymidine analogs as topical antiviral or chemotherapeutic agents.[13,14] We, therefore, measured the catabolism of 5-iododeoxyuridine (IUDR) as an example of a thymidine analog whose clinical efficacy might be limited by its catabolism within the epidermis.

MATERIALS AND METHODS

Cell Culture

Human neonatal foreskin keratinocytes were cultivated as previously described on a feeder layer of irradiated 3T3 cells in Dulbecco's modified Eagle's medium (DMEM) containing with 20% fetal calf serum, growth factors and antibiotics.[3,6] Secondary cultures became confluent within 2–3 weeks and formed a stratified tissue of 5–7 cell layers.

Tissues

Human neonatal foreskins were obtained from the Maternity Ward and specimens of adult human skin were obtained from the Skin Bank or Surgical Pathology at Yale New Haven Hospital in accordance with approved protocols. Guinea pig skin was obtained from Hartley guinea pigs (Camm Research, Wayne, NJ). Guinea pig skin was prepared by removing the hair with clippers and by removing the epidermis with a dermatome (Storz Instrument Co., St. Louis, MO) set at 300 μm.

Enzyme Activity in Extracts

Crude, soluble extracts were prepared from cultured keratinocytes and skin in phosphate-buffered saline as previously described.[6] The protein concentration was measured by the Bradford assay.[15]

Thymidine catabolizing activity in extracts was measured in two ways. Radioactive thymidine (100 μM, 4 μCi/ml) was incubated at 37° with crude extracts in phosphate-buffered saline to give a final volume of 100 μl. The percent of thymidine catabolized to thymine was determined by scintillation counting of fractions from a thin-layer chromatogram that separated the nucleoside from the pyrimidine base.[6,16] Alternatively, enzyme activity was measured by a spectrophotometeric assay in which the change in absorbance was determined as the nucleoside was catabolized to the pyrimidine base.[14] The nucleoside (100 μM) was incubated with extracts in phosphate-buffered saline in a 1-ml cuvette. The maximum change in absorbance, taken as the complete conversion of 100 μM nucleoside to its base, was determined using 5 units (5 μmole of thymine formed per min) of *E. coli* thymidine phosphorylase (Sigma Chemical Co., St. Louis, MO). The change in absorbance in 10 min in the presence of the extract as a percent of the maximum change in absorbance was used as a measure of the amount of nucleoside catabolized. Thymidine catabolizing activity measured spectrophotometrically was comparable to activity measured by the radioisotope assay.

Enzyme Activity in Tissues

A modified Franz diffusion chamber[17] was fabricated with a reservoir of 1.6 ml and an orifice 9 mm in diameter. A nylon screen made of 48-μm mesh Nitex fabric (Tetko, Inc., Elmsford, NY) supported the tissue and was positioned on the lower chamber that was filled with phosphate-buffered saline. The tissue was clamped in place with the upper chamber. The lower chamber was stirred constantly and the diffusion chamber was maintained at 37°.

The diffusion chamber was prepared either with human skin or with stratified tissue grown in culture. The skin was prepared by removing most of the stratum corneum by tape-stripping 15 times with cellophane tape, and then the epidermis was removed with a dermatome set at 300 μm. The cultured, stratified tissue of keratinocytes was removed from a 35-mm culture dish by incubating the culture with Dispase (Boehringer Mannheim Biochemicals, Indianapolis, IN), 2.5 mg/ml, for 20 min. The tissue was washed with phosphate-buffered saline and placed on Nitrex fabric with the basal cell layer against the screen. The tissue was examined under a dissecting microscope and only intact tissues without any evidence of holes were used.

Experiments were initiated by adding 100 μl of radioactive thymidine, 7.7 μCi, at a concentration of 1, 10 or 100 μM to the tissue supported in the Franz diffusion chamber. During the incubation, samples of buffer were removed from the side arm of the lower chamber for analysis. Samples were analyzed for total radioactivity and percent thymidine by thin-layer chromatography as described above.

Chemicals

Radioactive thymidine (77 Ci/mmole, 1 mCi/ml) was purchased from New England Nuclear (Boston, MA). Thymidine, thymine, 5-iododeoxyuridine and other chemicals were purchased from Sigma. 6-Aminothymine was synthesized in three steps from 2-bromopropionic acid.[18]

RESULTS

Catabolism of Thymidine by Stratified Cultures of Human Keratinocytes and by Skin in Diffusion Chambers

We have shown that extracts of cultured human keratinocytes and human skin catabolize thymidine (REF. 7 and TABLE 1). To measure the activity of intact keratinocytes, multilayered, cultured epithelial tissues were placed in a modified Franz diffusion chamber. When 1 or 10 μM radioactive thymidine was added to the surface of the tissue in the diffusion chamber, radioactivity diffused rapidly; within 1–3 min, more than 80% of the radioactivity applied to the tissue was found in the lower chamber (TABLE 2). Yet only 27–37% of the radioactivity in the lower chamber was thymidine. The capacity of the stratified cultured keratinocytes to catabolize thymidine based on surface area was proportional to the concentration of thymidine over the range of 1 to 100 μM, indicating that the extent of catabolism of thymidine at these concentrations was limited by diffusion through the epithelium, not by the capacity of the enzymatic activity.

When the Franz chamber was prepared with tape-stripped human skin, the diffusion of thymidine was much slower than the diffusion of thymidine through

TABLE 1. Catabolism of Thymidine by Extracts of Skin[a]

Source	Activity nmole Thymidine Degraded mg Protein · 30 Min
Human neonatal foreskin	59
Cadaver skin	70
Human skin biopsy	52
Mouse skin	6
Rabbit skin	9
Guinea pig skin	2

[a]Soluble extracts of skin were incubated with 100 μM radioactive thymidine in phosphate buffer at 37°. After 30 min, thin-layer chromatography was used to determine the amount of radioactive thymidine as a proportion of the total radioactivity in order to calculate the amount of thymidine that had been catabolized.

cultured tissue. After 3 hr, approximately 2% of total radioactivity applied to the upper chamber was recovered in the lower chamber. At this time, the first sample with sufficient radioactivity to analyze by thin-layer chromatography, only 5% of the radioactivity in the lower chamber was identified as thymidine.

Catabolism of 5-Iodo-2'-deoxyuridine (IUDR)

IUDR is a substrate for thymidine phosphorylase.[14] We, therefore, determined the rate of catabolism of IUDR and thymidine by extracts of human skin (TABLE 3). IUDR was catabolized 3 times faster than thymidine by crude, soluble extracts of human skin, a finding consistent with that of others using thymidine phosphorylase from human platelets.[14] The catabolism of 100 μM IUDR was inhibited by more than 95% in the presence of 100 μM 6-aminothymine, an inhibitor of thymidine phosphorylase.[14]

As a topical antiviral agent, IUDR is more effective in guinea pigs than in humans.[19] Since we have found that human skin contains significantly more thymidine phosphorylase activity than guinea pig skin, we compared the catabolism of IUDR in extracts of human skin to that in extracts of guinea pig skin. In contrast to extracts from human skin, crude soluble extracts of guinea pig skin contained little or no activity for catabolizing IUDR (TABLE 3).

TABLE 2. Catabolism of Thymidine by Stratified Cultures of Human Keratinocytes in a Diffusion Chamber[a]

^{3}H-Thymidine Applied (μM)	Total ^{3}H in Lower Chamber (% Applied)	^{3}H-Thymidine in Lower Chamber (% Recovered ^{3}H)	Thymidine Catabolized pmole Min · cm^2
1	88	37	98
10	90	27	1140

[a]Radioactive thymidine was added to the surface of a stratified tissue of cultivated human keratinocytes supported in a modified Franz diffusion chamber. After 1 min, the radioactivity that diffused through the tissue into the lower chamber was measured and the percent thymidine determined by thin-layer chromatography.

DISCUSSION

Effects of Keratinocyte Thymidine Phosphorylase on the Availability of Thymidine within the Epidermis

Human keratinocytes possess thymidine catabolizing activity that could regulate the amount of thymidine within the epidermis. The relevance of this activity can be evaluated by comparing the capacity of epidermal cells to catabolize thymidine to the availability of thymidine from the circulation and from degradation of DNA of terminally differentiating keratinocytes. The availability of thymidine to the epidermis from the systemic circulation can be estimated from the concentration of thymidine in serum, approximately 0.1 μM,[8] and the blood flow to the skin, 8% of cardiac output or about 400 ml/min.[20] Thus, a maximum of 40 nmole of thymidine per min is available to the epidermis. Based on surface area of the epidermis, 1.7×10^4 cm^2, the availability of thymidine from the systemic circulation can be estimated to be 2.4 pmole of thymidine/min/cm^2. The amount of thymidine available from degradation of keratinocyte DNA can be estimated from the daily turnover of cells in the epidermis, 1.5×10^5 cells/cm^2 per day,[21] and the assumption that thymidine comprises 25% of the DNA content of mammalian cells, 10 pg DNA/cell,[22] or 8.3×10^{-3} pmole thymidine/cell. Thus the maximum amount of thymidine available from DNA degradation of keratinocytes is approximately 0.8 pmole thymidine/min/cm^2.

At a physiological concentration of thymidine, 0.1 μM, the capacity of the epidermis to catabolize thymidine is approximately 10 pmole of thymidine/min/cm^2; this number is derived from a linear extrapolation of experimental data obtained at 1 and 10 μM thymidine, since thymidine catabolizing activity is proportional to the concentration of thymidine at these low concentrations (TABLE 2). Therefore, the capacity of the epidermis to degrade thymidine is some 3 to 4 times greater than the availability of thymidine to the epidermis from the circulation and from DNA degradation combined (FIG. 3). The high capacity of the epidermis to degrade thymidine is consistent with the reported low tissue concentration of thymidine in human skin, 0.53 nmole thymidine per gram wet weight of whole skin; this value is approximately 0.53 μM, similar to the concentration of thymidine in the circulation.[23]

The high capacity of the epidermis to catabolize thymidine relative to its availability does not preclude the utilization of thymidine for DNA synthesis as demonstrated by numerous studies using radioactive thymidine both *in vivo* and *in*

TABLE 3. Catabolism of Thymidine and 5-Iododeoxyuridine by Extracts of Skin[a]

Substrate 100 μM	6-Aminothymine 100 μM	Activity nmole deoxynucleoside degraded	
		mg protein · 10 min	
		Human Skin	Guinea Pig Skin
Thymidine	—	23	ND
	+	7.2	ND
Idodeoxyuridine	—	70	8.2
	+	5.3	ND

[a]Substrates in phosphate-buffered saline were incubated in the presence or absence of 6-aminothymine in a 1-ml cuvette at 37°. The change in UV absorbance 10 min after the addition of extract was measured at 280 nm for thymidine and at 300 nm for IUDR. ND, not determined.

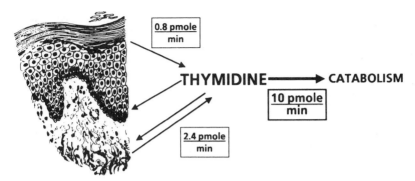

FIGURE 3. The capacity of the human epidermis to catabolize thymidine. The availability of thymidine was estimated from data in the literature (see text). Catabolism of thymidine was estimated from *in vitro* experiments (see TABLE 2). The capacity of epidermal cells to catabolize thymidine is 3 to 4 times greater than the availability of thymidine from the circulation and from DNA degradation.

vitro.[1-3] The level of thymidine available for DNA synthesis, however, may regulate the cellular demand for the *de novo* synthesis of this pyrimidine precursor (see FIG. 2). One hypothesis that is supported by the quantitative data presented here but that ignores the possibility of tissue or intracellular compartmentalization of the enzyme or its substrate is that thymidine levels within the epidermis are regulated by thymidine phosphorylase.

A more complete understanding of the significance of the high capacity of keratinocytes to catabolize thymidine will depend upon determining the localization of thymidine phosphorylase activity in the epidermis. The localization of thymidine catabolizing activity within the epidermis is a real possibility, since thymidine phosphorylase has been shown to be a marker of differentiation in lymphoid cells.[24,25] If, for example, thymidine phosphorylase is more active in well-differentiated keratinocytes than in basal keratinocytes, then thymidine catabolizing activity could affect the availability of thymidine from DNA degradation in granular cells to a greater extent than the availability of thymidine from the circulation.

Effects of Keratinocyte Thymidine Phosphorylase on the Systemic Circulation of Thymidine

The high capacity of the epidermis to catabolize thymidine relative to its availability from the circulation and from DNA degradation suggests that the epidermis might participate in the homeostatic regulation of the concentration of circulating thymidine. Zaharko *et al.*[26] and Covey and Straw[27] have reported that the liver cannot be solely responsible for the systemic catabolism of thymidine. These authors have suggested that spleen, intestine and colon must play an important role in the systemic catabolism of thymidine. These tissues catabolize thymidine with activity ranging from 58 to 89 nmole thymidine/mg protein in 30 mins, which is comparable to the thymidine catabolizing activity we report in human epidermis. These activities are about one-half to one-third the activity of highly active tissues such as liver and spleen which catabolize 113 to 155 nmole thymidine/mg protein in 30 min.[28] Our data suggest that the epidermis also may be an important tissue in the systemic catabolism of thymidine in man. The thymidine phosphorylase activity within the epidermis, 10 pmole of thymidine degraded per min. per cm^2 of tissue, has the potential to clear all the thymidine delivered to the skin by the circulation, estimated to be 2.4 pmole of thymidine per min per cm^2 of skin.

The degradation of DNA by terminally differentiating keratinocytes yields a significant amount of thymidine per day; based on the daily turnover of cells per day and the 8.3 fmoles of thymidine/cell that could be derived from DNA degradation, as mentioned above, approximately 20 μmole of thymidine or 5 mg, is generated daily by DNA catabolism in the epidermis. If this amount of thymidine entered the circulation in a bolus, it would raise the thymidine concentration in the circulation to 4 μM, a concentration some 10 times greater than the concentration of thymidine normally found in the blood. Thus, the high capacity of the epidermis for catabolizing thymidine might be important in limiting the amount of keratinocyte-derived thymidine that enters the circulation.

Effect of Epidermal Thymidine Phosphorylase on the Catabolism of 5'-Iododeoxyuridine (IUDR)

The high capacity of the epidermis to catabolize thymidine would be expected to affect local concentrations of thymidine analogs that are substrates for thymidine phosphorylase.[14] IUDR is a thymidine analog which is a potent antiviral agent *in vitro* but is not a very effective topical antiviral agent in man.[19] This lack of efficacy has been attributed to the poor penetration of the drug through the skin.[19] As demonstrated above, IUDR is rapidly catabolized by human skin. Although the penetration of IUDR to its active site could limit its efficacy, the catabolism of IUDR by epidermal thymidine phosphorylase certainly will reduce the local concentration of IUDR unless very high doses of the drug are applied.

The efficacy of IUDR as a topical therapy for herpes infections in the guinea pig,[19] in contrast to its lack of efficacy in humans, may be explained by differences in thymidine catabolizing activity in guinea pig and human skin. Guinea pig skin contains only 4% of the activity of human skin (REF. 6 and TABLE 1). Like thymidine, IUDR is poorly catabolized by extracts of guinea pig skin. Thus, the guinea pig model would not predict the lack of efficacy of IUDR in man if that ineffectiveness were due to the catabolism of IUDR. The high capacity of the human epidermis to catabolize thymidine indicates that the catabolism of new, promising antiviral 5-substituted thymidine analogs will be an important consideration in evaluating the efficacy of these drugs as topical therapies for cutaneous viral diseases in man.

SUMMARY

The human epidermis possesses a very active thymidine phosphorylase with the capacity to catabolize all of the thymidine available to the epidermis from the circulation and from the degradation of DNA in terminally differentiating keratinocytes. This high capacity of keratinocytes to catabolize thymidine could affect local levels of thymidine within the epidermis for DNA synthesis and could contribute to the regulation of the concentration of thymidine in the systemic circulation. Further work is needed to delineate the physiologic role of this keratinocyte enzyme. A practical consequence of the activity of epidermal thymidine phosphorylase is the role it may play in limiting the clinical efficacy of certain thymidine analogs, an important class of antiviral agents.

REFERENCES

1. GELFANT, S. 1982. On the existence of non-cycling germinative cells in human epidermis *in vivo* and cell cycle aspects of psoriasis. Cell Tissue Kinet. **15**: 393–397.
2. LACHAPELLE, J. M. & T. GILLMAN. 1969. Tritiated thymidine labelling of normal human epidermal cell nuclei. Br. J. Dermatol. **81**: 603–616.
3. MILSTONE, L. M. & J. F. LAVIGNE. 1985. Heterogeneity of basal keratinocytes: Nonrandom distribution of thymidine-labeled basal cells in confluent cultures is not a technical artifact. J. Invest. Dermatol. **84**: 504–507.
4. SANTOIANNI, P. & S. ROTHMAN. 1961. Nucleic acid-splitting enzymes in human epidermis and their possible role in keratinization. J. Invest. Dermatol. **37**: 489–495.
5. OGURA, R., T. HIDAKA & H. KOGA. 1977. Biochemistry of nucleases in relation to epidermal DNA catabolism. *In* Biochemistry of Cutaneous Epidermal Differentiation. M. Seiji & I. A. Bernstein, Eds.: 129–147. University Park Press. New York, NY.
6. SCHWARTZ, P. M., L. C. KUGELMAN, Y. COIFMAN, L. M. HOUGH & L. M. MILSTONE. 1988. Human keratinocytes catabolize thymidine. J. Invest. Dermatol. **90**: 8–12.
7. SCHWARTZ, P. M. & L. M. MILSTONE. 1988. Thymidine phosphorylase in human epidermal keratinocytes. Biochem. Pharmacol. **37**: 353–355.
8. HOWELL, S. B., S. J. MANSFIELD & R. TAETLE. 1981. Significance of variation in serum thymidine concentration for the marrow toxicity of methotrexate. Cancer Chemother. Pharmacol. **5**: 221–226.
9. AMANO, S., T. WATANABE, M. IKEUCHI, M. SASAHARA, E. YAMADA & F. HAZAMA. 1985. 5'-Nucleotidase activities in human fetus. Biol. Neonate **47**: 253–258.
10. RYAN, T. J. 1983. Cutaneous Circulation. *In* Biochemistry and Physiology of the Skin. L. Goldsmith, Ed. Ch. 35. Oxford University Press, New York, NY.
11. HODGSON, C. 1962. Nucleic acids and their decomposition products in normal and pathological horny layers. J. Invest. Dermatol. **39**: 69–78.
12. WHEATLEY, V. R. & E. M. FARBER. 1962. Chemistry of psoriatic scales. II. Further studies of the nucleic acids and their catabolites. J. Invest. Dermatol. **39**: 79–89.
13. KAHILAINEN, L., D. BERGSTROM, L. KANGAS & J. A. VILPO. 1986. *In vitro* and *in vivo* studies of a promising antileukemic thymidine analogue, 5-hydroxymethyl-2'-deoxyuridine. Biochem. Pharmacol. **35**: 4211–4215.
14. DESGRANGES, C., G. RAZAKA, M. RABAUD, H. BRICAUD, J. BALZARINI & E. DECLERCQ. 1983. Phosphorolysis of BVDU and other 5-substituted-2'-deoxyuridines by purified human thymidine phosphorylase and intact blood platelets. Biochem. Pharmacol. **32**: 3583–3590.
15. BRADFORD, M. M. 1976. A rapid and sensitive method for the quantitation of microgram quantities of protein utilizing the principle of protein-dye binding. Anal. Biochem. **72**: 248–254.
16. SCHWARTZ, P. M., R. D. MOIR, C. M. HYDE, P. J. TUREK & R. E. HANDSCHUMACHER. 1985. Role of uridine phosphorylase in the anabolism of 5-fluorouracil. Biochem. Pharmacol. **34**: 3583–3589.
17. FRANZ, T. J. 1975. Percutaneous absorption. On the relevance of *in vitro* data. J. Invest. Dermatol. **64**: 190–195.
18. BERGMANN, W. & T. B. JOHNSON. 1933. Researches on pyrimidines. CXXXII. A new synthesis of thymine. J. Am. Chem. Soc. **55**: 1733–1735.
19. FREEMAN, D. J. & S. L. SPRUANCE. 1986. Efficacy of topical treatment for herpes simplex virus infections: Predictions from an index of drug characteristics *in vitro*. J. Infect. Dis. **153**: 64–70.
20. TAN, O. T. & T. J. STAFFORD. 1987. Cutaneous circulation. *In* Dermatology and General Medicine. T. B. Fitzpatrick, A. Z. Eisen, K. Wolff, I. M. Freedberg & K. F. Austen, Eds.: 357–367. McGraw-Hill Book Co. New York, NY.
21. WEINSTEIN, G. D., J. L. MCCULLOUGH & P. ROSS. 1984. Cell proliferation in normal epidermis. J. Invest. Dermatol. **82**: 623–628.
22. PATTERSON, M. K. 1979. Measurements of growth and viability of cells in culture. Methods Enzymol. **58**: 141–152.
23. HARMENBERG, J., M. MALM & G. ABELE. 1985. Deoxythymidine pools of human skin and guinea pig organs. FEBS Lett. **188**: 219–221.

24. PALU, G. 1980. Thymidine phosphorylase: A possible marker of cell maturation in acute myeloid leukaemia. Tumori **66:** 35–42.
25. SRIVASTAVA, B. I. S. & T. HAN. 1984. Alterations in enzyme expression on 12-*O*-tetradecanoyl-phorbol-13-acetate-induced differentiation of chronic lymphocytic leukemia cells. FEBS Lett. **170:** 152–156.
26. ZAHARKO, D. S., B. J. BOLTEN, D. CHIUTEN & P. H. WIERNIK. 1979. Pharmacokinetic studies during phase I trials of high-dose thymidine infusions. Cancer Res. **39:** 4777–4781.
27. COVEY, J. M. & J. A. STRAW. 1983. Nonlinear pharmacokinetics of thymidine, thymine, and fluorouracil and their kinetic interactions in normal dogs. Cancer Res. **43:** 4587–4595.
28. MAEHARA, Y., H. NAKAMURA, Y. NAKANE, K. KAWAI, M. OKAMOTO, S. NAGAYAMA, T. SHIRASAKA & S. FUJII. 1982. Activities of various enzymes of pyrimidine nucleotide and DNA synthesis in normal and neoplastic human tissues. Gann **73:** 289–298.

Does Catabolism of Stratum Corneum Proteins Yield Functionally Active Molecules?

IAN R. SCOTT,[a] SUSAN RICHARDS,[b] CLIVE HARDING,[a]
J. ERYL LIDDELL,[b] AND C. GERALD CURTIS[b]

[a]*Unilever Research Colworth Laboratory*
Colworth House
Sharnbrook, Bedford MK44 1LQ, United Kingdom
and
[b]*Department of Biochemistry*
University College
P.O. Box 78
Cardiff CF1 1XL, United Kingdom

INTRODUCTION

The title of this paper is a rather broad and cryptic question, so let us start by decoding and answering it.

There are many stratum corneum proteins that are presumably degraded as the stratum corneum matures but one protein dominates all others on a quantitative basis. Filaggrin may comprise more than one quarter of the total protein complement of the newly formed stratum corneum cell but by the time that cell is 3–4 days old, virtually no filaggrin remains and instead the cell contains an enormously concentrated "soup" of amino acids and derivatives thereof.[1-3] These are the catabolised molecules—are they "functionally active?"

Functional Activity of Filaggrin Catabolites

The answer to that question is undoubtedly yes, and this knowledge predates by far the discovery of filaggrin. Two compounds which derive from filaggrin catabolism can be picked out as particularly important.

Urocanic acid, so named because it was first isolated from dogs' urine, is produced within the cells of the stratum corneum from histidine derived from filaggrin catabolism.[4] The enzyme involved, histidase, is found in only one other organ, the liver,[5] making its presence in the stratum corneum significant. Urocanic acid strongly absorbs ultraviolet light in the most damaging part of the solar spectrum and thus forms an important part of the skin's defences against UV-induced damage.[6]

The second compound, pyrrolidone carboxylic acid or PCA, is the major catabolite of filaggrin. It is produced, again within the stratum corneum, by a not completely defined route, from the glutamine that comprises about 30% of the filaggrin molecule.[2,3]

The salt forms of PCA have the important characteristic of being intensely hygroscopic so that a tissue containing relatively high concentrations of PCA (the stratum corneum contains as much as 10% by weight) remains hydrated even when its environment is rather dry as is normally the case with the stratum corneum.[7] As

125

hydration is vital to the maintenance of the elasticity and flexibility of the stratum corneum,[8,9] the action of this hygroscopic substance is vital to its proper function.

The importance of this hygroscopic material is further emphasised by the fact that the stratum corneum has evolved a simple but effective system to ensure that filaggrin catabolism does not commence until the cell containing it becomes partly dehydrated.[10] The kep steps in the process of synthesis and catabolism of filaggrin are summarised in FIGURE 1.

Other Functions for Filaggrin

So, having answered the question in the title of the paper without resorting to any new experimental findings, I will come to the main point of this paper which is to ask whether this function of filaggrin catabolites, together with the other proposed function of filaggrin as a keratin filament aggregating protein[11] is sufficient to account for the synthesis of filaggrin in all known circumstances. I hope to demonstrate that this is not so and to provide evidence for further previously undiscovered functions for filaggrin.

The first function proposed for filaggrin was as a matrix protein which would cause keratin filaments to take up a close packed structure in the cells of the stratum corneum.[11]

Doubts about the need for a specific keratin aggregating protein were first raised by the observation that close packed structures of keratin filaments could be produced just by adding inorganic salts to the keratin.[12] Filaggrin is a very basic protein, largely due to a high arginine content, and it was possible that its keratin aggregating ability

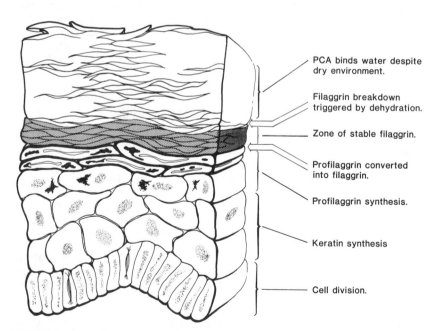

PCA binds water despite dry environment.

Filaggrin breakdown triggered by dehydration.

Zone of stable filaggrin.

Profilaggrin converted into filaggrin.

Profilaggrin synthesis.

Keratin synthesis

Cell division.

FIGURE 1. Schematic structure of the epidermis showing the major steps in filaggrin synthesis and catabolism.

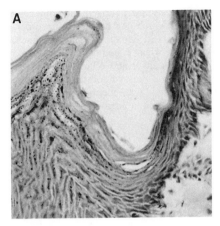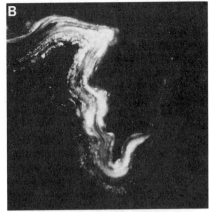

FIGURE 2. Indirect immunofluorescence micrograph of the junctional region between soft and hard palates of the adult rat. (**A**) H and E stained; (**B**) antifilaggrin stained.

was simply a reflection of its basicity. This possibility was apparently ruled out by the failure of the basic nucleoproteins to cause keratin aggregation.[11,13]

We have however observed that if histone H1 is prepared in a similar way to that used for filaggrin (critically, involving denaturation) it has very similar keratin aggregating properties to filaggrin (Scott, unpublished data). The key finding however casting doubt on the *in vivo* importance of keratin aggregation by filaggrin came from the same group that originally identified the keratin aggregating ability of filaggrin. The absence of keratohyalin granules in the affected epidermis of patients suffering from ichthyosis vulgaris is well known. In a recent paper[14] it was confirmed that this absence of keratohyalin is correlated with absence of filaggrin.

Surprisingly, however, the absence of filaggrin had no apparent effect on the keratin pattern seen in affected stratum corneum. It must be concluded therefore that, while filaggrin does interact with keratin *in vitro,* and may interact *in vivo,* the interaction is not vital to the aggregation of keratin seen in the stratum corneum.

What then of the function of filaggrin catabolites? While undoubtedly of importance for the dry surfaced and UV-exposed epidermis, one would predict that filaggrin would be absent from wet surfaced epithelia such as in the oral cavity. This however is not so. Some keratinised oral epithelia do lack filaggrin but in others it is clearly present, for example, in the junctional region between the hard and soft palates of the rat[15] (FIG. 2). There is no evidence for abnormal keratin packing in those tissues lacking filaggrin and no requirement for hygroscopic material in those wet surfaced epithelia in which it is found. Indeed the continuous high humidity of the mouth prevents filaggrin catabolism from even starting, as can be seen from the uniform staining for filaggrin throughout the stratum corneum in FIGURE 2. It seems therefore that we should seek a third function for filaggrin.

Filaggrin and Keratinocyte Envelope Formation

Indications of what that function might be came from experiments designed to identify putative precursors of the cornified, transglutaminase cross-linked, envelope that surrounds the fully differentiated keratinocyte.

Substrates for transglutaminase can be identified by incubating tissue homogenates containing the enzyme with dansyl cadaverine, which can be covalently cross-linked to the substrate protein by transglutaminase and detected by fluorescence. FIGURE 3 shows the results of a typical experiment when a homogenate of whole newborn rat epidermis was incubated with dansyl cadaverine. When analysed by SDS-polyacrylamide gel electrophoresis, there were two major fluorescent products, one of very high molecular weight and the other having an identical mobility to filaggrin. These results are similar to those of Hanigan and Goldsmith although those authors did not identify the protein as filaggrin.[16]

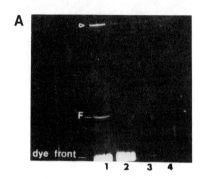

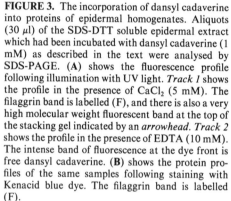

FIGURE 3. The incorporation of dansyl cadaverine into proteins of epidermal homogenates. Aliquots (30 μl) of the SDS-DTT soluble epidermal extract which had been incubated with dansyl cadaverine (1 mM) as described in the text were analysed by SDS-PAGE. (A) shows the fluorescence profile following illumination with UV light. *Track 1* shows the profile in the presence of $CaCl_2$ (5 mM). The filaggrin band is labelled (F), and there is also a very high molecular weight fluorescent band at the top of the stacking gel indicated by an *arrowhead. Track 2* shows the profile in the presence of EDTA (10 mM). The intense band of fluorescence at the dye front is free dansyl cadaverine. (B) shows the protein profiles of the same samples following staining with Kenacid blue dye. The filaggrin band is labelled (F).

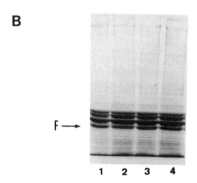

Analysis by two-dimensional electrophoresis confirmed the identity of this fluorescent product as filaggrin (not shown). On its own this was not a remarkable finding as it is known that filaggrin has the potential to act as a transglutaminase substrate. However, when the experiment was repeated using viable organ cultures of newborn rat skin, incubated with dansyl cadaverine, a similar result was obtained (FIG. 4). The viability of the organ culture was established by histology and by measuring the cultures continued ability to synthesise profilaggrin from [³H]histidine which is a particularly sensitive test of the health of a culture.

From this result we must conclude that not only does filaggrin have the *potential* to be a transglutaminase substrate but that in living skin it can be *and is* acted on by the

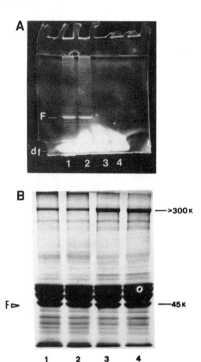

FIGURE 4. The incorporation of dansyl cadaverine into epidermal proteins in whole skin organ culture. Whole skin organ culture experiments were carried out in the presence of dansyl cadaverine (1 mM) as described in the text. Aliquots (30 μl) of the SDS-DTT soluble supernatant were analysed by SDS-PAGE. (A) shows the fluorescence profile under UV light of proteins labelled with dansyl cadaverine. *Tracks 1 and 2* show the profile in the presence of $CaCl_2$ (5 mM), whilst *tracks 3 and 4* show the profile in the presence of EDTA (10 mM). The filaggrin band is denoted by the letter (F). The intense band of fluorescence at the dye front (labelled df) is due to free dansyl cadaverine diffusing back into the gel during photography. (B) shows the protein profile of the same samples following staining with Kenacid blue dye. The filaggrin band is labelled (F) and has a molecular weight of 45 K. The band at approximately 300 K is profilaggrin.

endogenous transglutaminase. It seems likely therefore that at least some filaggrin must be incorporated into the keratinocyte envelope.

Keratinocyte envelopes are extremely resistant to solubilisation and it is therefore not possible to confirm the presence of filaggrin in the structure directly. We therefore adopted an immunological approach using an affinity purified polyclonal antiserum against rat filaggrin.

Indirect immunofluorescence microscopy generally shows a pattern of intense staining for filaggrin in the keratohyalin granule and lower parts of the stratum

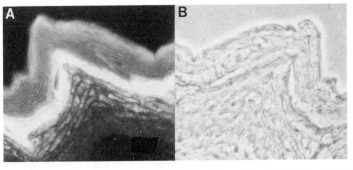

FIGURE 5. Immunofluorescence micrograph of 7-day-old rat skin stained with antirat filaggrin antibody. (A) Immunofluorescence; (B) same field by phase contrast (mirror image)

corneum with little staining in the upper stratum corneum.[10] When, however, favourable sections are studied (FIG. 5) specific staining of the cell margins in the upper stratum corneum is visible.

In an attempt to quantify the amount of filaggrin present in the cell envelope, an ELISA method was developed. This was based on the binding of purified cell envelopes to ELISA wells, followed by incubation with antifilaggrin antiserum and alkaline phosphatase conjugated second antibody.

Antigen and antibody dilution curves were prepared with both envelopes and

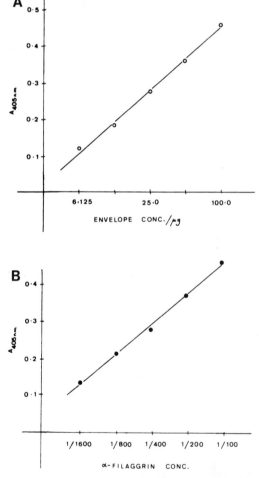

FIGURE 6. Cross-reactivity between purified epidermal cell envelopes and antifilaggrin. The antigen dilution curve (**A**) was obtained when ELISA plates were absorbed with purified cell envelopes (6.125–100 μg/well) and then reacted with affinity purified antifilaggrin at a dilution of 1/100 in PBS. The antibody dilution curve (**B**) was obtained when ELISA plates were absorbed with purified cell envelopes (100 μg/well) and reacted with increasing concentrations of affinity purified antifilaggrin in PBS. All values have been corrected for nonimmune serum.

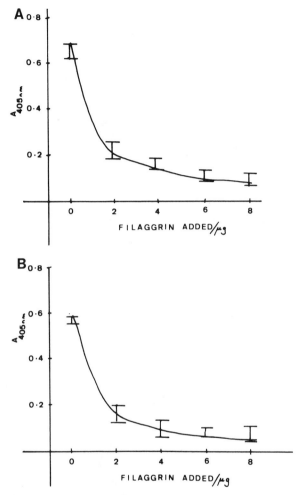

FIGURE 7. Inhibition of antifilaggrin binding by preincubation of the antibody with pure filaggrin. Inhibition curves were obtained following the application of an antifilaggrin: filaggrin complex containing varying amounts of filaggrin ($0-8$ μg) and a $1/1000$ dilution of antifilaggrin to ELISA plates absorbed with either (**A**) pure filaggrin (12.5 μg/well) or (**B**) purified cell envelopes (100 μg/well). The values have been corrected for nonimmune serum.

purified filaggrin bound on the plate. This gave the expected linear response (FIG. 6). The specificity of the antibody for filaggrin in envelopes was confirmed by an antigen competition experiment (FIG. 7) in which the competition between a solution of pure filaggrin and pure filaggrin bound to the ELISA plate was identical to that between a solution of pure filaggrin and envelopes bound to the plate.

The quantity of envelopes bound to the plate was estimated by recovering the envelopes from the ELISA plate in strong acid, hydrolysing them and assaying the free amino acids by the ninhydrin method. A standard curve of absorbance of p-nitrophenol vs filaggrin bound to the ELISA plate was constructed using [³H] histidine-labelled filaggrin with acid elution from the plate and scintillation counting. This method indicated that 13 ± 6% of the total envelope protein was represented by filaggrin (TABLE 1).

Naturally, this estimate of filaggrin content relies on the accuracy of certain assumptions. These are: (i) that filaggrin bound directly to the ELISA plate binds the same amount of antibodies as filaggrin covalently bound to the cell envelope. Steric constraints would most likely lead to reduced binding of antibody to the envelope-bound filaggrin leading to an underestimate of filaggrin content. (ii) The antigenic epitopes covalently bound to the cell envelopes represent the whole filaggrin molecule. If, for example, a strongly antigenic epitope representing only a small part of the filaggrin molecule was preferentially bound to the envelope, then the immunological method could seriously overestimate the filaggrin content.

TABLE 1. Filaggrin Content of Stratum Corneum Cell Envelopes

				% Filaggrin in Envelopes	
Expt. No.	Envelopes Applied to Elisa Well (μg)	Envelopes Bound (μg)	Filaggrin Detected (μg)	Means of Individual Expts.	Overall Mean
1	100	8.0	1.15	13 ± 2%	
		10.0	1.16		
2	100	5.0	0.60	9 ± 3%	
		7.5	0.61		
		10.0			13 ± 6%
3	50	2.5	0.68	22 ± 6%	
		4.0	0.672		
4	75	2.5	0.27	10 ± 1%	
		2.92	0.271		

CONCLUSIONS

Subject to the accuracy of the assumptions outlined above, it appears that filaggrin is a significant component of the cell envelope in the newborn rat stratum corneum. Filaggrin has a number of properties that make it unique among the cell envelope precursors so far identified. Firstly, it has a known affinity for keratin and could therefore be important in mediating the interaction between the envelope and the keratin fibres that fill the stratum corneum cell.

Secondly, it lacks the ability to self aggregate under the influence of transglutaminase as it substantially lacks lysine groups. It can only therefore become part of a highly cross-linked structure by acting as a backbone onto which other proteins with lysine donor groups are attached. In this it resembles involucrin which when present in cell free preparations catalyses the cross-linking of soluble proteins which would not be cross-linked in its absence.[17] Filaggrin represents a more extreme type of substrate in that it has an absolute requirement for other proteins to form cross-linked aggregates.[18] It is intriguing to speculate on whether filaggrin fulfills a similar function to involucrin. It may be significant that the species in which involucrin is found (human) has generally very much smaller quantities of filaggrin than the rodent species in which involucrin is absent.[19,20]

METHODS

Incorporation of Dansyl Cadaverine into Proteins of Epidermal Homogenates

Newborn (2–3-day-old) rats were killed by cervical dislocation. Epidermis was isolated either by heat separation (immersion in a water bath at 60°C for 1 min) or by freeze-scraping,[4] and homogenized at 4°C (1 epidermis/ml) in Tris-EDTA buffer pH 7.5 containing Tris-HCl (50 mM), EDTA (1 mM) and PMSF (2 mM), using a Kontes Duall 22 tissue grinder. Incubation mixtures comprised: epidermal homogenate (500 μl), dansyl cadaverine (1 mM), as the fumarate salt, and either $CaCl_2$ (5 mM) or EDTA (10 mM). Incubation mixtures were adjusted to a final volume (1 ml) with the Tris buffer and were incubated at 37°C for 2 h, after which time they were prepared for SDS-polyacrylamide gel electrophoresis (SDS-PAGE).

Incorporation of Dansyl Cadaverine into Epidermal Proteins
in Whole Skin Cultures

Pieces of full-thickness, newborn rat skin (1 cm × 1 cm) were floated dermis side down on standard Dulbecco's MEM (0.5 ml) containing dansyl cadaverine fumarate salt (1 mM), at 37°C in a humidified 5% CO_2 atmosphere for 1 h. Epidermis was then isolated by freeze-scraping and homogenized in SDS-lysis buffer (pH 8.0) comprising: Tris-HCl (100 mM), glycerol (10%), SDS (2%), EDTA (1 mM) and PMSF (2 mM), prior to electrophoresis.

The viability of this culture system was assessed by its ability to incorporate [³H]histidine (100 μl; 100 μCi) into profilaggrin as described by Harding and Scott.[19] In addition, pieces of cultured skin which had been incubated in the presence of dansyl cadaverine were fixed, embedded, stained with haematoxylin and eosin, and sectioned (8 μm thickness) for histology.

SDS-Polyacrylamide Gel Electrophoresis

Samples (except those from organ culture experiments), were adjusted to the following concentrations: SDS (2%), EDTA (1 mM), PMSF (2 mM) and glycerol (10%). All samples were made 20 mM with respect to DTT, incubated at 60°C for 30 min and centrifuged (10,000 g for 5 min).

Bromophenol blue tracking dye was added to the supernatant and aliquots (25 μl) were analysed on mini gradient slab gels.[19]

Gels were photographed under ultraviolet light to detect fluorescent bands and then stained with coomassie blue by standard procedures.

Preparation of [³H] Filaggrin

Newborn (1-day-old) rats were each injected subcutaneously with [³H]histidine (20 μl, 20 μCi), using a Hamilton PB600 repeating dispenser. Animals were sacrificed 24 h later and the epidermis was isolated by freeze-scraping, homogenized (1 epidermis/ml), in a urea buffer comprising: urea (8 M), Tris-HCl (10 mM), EDTA (1 mM), PMSF (2 mM) at pH 8.0, and incubated at 37°C for 30 min to extract the urea soluble proteins.

The extract was passed through a small column (1.5 × 10 cm) of DE52 cellulose in the same buffer to remove keratin, nucleic acids, and profilaggrin. The resulting partially purified filaggrin was stored at $-70°C$ in the urea buffer, and desalted immediately prior to use into 10 mM Tris-HCl (pH 7.5), using a centrifugal desalting method.[21]

The purity of the prepared filaggrin was determined by polyacrylamide gel electrophoresis followed by gel slicing and scintillation counting to quantitate the fraction of the total radioactivity accounted for by filaggrin. Typically this was greater then 70%. The protein content of filaggrin preparations was assayed by the ninhydrin method, following complete acid hydrolysis in 6 M HCl under N_2 at 110°C for 18 h, using glutamic acid as standard.

Preparation of Cell Envelopes

Newborn rats were killed and whole dorsal and ventral skin was removed and floated dermis side down for 90 min at 37°C on a solution of phosphate buffered saline containing Dispase (0.25%). The epidermis was then peeled away from the dermis.

The epidermis was rinsed in SDS-lysis buffer and homogenized in the same buffer. The homogenate was incubated at 60°C for 30 min and then centrifuged (10,000 g for 5 min), and the pellet was washed exhaustively with the same buffer.

The pellet was lyophilized, and the SDS removed by ion-pair extraction.[22] Protein was determined by dry weight.

Preparation of Antisera and Indirect Immunofluorescence Microscopy

The preparation and purification of rabbit antirat filaggrin antiserum and its use for indirect immunofluorescence microscopy has been described previously.[10]

Detection of Filaggrin in Purified Cell Envelopes
Using a Noncompetitive ELISA Technique

Cell envelopes were resuspended (1 mg/ml) in carbonate/bicarbonate coating buffer (15 mM, pH 9.6) containing Thimerosal (0.02%, Sigma).

The ELISA protocol was as follows:

1. Cell envelopes (100 μl) were absorbed onto 96-well microtitration plates (Falcon 3915, Becton Dickinson Labware, Oxnard, USA), at 4°C overnight.
2. Plate was washed ×3 with PBS containing Tween-20 (0.05%, Bio-Rad).
3. Each well was blocked with BSA (1%) in PBS (100 μl) for 1 h at room temperature.
4. As (2) above.
5. Affinity purified rabbit antirat filaggrin (100 μl) diluted in PBS was applied to each well and the plate was incubated for 1 h at room temperature.
6. As (2) above.
7. Antirabbit IgG alkaline phosphatase conjugate (Sigma), diluted 1/1000 in PBS was applied to each well (100 μl) and incubated at room temperature for 1 h.
8. As (2) above followed by one wash in distilled water.
9. Aliquots (100 μl) of disodium p-nitrophenyl phosphate (1 mg/ml) in glycine buffer (0.1 M, pH 10.4) containing $MgCl_2$ (1 mM) and $ZnCl_2$ (1 mM) were added and the colour was allowed to develop for 30 min.

10. Reaction was terminated by the addition of an aliquot (30 μl) of NaOH (1 M) to each well.

11. The absorbance was measured at 405 nm using an EIA reader (Bio-Rad).

Inhibition of Antifilaggrin Binding to Cell Envelopes Following Absorption of the Antibody with Purified Filaggrin

Plates were coated with cell envelopes (100 μg/100μl), overnight at 4°C. Antifilaggrin (1/4000 in PBS) was preincubated for several hours at 4°C with varying concentrations (2–8 μg/100 μl) of purified filaggrin, and this mixture was applied to the coated plates instead of the antibody alone. The remainder of the protocol was carried out as above.

Quantitation of Filaggrin in Cell Envelopes

FIGURE 6 shows the linear relationship between envelope concentration and absorbance at 405 nm, showing that envelopes contain a measurable amount of filaggrin. However, in order to quantify the amount of filaggrin in envelopes it was necessary to determine (a) the actual amount of filaggrin bound to the ELISA plate when carrying out the standard filaggrin calibration curve, and (b) the actual amount of cell envelope protein bound.

The relationship between the amount of filaggrin bound to the plate and the optical density at 405 nm was determined using the [^{3}H]histidine-labelled filaggrin. ELISA plates were coated with [^{3}H]-labelled filaggrin (100 μl: 1.79×10^3 dpm/μg) at concentrations ranging from (10–25 μg/well) at 4°C overnight in bicarbonate coating buffer. The plates were washed $\times 3$ with PBS/Tween, to remove the unbound [^{3}H] filaggrin and aliquots (200 μl) of 2 M HCl were applied to each well to remove the bound filaggrin. The plate was incubated at 37°C for 60 min and the HCl was then transferred into a scintillation vial. The wells were washed with 2×250 μl aliquots of 2 M HCl, and the combined washes (700 μl total) were neutralized with 2 M NaOH, and assayed for radioactivity by scintillation counting in Lumagel scintillant (10 ml). On the same plates, duplicate samples of bound [^{3}H] filaggrin were assayed using the ELISA protocol outlined above. The relationship between the amount of [^{3}H] filaggrin absorbed and the absorbance at 405 nm was found to be linear between OD of O and 1.0 (0 to 2 μg bound filaggrin).

The binding of epidermal cell envelopes was quantified by removing the bound protein from the washed plate by incubation at 37°C for 60 min, with 200 μl aliquots of 2 M Aristar HCl. The acid and subsequent washes (2×250 μl aliquots of 2 M Aristar HCl) were decanted into acid washed test tubes. The HCl was removed by freeze-drying and samples were redissolved in 6 M Aristar HCl (500 μl), sealed under vacuum, and hydrolysed at 110°C for 18 h.

The HCl was removed under vacuum, and samples were assayed for protein content using the ninhydrin method with glutamic acid as standard. Wells which had been coated with bicarbonate coating buffer only were processed in an identical manner and served as blanks in each assay.

The binding capacity of the ELISA plates varied from one batch to another, and values ranged from 6–8% for filaggrin and 4–20% for cell envelopes. Because of this variability it was essential that measurements of filaggrin binding, envelope binding, and the measurements of filaggrin in cell envelopes (using antifilaggrin) be carried out on the same ELISA plates. In this way the filaggrin concentration could be expressed as "% filaggrin in the bound envelope fraction" (see TABLE 1).

REFERENCES

1. SCOTT, I. R., C. R. HARDING & J. G. BARRETT. 1982. Histidine rich protein of the keratohyalin granules: Source of the free amino acids, urocanic acid and pyrrolidone carboxylic acid in the stratum corneum. Biochim. Biophys. Acta **719:** 110–117.
2. BARRETT, J. G. & I. R. SCOTT. 1983. Pyrrolidone carboxylic acid synthesis in guinea pig epidermis. J. Invest. Dermatol. **81:** 122–124.
3. SCOTT, I. R. & C. R. HARDING. 1981. Studies on the synthesis and degradation of a histidine rich phosphoprotein from mammalian epidermis. Biochim. Biophys. Acta **669:** 65–78.
4. SCOTT, I. R. 1981. Factors controlling the expressed activity of histidine-ammonia lyase in the epidermis and the resulting accumulation of urocanic acid. Biochem. J. **194:** 829–838.
5. ZANNONI, V. G. & B. N. LA DU. 1963. Biochem J. **88:** 160–162.
6. BADEN, H. P. & M. A. PATHAK. 1967. J. Invest. Dermatol. **48:** 11–17.
7. LADEN, K. & R. SPITZER. 1967. J. Soc. Cosmet. Chem. **18:** 351–360.
8. BLANK, I. H. 1952. J. Invest. Dermatol. **18:** 433.
9. BLANK, I. H. 1953. J. Invest. Dermatol. **21:** 259.
10. SCOTT, I. R. & C. R. HARDING. 1986. Filaggrin breakdown to water binding compounds during development of the rat stratum corneum is controlled by the water activity of the environment. Dev. Biol. **115:** 84–92.
11. DALE, B. A., K. A. HOLBROOK & P. M. STEINERT. 1978. Assembly of stratum corneum basic protein and keratin filaments in macrofibrils. Nature (London) **276:** 729–731.
12. FUKUYAMA, K., T. MUROZUKA, R. CALDWELL & W. L. EPSTEIN. 1978. Divalent cation stimulation of *in vitro* fibre assembly from epidermal keratin protein. J. Cell Sci. **33:** 255–263.
13. STEINERT, P. M., J. S. CANTIERI, D. C. TELLER, J. D. LONSDALE-ECCLES & B. A. DALE. 1981. Characterisation of a class of cationic proteins that specifically interact with intermediate filaments. Proc. Natl. Acad. Sci. USA **78:** 4097–4101.
14. SYBERT, V. P., B. A. DALE & K. A. HOLBROOK. 1985. Ichthyosis vulgaris: Identification of a defect in the synthesis of filaggrin correlated with an absence of keratohyalin granules. J. Invest. Dermatol **84:** 191–194.
15. DALE, B. A., W. B. THOMPSON & I. B. STERN. 1982. Arch. Oral Biol. **27:** 535–545.
16. HANIGAN, H. & L. A. GOLDSMITH. 1978. Biochim. Biophys. Acta **922:** 589–601.
17. SIMON, M. & H. GREEN. 1985. Cell **40:** 677–683.
18. BADEN, H. P. & J. KUBILUS. 1983. Cross-linking of epidermal fibrous protein. *In* Stratum Corneum. R. Marks & G. Plewig, Eds. Springer Verlag. Berlin.
19. HARDING, C. R. & I. R. SCOTT. 1983. Histidine-rich proteins (filaggrins). Structural and functional heterogeneity during epidermal differentiation. J. Mol. Biol. **170:** 651–673.
20. RICE, R. H. & S. M . THACHER. 1986. Involucrin, a constituent of cross-linked envelopes and marker of squamous maturation. *In* Biology of the Integument. Vol. 2. J. Bereiter-Hahn, A. G. Maltoltsy & K. S. Richard, Eds.: 752–761. Springer-Verlag. Berlin.
21. HELMERHOLST, E. & G. B. STOKES. 1980. Anal. Biochem. **104:** 130–135.
22. HENDERSON, L. E., S. OROSZLAN & W. KONIGSBERG. 1979. Anal. Biochem. **93:** 153–157.

A Tumor-Secreted Protein Associated with Human Hypercalcemia of Malignancy

Biology and Molecular Biology[a]

MICHAEL ROSENBLATT,[b] MICHAEL P. CAULFIELD,[b]
JOHN E. FISHER,[b] NOBORU HORIUCHI,[c]
ROBERTA L. MCKEE,[b] SEVGI B. RODAN,[b]
MARK A. THIEDE,[b]DAVID D. THOMPSON,[b]
J. GREGORY SEEDOR,[b] RUTH E. NUTT,[b]
MARK E. GOLDMAN,[b] JANE E. REAGAN,[b]JAY J. LEVY,[b]
C. THOMAS GAY,[b] PATRICIA A. DEHAVEN,[b]
GORDON J. STREWLER,[d] ROBERT A. NISSENSON,[d]
THOMAS L. CLEMENS,[c] AND GIDEON A. RODAN[b]

[b]Department of Biological Research and Molecular Biology
Merck Sharp & Dohme Research Laboratories
West Point, Pennsylvania 19486

[c]Regional Bone Center
Helen Hayes Hospital
(New York State Department of Health)
West Haverstraw, New York 10993

[d]Veterans Administration Medical Center
and
University of California at San Francisco
San Francisco, California 94121

INTRODUCTION

In 1941 Albright[1] proposed that tumors could secrete parathyroid hormone (PTH) ectopically and cause the paraneoplastic syndrome of humoral hypercalcemia of malignancy (HHM). Since that time, this aspect of tumor biology has been the subject of intensive investigation. Factors secreted by tumors and postulated to produce hypercalcemia include: prostaglandins, growth factors, and peptides or proteins that circulate in blood and act on bone or kidney in an endocrine fashion, even in the absence of tumor metastasis to bone.

Several studies have demonstrated that a PTH-like factor, physicochemically and immunologically distinct from PTH, is secreted by tumor cells.[2-5] Furthermore, messenger RNA (mRNA) for PTH is not found in such tumors.[6] However, like PTH, the factor does stimulate adenylate cyclase *in vitro* in PTH target cells, and this

[a]This work was supported in part by National Institutes of Health Grants AR 36446, AR 39191, AM-35323, AM-11262, and CA-34738, by American Cancer Society Grant PDT-229, and by funds from the Medical Research Service of the Veterans Administration, and Merck Sharp & Dohme Research Laboratories.

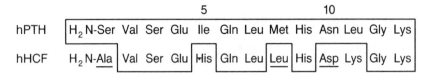

FIGURE 1. NH$_2$-terminal amino acid sequence of human PTH-(1—34) and hHCF-(1—34)NH$_2$. Identical amino acids are enclosed in the *boxed area*. *Underlined* residues differ by a single nucleotide in their trinucleotide codons.

activity can be inhibited by PTH antagonists.[2–5] Three groups have recently isolated peptides derived from several different human tumors (lung squamous carcinoma, renal cell carcinoma, breast carcinoma) and obtained partial amino acid sequences of these peptides.[7–9] The putative full-length peptide structures of the human humoral hypercalcemia factor (hHCF) based on the nucleotide sequence of the cloned complementary DNAs (cDNA) have also been elucidated.[10–12] In each case, within the NH$_2$-terminal 13 residues there is considerable homology between hHCF and the biologically critical NH$_2$-terminal region of PTH (FIG. 1), although the tumor factor appears to be the product of a different gene than that encoding PTH.[10,12]

An NH$_2$-terminal fragment of the tumor-secreted protein, hHCF-(1—34)NH$_2$, has been chemically synthesized and its biological properties evaluated.[13,14] This peptide displays multiple actions similar to PTH *in vivo* and can produce components of the clinical syndrome of HHM.[14] Like PTH, it increases levels of 1,25-dihydroxyvitamin D$_3$.[14] It acts directly on bone, binds to PTH receptors, and stimulates multiple PTH-like postreceptor responses.[13–15] Furthermore, the PTH antagonist [Tyr-34]bPTH-(7—34)NH$_2$ blocks its actions.[16] Cloning revealed the presence of two forms of mRNA encoding the tumor protein, originating from a single gene by an alternative splicing mechanism.[12] The 3'-untranslated region is homologous to the corresponding domain of the c-myc proto-oncogene. These studies address issues in tumor biology and establish a potential, novel approach using PTH antagonists for treatment of tumor-associated hypercalcemia.

Cloning and Structure of the Tumor Factor

A peptide secreted by tumors associated with the clinical syndrome of HHM was recently purified from a human renal carcinoma cell line (786-0).[9] Its NH$_2$-terminal amino acid sequence has considerable similarity with that of PTH and peptides isolated from human breast and lung carcinomas (BEN).[7,8,10–12]

A cDNA complementary to the 1800-base mRNA encoding a PTH-like peptide expressed by the human renal carcinoma cell 786-0 was isoalted and sequenced. The cDNA contains an open reading frame encoding a leader sequence of 36 amino acids and a 139-residue peptide. Eight of the first 13 amino acid residues are identical to those of the corresponding region of PTH. Through the first 828 bases the sequence of this cDNA is identical with one recently isoalted from a BEN cell library.[10] However, beginning with base 829 the sequences diverge and the open reading frame is shortened by two amino acids (FIG. 2).

Our findings reveal that renal 786-0 cells are making two major hHCF-related mRNAs of 1.8 kb and 1.55 kb, respectively, which differ in their 3'-untranslated sequences as detected by differential RNA blot analysis. The 1.8 kb mRNA corresponds to the cDNA described above and the 1.55 kb mRNA to the cDNA first isolated from BEN cells.[10] The evidence obtained shows that these two mRNAs are present in approximately equal abundance in the 786-0 cells and are generated by alternative splicing of sequences at the 3' end of the coding sequence. Primer-extension analysis of 786-0 cell poly (A)+ RNA together with Southern blot analysis of human DNA demonstrated the presence of a single copy gene coding for multiple mRNAs which result from alternative splicing. Such a minor alteration in the structure of a peptide as a result of alternative splicing is rare. The functional consequences of such a change are unknown, but could potentially have influence on secretion of the protein or stability of its mRNA.

The 1.8 kb mRNA also contains homology to another tumor-associated gene product, the c-myc proto-oncogene. Sequences similarities shared by the 3'-untranslated sequence of this cDNA and the c-myc proto-oncogene may be relevant to the expression of the two genes in tumor cells. Elevated levels of c-myc have been correlated with cellular transformation[17] and may play a role in the phenotype of these tumor cells, or the AU-rich sequences shared by the hHCF and c-myc mRNAs may represent a common signal influencing mRNA turnover.[18]

Little is known about the physiological regulation of hHCF gene expression. The hHCF could be the product of a developmentally regulated gene whose transcription is promoted as a consequence of cell transformation. However, it may normally be expressed in lower levels in nontumor cells, such as keratinocytes,[19] serving a physiological function yet to be elucidated. Further studies are needed to elucidate the pattern of hHCF gene expression in normal tissues and in tumors.

Biological Properties In Vivo *and* In Vitro

In order to determine if hHCF alone can, as implied, produce hypercalcemia and other components of the HMM syndrome *in vivo*, and to compare its biological profile and potency to PTH, we chemically synthesized[14] an NH_2-terminal fragment of the tumor peptide, hHCF-(1—34)NH_2.

Using thyroparathyroidectomized rats,[20] we compared the activity of hHCF-(1—34)NH_2 with that of bovine PTH, bPTH-(1—84). In this system, infusion of hHCF-(1—34)NH_2 produced hypercalcemia with an apparent potency 6 to 10 times greater than bPTH-(1—84) (FIG. 3).[14] The hypercalcemia could result from direct actions on bone or kidney, or via 1,25-dihydroxyvitamin D_3 action on gut, or any combination of these activities.

The hHCF-(1—34) produces several PTH-like effects on kidney.[14] It reduces serum phosphate and elevates circulating 1,25-dihydroxyvitamin D_3, the active form of vitamin D which is physiologically regulated by PTH. It also increases excretion of

```
        ↓
CTCGATTCAC GGTAACAGGCTTCTCGGCCCGTAGCCTCAGCGGGGTGC
L   D   S   R   - - -

CTCGATTCAC GGAGGCATTGAAATTTTCAGCAGAGACCTTCCAAGGAC
L   D   S   R   R   H   - - -
```

FIGURE 2. Nucleotide sequences of renal cell-carcinoma (10B5) cDNA (*above*) compared to lung (BEN) cell (BRF .50) (*below*) at the presumptive site of alternative splicing. The *arrow* represents the point at which the sequences diverge. In-frame stop codons present in the sequences of 10B5 or BRF.50 are *underlined*.[12]

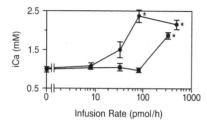

FIGURE 3. Effects of bPTH-(1—84) and hHCF-(1—34)NH$_2$ on serum concentrations of ionized calcium in thyroparathyroidectomized rats. bPTH-(1—84) (■); hHCF-(1—34)NH$_2$ (●) were continuously infused (8–480 pmol/h) into rats for 16 h.[16] Values represent the mean ±SEM for each group (n = 3–5). Values that are significantly different from control are indicated by * ($p < 0.05$).[14]

adenosine 3',5'-monophosphate (cAMP) and phosphate. Again, hHCF-(1—34)NH$_2$ was more potent than bPTH-(1—84).

The action of hHCF-(1—34)NH$_2$ on vitamin D metabolism was particularly pronounced relative to bPTH and important with regard to its implication for promoting long-term hypercalcemia clinically. In addition to mobilizing calcium release from bone and having PTH-like actions on kidney, hHCF could contribute to HHM by increasing levels of 1,25-dihydroxyvitamin D$_3$ that then promote the gastrointestinal absorption of dietary calcium. Elevated levels of 1,25-dihydroxyvitamin D$_3$ are inconsistently found in human hyperparathyroidism and usually not found in HHM, perhaps because 1,25-dihydroxyvitamin D$_3$ production is suppressed by elevated calcium levels or because renal function is compromised in HHM.[21–23] However, in animal models of HHM, including the transplantation of human tumor tissues into nude mice,[24–26] levels of 1,25-dihydroxyvitamin D$_3$ are consistently elevated.

To determine whether bone is a target organ for hHCF-(1—34)NH$_2$ and whether the action on bone per se can produce hypercalcemia, we used thyroparathyroidectomized rats fed a low calcium diet.[27] When the tumor peptide was infused at dose-rates sufficient to maintain normal levels of calcemia chronically (48 h), bone underwent histological changes similar to those observed when PTH was administered, namely an absolute increase in numbers of osteoclasts and an increase in osteoclasts apposed to trabecular surfaces (FIG. 4). These dose-dependent histological findings provide direct evidence for hHCF-(1—34)NH$_2$ action on bone and suggest that the blood calcemic response reflects hHCF-stimulated increases in bone resorption and accompanying calcium release with potency comparable to PTH.

Further evidence that bone is a direct target tissue for hHCF-(1—34)NH$_2$ action was obtained in studies in which the peptide was infused in thyroparathyroidectomized and nephrectomized rats fed a low calcium diet.[14] The low calcium diet reduces the contribution of dietary calcium; nephrectomy eliminates an action on alteration of

FIGURE 4. Histomorphometric quantification of the number of osteoclasts/mm^2 (mean ±SE) or proximal tibial metaphysis after 48 h of infusion with the indicated concentrations of hHCF-(1—34)NH$_2$ or bPTH-(1—34).[25]

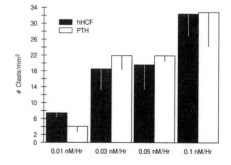

renal clearance of calcium and minimizes the effect of generating 1,25-dihydroxyvitamin D_3 on overall calcium metabolism. Again in this assay system, hypercalcemia is skeletal in origin. The hHCF-(1—34)NH_2 increased calcium to an extent comparable to that obtained with bPTH-(1—34).

Receptor Interaction and Postreceptor Effects

We also evaluated the interaction of hHCF-(1—34)NH_2 with PTH receptors *in vitro*. Binding affinity comparable to PTH for renal PTH receptors was observed (TABLE 1). In an adenylate cyclase assay utilizing bovine renal cortical membranes,[14] hHCF-(1—34)NH_2 had potency comparable to PTH. A close correspondence between renal binding (K_b) and activation constants (adenylate cyclase stimulation, K_m) was observed for all peptides tested (TABLE 1).

In another assay for adenylate cyclase, intact bone-derived osteosarcoma (ROS 17/2.8) cells were employed and cAMP levels measured.[15] The hHCF-(1—34)NH_2, unlike the bovine and human 1—34 and 1—84 PTH peptides, was more potent, approximately 100-fold more potent than hPTH-(1—84) (TABLE 1) in stimulating bone than renal adenylate cyclase, a finding similar to that observed by others using tumor extracts, conditioned media, or other bone-derived cells.[4,24]

TABLE 1. hHCF and PTH Compared *In Vitro*

Peptide	Renal Membranes		Bone (ROS Cells)
	Binding K_b (nM)	Adenylate Cyclase K_m (nM)	Adenylate Cyclase K_m (nM)
hHCF-(1—34)NH_2	13.3 ± 2.6	36.6 ± 4.3	1.0 ± 0.1
bPTH-(1—84)	4.9 ± 0.2	14.3 ± 8.7	4.8 ± 0.1
bPTH-(1—34)	0.8 ± 0.2	1.8 ± 0.1	1.1 ± 0.1
hPTH-(1—84)	63.7 ± 19.4	116.0 ± 32.0	91.5 ± 30.5
hPTH-(1—34)	1.2 ± 0.2	5.4 ± 0.2	2.0 ± 0.5

Both HCF and PTH produced identical postreceptor effects in bone-derived ROS 17/2.8 cells.[15] In addition to stimulation of adenylate cyclase, the extent of cyclase stimulation was enhanced equally by dexamethasone in the case of both peptides. The hHCF-(1—34)NH_2 also inhibited growth with an ED_{50} of $\simeq$0.1 nM in the presence of dexamethasone, but (like PTH) had no effect in the absence of dexamethasone. Both hormones also reduced alkaline phosphatase activity (FIG. 5), probably due to the corresponding reduction in mRNA for alkaline phosphatase. Taken together, these effects strongly suggest that hHCF mediates its actions through the PTH receptor.

The data obtained from bone cells may explain in part the enhanced calcemic response to hHCF-(1—34)NH_2 versus bPTH-(1—84) *in vivo*. Alternatively, the finding that PTH receptor affinity is closely similar for hHCF-(1—34)NH_2 and several forms of PTH suggests that hHCF-(1—34)NH_2 might possess increased stability *in vivo*. The increased potency of hHCF-(1—34)NH_2 relative to PTH may also indicate differences between bone and renal PTH receptors (suggesting possible receptor subtypes), or differences in receptor coupling to second messenger systems, or even the presence of a new class for receptors of hHCF with which both PTH and hHCF might interact.

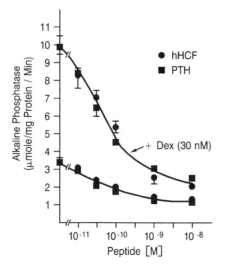

FIGURE 5. Dose-dependent inhibition of alkaline phosphatase activity (AP) by hHCF and PTH in the presence and absence of dexamethasone. ROS 17/2.8 cells were plated in F-12 medium with 5% FBS in 24-well multidishes. Four days later dexamethasone (30 nM), or ethanol (0.0025%) were added either with hHCF (●) or PTH (■) at the indicated concentrations. Alkaline phosphatase activity and protein were determined after 3 days of treatment. Data are the mean ± SD of determinations from triplicate wells from one of two similar experiments.[15]

A PTH Antagonist Inhibits hHCF Actions

Peptide hormone antagonists that are effective *in vivo* are uniquely precise tools for biomedical research. They can be used to determine how peptide hormones act, what their role it is normal physiologic processes, and how they contribute to pathophysiological states. In addition, they have potential clinical utility in the diagnosis and treatment of syndromes of hormone excess.

The biological profile displayed by hHCF-(1—34)NH$_2$ suggests that PTH antagonists could inhibit all or most of its actions. Previous structure-function studies of PTH analogs led to the design of a PTH antagonist effective *in vitro* and *in vivo*.[28] This analog, [Tyr-34]bPTH-(7—34)NH$_2$, is a competitive antagonist of PTH designed to reversibly occupy PTH receptors. The analog inhibits major actions of PTH, such as the PTH-stimulated calcemic response,[29] phosphaturia,[28] urinary cAMP excretion,[30] and stimulation of renal 1α-hydroxylase activity;[20] the analog is devoid of PTH-like agonist properties.

Using this PTH antagonist, we performed the biochemical equivalent of the classic "endocrine ablation" experiment, namely, we attempted to "remove" tumor by blocking the action of its secretory product.[16] A rat bioassay was devised: rats were

FIGURE 6. Preliminary studies of the antagonism of the hHCF-(1—34)NH$_2$-stimulated calcemic response by the PTH antagonist [Tyr-34]bPTH-(7—34)NH$_2$. Rats were TPTX'd at −16 h, placed on a low calcium diet at −12 h, and administered an infusion of hHCF-(1—34)NH$_2$ (0.06 nmoles/h) beginning at time 0. At 6 h, antagonist (at 200:1 molar-dose ratio over agonist) was infused continuously for the next 18 h. Animals receiving agonist + vehicle (*open symbols*) (n = 6); animals receiving agonist + antagonist (*closed symbols*) (n = 5).[16]

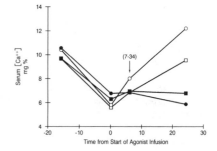

thyroparathyroidectomized, cannulated bilaterally, and then placed on a low calcium diet.

Calcium levels declined markedly. After 16 h on the low calcium diet, infusion of hHCF-(1—34)amide was begun and continued for 24 h. Six hours after beginning hHCF-(1—34)NH_2 infusion, one-half of the group of animals receiving the tumor peptide were also given [Tyr-34]bPTH-(7—34)NH_2 (12 nmol/h) intravenously (peptide purchased from Bachem, Torrance, CA or Protein Research Foundation, Osaka, Japan). In preliminary experiments, despite administration of hHCF-(1—34)NH_2 in advance of the PTH antagonist, marked inhibition of the calcemic response was achieved (FIG. 6). This occurred at a molar dose-ratio of antagonist to tumor peptide (200:1) comparable to that used to antagonize PTH action *in vivo* previously.[20,28-30]

The promotion of urinary phosphate and cAMP excretion by hHCF-(1—34)NH_2 also was antagonized by [Tyr-34]bPTH-(7—34)NH_2, using a previously described rat assay system.[20,28] In addition, the increases in 1,25-dihydroxyvitamin D_3 levels stimulated by hHCF-(1—34)NH_2 were inhibited by the PTH antagonist.

In vitro, hHCF-(1—34)NH_2-stimulated adenylate cyclase activity in both renal- and skeletal-derived tissue was inhibited by [Tyr-34]bPTH-(7—34)NH_2 (TABLE 2).

TABLE 2. Antagonists of hHCF and PTH Tested against the Agonist Peptides

		Blockade of:	
Analog	Binding K_b (nM)	3 nM [Nle^{8,18}, Tyr^{34}] bPTH(1—34)NH_2-Stimulated Adenylate Cyclase K_i (nM)	30 nM hHCF(1—34)NH_2-Stimulated Adenylate Cyclase K_i (nM)
[Tyr^{34}]bPTH(7—34)NH_2	133 ± 30	338 ± 104	430 ± 170
hHCF(7—34)NH_2	242 ± 25	801 ± 119	411 ± 52

SUMMARY AND CONCLUSIONS

This investigation addresses a theoretical concept of tumor pathogenesis proposed over 40 years ago, namely that malignancy-associated hypercalcemia can result from endocrine secretion by tumors of a PTH-like factor. These studies demonstrate that a fragment of hHCF alone, without added or tumor-secreted cofactors or hormones, can produce hypercalcemia and other biochemical abnormalities associated with HHM. The hypercalcemia can be generated by hHCF-(1—34)NH_2 action on bone, although kidney and gut could contribute to the HHM syndrome when it occurs naturally. No other tumor-secreted peptide displays this biological profile. These studies establish one (PTH-like) mechanism by which human tumors could produce hypercalcemia. Furthermore, the finding that hHCF-(1—34)NH_2 is more potent than PTH in some systems is of considerable interest for the future design of hormone analogs.

A broad spectrum of biological properties of hHCF-(1—34)NH_2, including production of components of the HHM syndrome, can be inhibited by a PTH antagonist. Because [Tyr-34]bPTH-(7—34)NH_2 selectivly and competitively occupies PTH receptors, our studies demonstrate formally that hHCF-(1—34)NH_2 mediates some (and perhaps all) of its actions via receptors conventionally regarded as intended for interaction with PTH, but which actually may be present to allow for expression of bioactivity of both secreted proteins. Although some structural homology is shared by the two hormones and many contribute to interaction with receptors, the disparity in structure, especially within the 1—34 domains responsible for bioactivity in both hormones, is more pronounced. The similarity in biological profiles despite structural differences between hHCF and PTH is emphasized by the inhibitory action of [Tyr-34]bPTH-(7—34)NH$_2$ against the tumor peptide even in the absence of much of the homologous region in the PTH antagonist. This investigation provides impetus for designing more potent antagonists, which must now be regarded more appropriately as inhibitors of both PTH and hHCF. Such antagonists may best be generated from hybrid structures of the two hormones. In any case, these studies establish a promising new approach to therapy of tumor-associated hypercalcemia.

ACKNOWLEDGMENT

We are grateful to S. Camburn for expert secretarial assistance.

REFERENCES

1. Case records of the Massachusetts General Hospital. Case 27461. 1941. N. Engl. J. Med **225:** 789–791.
2. STEWART, A. F., K. L. INSOGNA, D. GOLTZMAN & A. E. BROADUS. 1983. Proc. Natl. Acad. Sci. USA **80:** 1454–1458.
3. STREWLER, G. J., R. D. WILLIAMS & R. A. NISSENSON. 1983. J. Clin. Invest. **71:** 769–774.
4. NISSENSON, R. A., G. J. STREWLER, R. D. WILLIAMS & S. C. LEUNG. 1985. Cancer Res. **45:** 5358–5363.
5. RODAN, S. B., K. L. INSOGNA, A. VIGNERY, A. F. STEWART, A. E. BROADUS, S. M. D'SOUZA, D. R. BERTOLINI, G. R. MUNDY & G. A. RODAN. 1983. J. Clin. Invest. **72:** 1511–1515.
6. SIMPSON, E. L., G. R. MUNDY, S. M. D'SOUZA, K. J. IBBOTSON, R. BOCKMAN & J. W. JACOBS. 1983. N. Engl. J. Med. **309:** 325–330.
7. MOSELEY, J. M., M. KUBOTA, H. D. DIEFENBACH-JAGGER, R. E. H. WETTENHALL, B. E. KEMP, L. J. SUVA, C. P. RODDA, P. R. EBELING, P. J. HUDSON, J. D. ZAJAC & T. J. MARTIN. 1987. Proc. Natl. Acad. Sci. USA **84:** 5048–5052.
8. STEWART, A. F., T. WU, D. GOUMAS, W. J. BURTIS & A. E. BROADUS. 1987. Biochem. Biophys. Res. Commun. **146:** 672–678.
9. STREWLER, G. J., P. H. STERN, J. W. JACOBS, J. EVELOFF, R. F. KLEIN, S. C. LEUNG, M. ROSENBLATT & R. A. NISSENSON. 1987. J. Clin. Invest. **80:** 1803–1807.
10. SUVA, L. J., G. A. WINSLOW, R. E. H. WETTENHALL, R. G. HAMMONDS, J. M. MOSELEY, H. DIEFENBACH-JAGGER, C. P. RODDA, B. E. KEMP, H. RODRIGUEZ, E. Y. CHEN, P. J. HUDSON, T. J. MARTIN & W. I. WOOD. 1987. Science **237:** 893–896.
11. MANGIN, M. *et al.* 1988. Proc. Natl. Acad. Sci. USA **85:** 597–601.
12. THIEDE, M. A., G. J. STREWLER, R. A. NISSENSON, M. ROSENBLATT & G. A. RODAN. 1988. Proc. Natl. Acad. Sci. USA **85:** 4605–4609.
13. KEMP, B. E., J. M. MOSELEY, C. P. RODDA, P. R. EBELING, E. H. WETTENHALL, D. STAPLETON, H. DIEFENBACH-JAGGER, F. URE, V. P. MICHELANGELI, H. A. SIMMONS, L. G. RAISZ & T. J. MARTIN. 1987. Science **238:** 1568–1569.
14. HORIUCHI, N., M. A. CAULFIELD, J. E. FISHER, M. E. GOLDMAN, R. L. MCKEE, J. E. REAGAN, J. J. LEVY, R. F. NUTT, S. B. RODAN, T. L. SCHOFIELD, T. L. CLEMENS & M. ROSENBLATT. 1987. Science **238:** 1566–1568.
15. RODAN, S. B., M. NODA, G. WESOLOWSKI, M. ROSENBLATT & G. A. RODAN. 1988. J. Clin. Invest. **81:** 924–927.

16. HORIUCHI, N., J. E. FISHER, M. P. CAULFIELD, D. THOMPSON, J. G. SEEDOR, J. J. LEVY, R. F. NUTT, G. A. RODAN, T. L. CLEMENS & M. ROSENBLATT. 1988. Abstract. J. Bone Miner. Res.
17. SHAW, G. & R. KAMEN. 1986. Cell **46:** 659–667.
18. RABBITS, T. H. 1985. Trends Gen. **1:** 327–331.
19. MERENDINO, J. J., JR., K. L. INSOGNA, L. M. MILSTONE, A. E. BROADUS & A. F. STEWART. 1986. Science **231:** 388–390.
20. HORIUCHI, N. & M. ROSENBLATT. 1987. Am. J. Physiol. **253:** E187–E192.
21. STEWART, A. F., R. HORST, L. J. DEFTOS, E. C. CADMAN, R. LANG & A. E. BROADUS. 1980. N. Engl. J. Med **303:** 1377–1383.
22. BUSHINSKY, D. A., G. S. RIERA, M. J. FAVUS & F. L. COE. 1985. J. Clin. Invest. **76:** 1599–1604.
23. HULTER, H. N., B. P. HALLORAN, R. D. TOTO & J. C. PETERSON. 1985. J. Clin. Invest. **76:** 695–702.
24. STREWLER, G. J., T. J. WRONSKI, B. P. HALLORAN, S. C. MILLER, S. C. LEUNG, R. D. WILLIAMS & R. A. NISSENSON. 1986. Endocrinology **119:** 303–310.
25. INSOGNA, K. L., A. F. STEWART, A. M.-C. VINGERY, E. C. WEIR, P. A. NAMNUM, R. E. BARON, J. M. KIRKWOOD, L. M. DEFTOS & A. E. BROADUS. 1984. Endocrinology **114:** 888–896.
26. GKONOS, P. J., T. HAYES, W. BURTIS, R. JACOBY, J. MCGUIRE, R. BARON & A. F. STEWART. 1984. Endocrinology **115:** 2384–2390.
27. THOMPSON, D. D., J. G. SEEDOR, J. E. FISHER, M. ROSENBLATT & G. A. RODAN. 1988. Proc. Natl. Acad. Sci. USA **85:** 5673–5677.
28. ROSENBLATT, M. 1986. N. Engl. J. Med **315:** 1004–1013.
29. DOPPELT, S. H., R. M. NEER, S. R. NUSSBAUM, P. FEDERICO, J. T. POTTS, JR. & M. ROSENBLATT. 1986. Proc. Natl. Acad. Sci. USA **83:** 7557–7560.
30. HORIUCHI, N., M. F. HOLICK, J. T. POTTS, JR. & M. ROSENBLATT. 1983. Science **220:** 1053–1055.

Characterization of a Parathyroid Hormonelike Peptide Secreted by Human Keratinocytes

KARL L. INSOGNA,[a,b,c] ANDREW F. STEWART,[b,c]
KYOJI IKEDA,[b] MICHAEL CENTRELLA,[d]
AND LEONARD M. MILSTONE[b,c]

[b]Yale University School of Medicine
P.O. Box 3333
333 Cedar Street
New Haven, Connecticut 06510-8059

[c]Veterans Administration Medical Center
West Spring Street
West Haven, Connecticut 06516

[d]University of Connecticut Health Sciences Center
Saint Francis Hospital
114 Woodland Street
Hartford, Connecticut 06105

INTRODUCTION

Malignant squamous tumors are often complicated by hypercalcemia.[1,2] The mechanisms which underlie this common and important paraneoplastic syndrome have been the focus of investigative interest for more than 60 years.[1,3] In the past decade, several laboratories have attempted to better delineate this clinical syndrome and to undertake the identification of potential mediators.[2,4–8]

Initial clinical studies indicated that patients with squamous malignancies and hypercalcemia were characterized by a uniform clinical and biochemical phenotype, so called humoral hypercalcemia of malignancy (HHM).[2,4] These patients typically had large, clinically evident tumors, with little or no evidence of skeletal metastases. In addition to hypercalcemia, hypophosphatemia and a reduction in renal phosphate threshold, affected patients also had low levels of circulating $1,25(OH)_2$ vitamin D and elevations in nephrogenous cyclic AMP excretion. However, blood immunoreactive PTH values were not elevated in this group.[4] Bone histomorphometry demonstrated a striking increase in bone resorption with suppressed bone formation, but no evidence of marrow involvement by tumor.[9] Taken together, these data suggested the presence of a circulating, bone-resorbing factor which had some of the characteristics of parathyroid hormone, but lacked others.

These observations were extended by Gkonos et al.[10] and Burtis et al.[11] who characterized an inducible, localized squamous cell carcinoma of the skin in mice

[a]Address for correspondence: Karl L. Insogna, M.D., Department of Endocrinology Research/151, Veterans Administration Medical Center, West Spring Street, West Haven, CT 06516.

146

which produced nearly the identical biochemical and histomorphometric phenotype observed in patients with HHM.

Subsequent *in vitro* studies demonstrated that media conditioned by cultured HHM-associated tumors and extracts of these tumor tissues contained a PTH-like adenylate cyclase-stimulating protein which appeared to be intereacting specifically with the PTH receptor.[7,12] However, immunologic and molecular biological data clearly indicated that this material was not native parathyroid hormone.[2,13] This protein has now been isolated and has been shown to be a 16,000–17,000 dalton protein which shares considerable amino-terminal homology with parathyroid hormone, but has a unique primary structure thereafter.[14-18] A smaller 6,000–9,000 dalton form of this protein has also been isolated.[16,19] These latter species share the same amino-terminal sequence as the 16 Kd protein and presumably result from processing at the carboxy-terminus. Full-length cDNAs encoding this protein have been identified by two groups and the gene for this protein mapped to chromosome 12.[17,18]

The possibility that this PTH-like protein was not simply a manifestation of the malignant phenotype, but might reflect deranged production of a normal secretory product has intrigued us for some time. The striking frequency with which squamous malignancies caused HHM led us to consider the possibility that nonmalignant squamous tissue might also produce this or a similar biological activity. What follows is a summary of our efforts to identify and characterize parathyroid hormonelike adenylate cyclase-stimulating activity derived from normal cultured human keratinocytes.

MATERIALS AND METHODS

Materials

Synthetic bovine (1–34)PTH and the competitive PTH-receptor antagonist Nle[8,18]Tyr[34]bPTH(3–34)amide were purchased from Bachem, Inc. (Torrance, CA). Dialysis was performed against distilled water using Spectrapor membranes (Spectrum Industries, Los Angeles, CA) with a MCO of 3,000 daltons.

G-5, an antisera raised in goats to human PTH, which has midregion specificity, was kindly provided by Dr. L. Mallette (VA Medical Center, Houston, TX).[20] CK-67, an antisera raised in chickens to human PTH, which has amino-terminus specificity, was kindly provided by Dr. G. Segre (Endocrine Unit, Massachusetts General Hospital, Boston, MA).[20]

Cell Cultures

Human keratinocytes. Neonatal foreskin keratinocytes were grown in dishes preconditioned with irradiated 3T3 fibroblasts.[21] For experiments examining PTH-like factor production, cells were studied one week past confluence in Dulbecco's Modified Eagle's Medium containing antibiotics and additions or modifications as detailed below. For purification studies second passage keratinocytes were grown to confluence in T-150 collagen coated (Vitrogen, Collagen Corp., Palo Alto, CA) flasks in calcium-free DMEM containing 10% DMEM supplemented with 1% fetal bovine serum and 0.3 mM $CaCl_2$.

Malignant squamous cell clines. Four malignant human squamous cell lines were examined for the production of the PTH-like factor. These included A431 (ATCC# CRL 1555) a vulvar squamous carcinoma, C-4I (ATCC# CRL 1594) a cervical squamous carcinoma, A253 (ATCC# HTB 41) an epidermal carcinoma of the submaxillary gland and Sq CC/Y1 a squamous carcinoma of the oral mucosa.[22] These cell lines were grown in DMEM supplemented with 10% fetal bovine serum and antibiotics. Media were harvested at 72 hr and assayed for biological activity.

ROS cells. The clonal rat osteocarcinoma line, cell line 17/2.8, was grown in modified HAMS F12 medium with supplemental calcium, glutamine, HEPES buffer, and antibiotics.[20]

Human dermal fibroblasts. Foreskins were obtained by routine circumcision of normal, neonatal infants and fibroblast cultures were initiated by explantation of minced skin fragments under cover slips. Two established human fibroblast cell lines, ATCC# CRL 1457 and CRL 1564 were also studied. Cells were grown in DMEM supplemented with 10% bovine serum and antibiotics.[23]

Adenylate Cyclase Assays

ROS assay. Using ROS 17/2.8 cells, PTH-like biological activity was detected by measuring the conversion of [^{3}H]-ATP to [^{3}H]-cyclic AMP induced by the occupancy of cell-surface PTH receptors.[20] Briefly, the cells are preincubated for two hr with [^{3}H]-adenine and then stimulated for 10 min with agonist. The reaction is stopped and the cells extracted by adding trichloroacetic acid. The [^{3}H]-cyclic AMP formed is separated from other nucleotides by affinity chromatography and quantitated by liquid scintillation counting. The lower limit of detection in this assay is 5×10^{-11} M PTH.

Dermal fibroblast adenylate cyclase assay. The adenylate cyclase activity in fibroblast lines was assayed using the same methods employed in the ROS assay.[23]

Bone-resorbing Assay

This assay measures the release of ^{45}Ca from prelabeled fetal rat long bones.[24] Results are expressed as the ratio of ^{45}Ca released from treated versus control bones. In this assay, PTH at a concentration of 4×10^{-9} M induces 1.69 ± 0.06-fold stimulation (mean of 16 determinations).

Growth Factor Assay

Transforming growth factor activity was detected using anchorage independent growth of NRK 49-F cells (American Type Tissue Culture) in soft agar.[24] In this assay, epidermal growth factor alone induces small colony formation (less than 100 μM in diameter) and TGF β alone has no activity. The combination of epidermal growth factor and TGF β induces large colony formation (greater than 200 μM in diameter).

Chromatography

Gel-permeation chromatography was performed on a 2.5×125 cm column using G-100 Sephadex (10–50 μ mesh, Pharmacia, Piscataway, NJ) in 100 mM acetic acid at 4°C. Six ml fractions were collected and aliquots assayed for biological activity.

Reverse-phase HPLC was performed on a Waters System[24] and employed a Vydac

C18 column 0.46 × 25 cm (Separations Group, Hesperia, CA). Gel-filtered keratinocyte-conditioned medium was applied to the column in 0.1% trifluoroacetic acid (TFA) and the column developed with a linear gradient of acetonitrile in 0.1% TFA at a flow rate of 1 ml per min.

Isoelectric Focusing

Flat-bed isoelectric focusing was performed using an LKB apparatus and Ultradex gel (LKB Instruments, Gaithersburg, MD). Ampholines (LKB) pH range 7–9 were used to establish the pH gradient. Samples were electrophoresed at 8 watts constant power for 18 hr at 4°C. The gel was then sliced, the fractions eluted with water, the pH measured and aliquots taken for biological activity.

FIGURE 1. Dose-response curves of KCM and (1–34)bPTH in the ROS assay. (From Merendino *et al.*[20] Reprinted by permission from *Science.*)

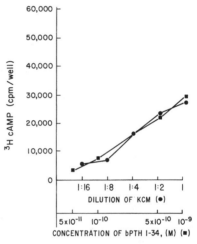

RESULTS

Production of PTH-like Adenylate Cyclase-stimulating Protein by Keratinocytes and Malignant Cell Lines

Conditioned medium (KCM) harvested from confluent keratinocyte cultures consistently demonstrated adenylate-cyclase-stimulating activity. Since our initial report,[20] we have examined over 50 primary keratinocyte cultures and found this activity to be uniformly present. On average, the amount of biological activity present in these cultures corresponds to the degree of stimulation induced in ROS cells by PTH in the concentration range 10^{-10} to 10^{-9} M. Growth on collagen did not seem to effect the elaboration of this biological activity. FIGURE 1 shows the dose-response relationship of keratinocyte-conditioned medium in the ROS assay. KCM induced a dose-dependent increase in [³H] cyclic AMP accumulation by the ROS cells in a manner completely analogous to that seen with bovine parathyroid hormone.

In contrast, conditioned media obtained from four cultures of irradiated murine 3T3 fibroblasts were devoid of this biological activity. Finally, on repeated assay, none

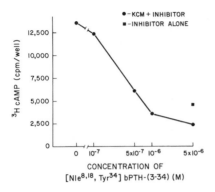

FIGURE 2. Effect of increasing concentrations of the synthetic PTH-receptor competitive antagonist Nle[8,18]Tyr[34]bPTH(3–34)amide on KCM-induced activity in the ROS assay. Note that inhibitor alone is a weak partial agonist in the assay. (From Merendino et al.[20] Reprinted by permission from Science.)

of the malignant squamous cell lines elaborated any adenylate cyclase-stimulating activity.

To confirm that the biological activity detected in KCM was specifically interacting with the PTH receptor on the ROS cells, the effect of the synthetic receptor-antagonist Nle[8,18]Tyr[34]bPTH(3–34)amide on KCM-induced PTH-like bioactivity was examined (FIG. 2). Increasing concentrations of the competitive receptor-antagonist led to a progressive inhibition of KCM-induced biological activity such that at an inhibitor concentration of 5×10^{-6} M biological activity was completely extinguished. To examine whether this biological activity was due to native parathyroid hormone itself, KCM was incubated with two antisera raised to human PTH. As shown in FIGURE 3, incubation of 7.5×10^{-10} M (1–34) bPTH for one hour at 27°C with 1:160 dilution of G-5 or 1:1000 dilution of CK-67 completely inhibited subsequent cyclic AMP accumulation by ROS cells. However, incubation of KCM with the same two antisera did not effect subsequent cyclic AMP production induced by these samples.

Regulation of production. FIGURE 4 summarizes the effects of variations in serum, calcium, and 1,25(OH)$_2$ vitamin D concentrations on elaboration of the PTH-like

FIGURE 3. Effect of preincubation with two PTH antisera on the bioactivity in the ROS assay induced by parathyroid hormone and KCM. Absolute [^{3}H] cyclic AMP accumulation (cpm/well) is given *above the bar* in each instance. Experimental details are provided in the text.

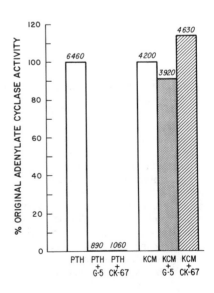

biological activity from keratinocytes. In the absence of serum, considerably less PTH-like biological activity was elaborated by the keratinocytes although activity was still detectable after 48 hr of culture. Increasing the concentration of serum up to 10% resulted in a 2- to 5-fold increase in elaboration of the PTH-like biological activity. Cultivation of keratinocytes for 2–14 days in calcium concentrations ranging from 0.3 to 3.0 mM had no effect on production of the PTH-like factor. Similarly, 72-hr incubation in varying concentrations of 1,25(OH)$_2$D ranging from 2 to 200 ng/ml did not effect elaboration of the PTH-like peptide by the keratinocytes. Finally, polypeptide growth factors such as insulin and EGF in concentrations ranging from 1 to 10 μg per ml had no significant effect on production of the keratinocyte-derived biological activity.

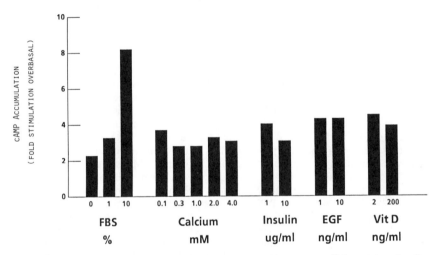

FIGURE 4. Effect of varying concentrations of fetal bovine serum, extracellular calcium, insulin, epidermal growth factor and 1,25(OH)$_2$ vitamin D on secretion of PTH-like bioactivity by keratinocytes. For these experiments cultures were grown to confluence in DMEM supplemented with 20% fetal bovine serum. At one week postconfluence cells were switched to either DMEM with 0, 1 or 10% FBS; calcium-free, serum-free DMEM with 0.1–4.0 mM calcium; or serum-free DMEM containing 0.2 mg/ml BSA and varying concentrations of insulin, EGF or 1,25(OH)$_2$ vitamin D.

Relative potency of the keratinocyte-derived PTH-like factor. Several groups including our own have recently reported that human dermal fibroblasts (HDFs) respond to parathyroid hormone with an increase in cyclic AMP production and have receptors that specifically bind PTH.[23,26–29] We, therefore, examined whether the KCM-derived PTH-like bioactivity would interact with these receptors. We initially surveyed 10 HDF lines for cyclic AMP accumulation in response to isoproterenol or parathyroid hormone. All 10 lines responded to isoproterenol while 6 of the 10 lines also responded to parathyroid hormone with 2.4-to 3.8-fold increases in cyclic AMP accumulation.[23] This response required a (1–34) bPTH concentration of 10^{-7} M. This relative insensitivity vis-à-vis the ROS cells was not due to fibroblast-derived proteolytic activity or other deleterious effects of fibroblasts of PTH, since after exposure to fibroblasts the hormone retained its bioactivity as measured in other assay systems.[23]

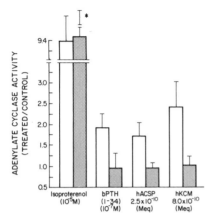

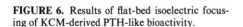

FIGURE 5. The mean effects of four agonists (□) and agonists plus Nle^{8,18}Tyr^{34}bPTH(3–34)amide 10^{-6} M (■) on the PTH-responsive HDF cell line CRL 1564. The values are the mean ± SEM of at least duplicate assays. hACSP refers to the human adenylate cyclase-stimulating protein or PTH-like peptide. The concentrations of hACSP and hKCM are expressed as molar equivalents (Meq) of (1–34)bPTH as defined in the ROS assay. The asterisk indicates SEM = 1.23. (From Wu et al.[23] Reprinted by permission from the *Journal of Clinical Endocrinology and Metabolism*.)

One of the 10 lines studied was examined further for its responsiveness to isoproterenol, PTH, KCM, and partially purified tumor-derived PTH-like peptide. As shown in FIGURE 5, isoproterenol at 10^{-6} M stimulated significant accumulation of cyclic AMP in these cells which was unaffected by coincubation with the PTH-receptor antagonist Nle^{8,18}Tyr^{34}bPTH(3—34)amide. Bovine parathyroid hormone at a concentration of 10^{-7} M also stimulated significant cyclic AMP accumulation in these cells which was completely extinguished by the competitive receptor-antagonist. The KCM induced a significant increase in cyclic AMP accumulation in the fibroblasts. This response was at least as great as that induced by parathyroid hormone, yet the amount of bioactivity present in this preparation corresponded to only 8×10^{-10} M equivalents of PTH as measured in the ROS assay. The human tumor-derived PTH-like bioactivity also produced a definite increase in cyclic AMP accumulation in these cells. However, in the ROS assay, the tumor-derived material corresponded to only a molar equivalent concentration of 2.5×10^{-10} M PTH. The bioactivity of both the tumor-derived and KCM-derived PTH-like activity in fibroblasts was completely abolished by the competitive receptor-antagonist.

Physical Characteristics of the KCM-derived PTH-like Factor

The KCM PTH-like biological activity was stable to heating at 100°C for 10 min and to treatment with 500 mM acetic acid. Incubation of KCM with trypsin (0.25 gm

FIGURE 6. Results of flat-bed isoelectric focusing of KCM-derived PTH-like bioactivity.

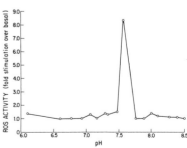

per liter at 27°C for 30 min) completely eliminated the ability of this preparation to stimulate cyclic AMP accumulation in the ROS cells. Crude KCM retained its activity when frozen at $-20°C$ for up to 3 months and when lyophilized and stored at $-20°C$, for up to 6 months.

Flatbed isoelectric focusing of KCM indicated a single discrete peak of biological activity eluting at pH 7.6 (FIG. 6). Gel-permeation chromatography also indicated a single peak of biological activity with an estimated molecular weight of 40 Kd (FIG. 7).

Association of Bone-Resorbing and Growth Factorlike Activities with PTH-like Bioactivity in Partially Purified KCM

Since the HHM-associated PTH-like peptide induces potent bone resorption *in vitro,* and since highly purified preparations of this peptide have been associated with TGF β-like activity,[24,30] we examined partially purified KCM preparations for these

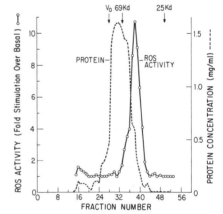

FIGURE 7. Gel-permeation chromatography of concentrated KCM using Sephadex G100.

two biological properties. Seven hundred ml of KCM was pooled and concentrated 10-fold using an Amicon D10 hollow-fiber dialysis/concentrator (Amicon Corp., Danvers, MA) using an HP10-20 cartridge with an approximate molecular weight cut off of 10,000 daltons. This material was dialyzed against distilled water, lyophilized and purified by gel-permeation chromatography followed by reverse-phase HPLC. For this latter step a linear (20→40%) gradient of acetonitrile was employed over 30 min. Although recovery of biological activity was poor from the HPLC step, a combination of these two physical separation techniques allowed separation of the PTH-like bioactivity from the vast majority of the protein. As shown in FIGURE 8, the PTH-like biological activity eluted as a discrete peak at 33% acetonitrile. Volumetrically equivalent pools were made across the chromatogram and assayed for bone-resorbing and growth factorlike activity.

Only the pool with the PTH-like biological activity contained significant amounts of bone-resorbing and growth-promoting activity. This pool induced 67% large colony formation in the transforming assay at a final concentration of 4 μg of protein/ml. Importantly, this activity could be demonstrated only in the presence of EGF, a finding

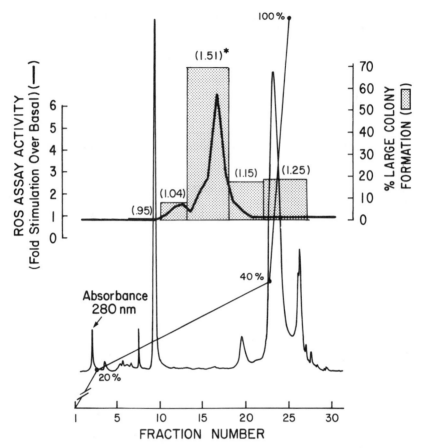

FIGURE 8. PTH-like, growth factorlike and bone-resorbing activites in purified KCM. In the *upper panel* PTH-like activity is represented by the *solid line,* growthlike activity by the *shaded bars,* and bone-resorbing activity by the *numbers in parentheses* above each pool. *Asterisk* indicates $p < 0.05$ for bone-resorbing activity compared with other pools. In the *lower panel* percentages refer to the gradient of acetonitrile.

similar to that seen with TGF β. Furthermore, this pool was the only one to demonstrate significant bone-resorbing activity at 1.5-fold over basal.

  RNA blot analysis. Using a full-length cRNA probe based on the coding region for the HHM-associated PTH-like peptide, mRNA from keratinocytes demonstrated a complex pattern of hybridizing transcripts, with two major species of about 1.6 and 2.1 Kb (FIG. 9, right panel). The three major transcripts appeared to comigrate with three transcripts of similar size in the HHM tumor mRNA (FIG. 9, left panel). The keratinocyte transcripts were roughly one-third to one-fifth the intensity of the transcripts seen in the tumor-derived mRNA. There were three keratinocyte-derived, less abundant, larger, mRNA species also noted.

DISCUSSION

The data we have presented are consistent with the hypothesis that the tumor-derived PTH-like protein, responsible for the syndrome of humoral hypercalcemia of malignancy, is also produced by normal human keratinocytes. Thus, in over fifty primary cultures examined, this activity is uniformly present. This biological activity is not invariably present in all cultured epithelial cells as evidenced by the absence of PTH-like activity in media cultured by the four malignant squamous tumors that we examined. The biological activity we have identified in keratinocyte-conditioned media interacts specifically with the PTH receptor since it is completely extinguished with the competitive, PTH-receptor antagonist $Nle^{8,18Tyr34}bPTH(3–34)$amide. Further, its dose-response curve parallels that of parathyroid hormone. The fact that it is heat-stable is also reminiscent of the behavior of the HHM-related PTH-like peptide.[7] That it is not parathyroid hormone itself is clearly indicated by the immunologic data which demonstrate no cross-reactivity with two antisera directed against two different epitopes of native parathyroid hormone.

The factors that regulate the production and secretion of this factor *in vivo* are not known. *In vitro,* factors known to regulate secretion of parathyroid hormone from cultured parathyroid cells had no effect on the production of the keratinocyte peptide. Specifically, $1,25(OH)_2$ vitamin D and calcium had no effect on the amount of the

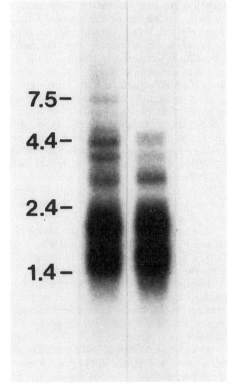

FIGURE 9. Blot-hybridization analysis of mRNA prepared from: *left panel:* human renal carcinoma causing HHM (line SKRC-52); *right panel:* human keratinocytes.

biological activity secreted by keratinocytes. However, the amount of serum present in the culture media did have a profound effect on the production and/or secretion of this factor. The serum component responsible for this effect is not known.

The molecular weight of the KCM-derived biological activity seems considerably higher than that of the recently purified HHM-derived PTH-like peptide.[14] However, our result was obtained with crude conditioned medium; and similar results have been reported with crude preparations of the HHM-associated peptide.[31] This may reflect binding to carrier proteins or nonspecific interactions with the chromatography gel. The pI of the HHM-associated peptide is greater than pH 9.0 which differs considerably from the pI obtained with crude keratinocyte-conditioned medium. While it is possible that this reflects the impure nature of the preparation, it also is consistent with some divergence in amino acid sequence or a size difference. Finally, the Northern blot analysis indicates that mRNA transcripts of similar size and number to those seen in humoral hypercalcemia of malignancy tumor-derived mRNA are present in human keratinocytes.

Our finding that bone-resorbing activity and TGF β-like activity copurify with the PTH-like bioactivity in keratinocyte conditioned media is intriguing. Since the HHM-associated peptide has potent bone-resorbing activity *in vivo* and *in vitro,* it is likely that the KCM-derived peptide will also show this biological activity. Whether it will also possess growth factorlike properties remains to be determined and must await the availability of homogeneously pure material.

The physiological role of this peptide remains to be defined. A particularly appealing possibility is that it has a paracrine or autocrine role in regulating epidermal function. Since it clearly has the capability to alter intracellular cyclic AMP levels in some cells, this is one potential mechanism of action for this peptide. It has been well documented that elevations in intracellular cyclic AMP stimulate proliferation in subconfluent keratinocyte cultures while in confluent cultures elevations in cAMP tend to inhibit proliferation.[32,33] Whether the PTH-like peptide can alter intracellular levels of cAMP in keratinocytes remains to be determined.

This factor, instead, may have an indirect role in controlling epithelial homeostasis by virtue of its effects on calcium or vitamin D metabolism. While external calcium concentrations have dramatic effects on epithelial stratification and differentiation,[34,35] we have no direct evidence that our factor alters either intra- or extracellular calcium. Some evidence supports the possibility of an influence of the keratinocyte peptide in $1,25(OH)_2$ metabolism. This vitamin D metabolite has been shown to alter keratinocyte differentiation.[36] Recently, the conversion of $25(OH)_2D$ to $1,25(OH)_2D$ has been shown to be stimulated by parathyroid hormone in keratinocytes.[37] Thus, it is possible that the keratinocyte-derived PTH-like peptide could influence levels of this vitamin D metabolite in skin.

The apparent preferential sensitivity of human dermal fibroblasts to the KCM-derived PTH-like peptide, as compared to parathyroid hormone, suggests that there may be a separate class of PTH receptors for this peptide in dermis. Therefore, a paracrine role for this hormone is possible.

Finally, the observation that this factor has TGF β-like properties offers a third potential mechanism for its action in skin. TGF β has been reported to stimulate fibronectin secretion by human keratinocytes and to induce the synthesis of plasminogen activator inhibitor by these cells.[38,39] In addition, it has been shown to inhibit keratinocyte proliferation.[40]

The characteristics of this bioactivity in skin support our initial speculation that secretion of the PTH-like peptide is not simply dependent upon the malignant phenotype, but has yet to be defined actions in normal skin. Elucidating its primary structure and defining its physiologic role remain important and exciting areas of investigation.

ACKNOWLEDGMENTS

We wish to thank Dr. Michael Reiss for providing the malignant epithelial cell lines to us and Mrs. Nancy Canetti for preparing the manuscript.

REFERENCES

1. RODMAN, J. S. & L. M. SHERWOOD. 1978. Disorders of mineral metabolism in malignancy. *In* Metabolic Bone Disease. Vol. 2. L. V. Avioli, & S. M. Krane, Eds.: 555–631.
2. GODSALL, J. W., W. J. BURTIS, K. L. INSOGNA, A. E. BROADUS & A. F. STEWART. 1986. Nephrogenous cyclic AMP, adenylate cyclase-stimulating activity, and the humoral hypercalcemia of malignancy. *In* Recent Progress in Hormone Research. R. O. Greep, Ed. **40:** 705–750. Academic Press. New York, NY.
3. ZONDEK, H., H. PETOW & W. SIEBERT. 1923. Die bedeutung der calciumbestimmung im blute für die diagnose der niereninsuffizientz. Z. Klin. Med **99:** 129–138.
4. STEWART, A. F., R. H. HORST, L. J. DEFTOS, E. C. CADMAN, R. LANG & A. E. BROADUS. 1980. Biochemical evaluation of patients with cancer-associated hypercalcemia. N Engl. J. Med. **303:** 1377–1383.
5. MUNDY, G. R., K. J. IBBOTSON, S. M. D'SOUZA, E. L. SIMPSON, J. W. JACOBS, & T. J. MARTIN. 1984. The hypercalcemia of cancer. N. Engl. J. Med. **310:** 1718–1727.
6. RUDE, R. K., C. F. SHARP, R. S. FREDERICKS, S. B. OLDHAM, N. ELBAUM, J. LINK, L. IRWIN & F. R. SINGER. 1981. Urinary and nephrogenous adenosine 3′,5′-monophosphate in the hypercalcemia of malignancy. J. Clin. Endocrinol. Metab. **52:** 765–771.
7. STEWART, A. F., K. L. INSOGNA, D. GOLTZMAN & A. E. BROADUS. 1983. Identification of adenylate cyclase-stimulating activity and cytochemical bioactivity in extracts of tumors from patients with humoral hypercalcemia of malignancy. Proc. Natl. Acad. Sci. USA **80:** 1454–1462.
8. STREWLER, G. J., R. D. WILLIAMS & R. A. NISSENSON. 1983. Human renal carcinoma cells produce hypercalcemia in the nude mouse and a novel protein recognized by parathyroid hormone receptors. J. Clin. Invest. **71:** 769–774.
9. STEWART, A. F., A. VIGNERY, A. SILVERGATE, N. D. RAVIN, L. LIVOLSI, A. E. BROADUS & R. BARON. 1982. Quantitative bone histomorphometry in humoral hypercalcemia of malignancy: Uncoupling of bone cell activity. J. Clin. Endocrinol. Metab. **55:** 219–227.
10. GKONOS, P. J., T. HAYES, W. BURTIS, J. MCGUIRE, R. JACOBY & A. STEWART. 1984. Squamous carcinoma model of humoral hypercalcemia of malignancy. Endocrinology **115:**2384–2390.
11. BURTIS, W. J., K. L. INSOGNA, A. E. BROADUS & A. F. STEWART. 1986. Two species of adenylate cyclase-stimulating activity in the murine squamous carcinoma model of humoral hypercalcemia of malignancy. Endocrinology **118:**1982–1988.
12. RODAN, S. B., K. L. INSOGNA, A.-M. C. VIGNERY, A. STEWART, A. BROADUS, S. D'SOUZA, M. BERTOLINI, G. MUNDY, & G. RODAN. 1983. Factors associated with humoral hypercalcemia of malignancy stimulate adenylate cyclase in osteoblastic cells. J. Clin. Invest. **72:** 1511–1515.
13. SIMPSON, E. L., G. R. MUNDY, S. M. D'SOUZA, K. J. IBBOTSON, R. BOCKMAN & J. R. JACOBS. 1983. Absence of parathyroid hormone messenger RNA in nonparathyroid tumors associated with hypercalcemia. N. Engl. J. Med. **309:** 325–330.
14. BURTIS, W., T. WU, C. BUNCH, J. WYSOLMERSKI, K. INSOGNA, A. BROADUS & A. STEWART. 1987. Identification of a novel 17,000 dalton PTH-like adenylate cyclase-stimulating protein from a tumor associated with humoral hypercalcemia of malignancy. J. Biol. Chem. **262:** 7151–7156.

15. Moseley, J. M., M. Kubota, H. Dieffenbach-Jagger, R. E. H. Wettenhall, B. E. Kemp, L. J. Suva, C. P. Rodda, P. R. Ebeling, P. J. Hudson, J. D. Zajac & T. J. Martin. 1987. Parathyroid hormone-related protein purified from a human lung cancer line. Proc. Natl. Acad. Sci. USA **84:** 5048–5052.

16. Strewler, G. J., P. H. Stern, J. W. Jacobs, J. Eveloff, R. D. Klein, S. C. Leung, M. Rosenblatt & R. A. Nissenson. 1987. Parathyroid hormone-like protein from human renal carcinoma cells: Structural and functional homology with parathyroid hormone. J. Clin. Invest. **80:** 1803–1807.

17. Suva, L. J., G. A. Winslow, R. Wettenhall, R. G. Hammonds, J. M. Moseley, H. Diefenbach-Jagger, C. P. Rodda, B. E. Kemp, H. Rodriguez, E. Y. Chen, P. J. Hudson, T. J. Martin & W. I. Wood. 1987. A parathyroid hormone-related protein implicated in malignant hypercalcemia: Cloning and expression. Science **237:** 893–896.

18. Mangin, M., A. C. Webb, B. . Dreyer, J. T. Posillico, K Ikeda, E. C. Weir, A. F. Stewart, N. H. Bander, L. Milstone, D. E. Barton, U. Francke & A. E. Broadus. 1988. Identification of a complementary DNA encoding a parathyroid hormone-like peptide from a human tumor associated with humoral hypercalcemia of malignancy. Proc. Natl. Acad. Sci. USA **85:** 597–607.

19. Stewart, A. F., W. J. Burtis, T. Wu, D. Goumas & A. E. Broadus. 1987. Two forms of parathyroid hormone-like adenylate cyclase-stimulating protein derived from tumors associated with humoral hypercalcemia of malignancy. J. Bone Miner. Res. **2:** 586–593.

20. Merendino, J., K. Insogna, L. Milstone, A. Broadus & A. Stewart. 1986. Cultured human keratinocytes produce a parathyroid hormone-like protein. Science **231:** 388–390.

21. Schwartz, P. M., L. C. Kugelman, Y. Coifman, L. M. Hough, & L. M. Milstone. 1988. Human keratinocytes catabolize thymidine. J. Invest. Dermatol. **90:** 8–12.

22. Ress, M., S. W. Pitman & A. C. Sartorelli. 1985. Modulation of the terminal differentiation of human squamous carcinoma cells *in vitro* by all-trans-retinoic acid. J. Natl. Cancer Inst. **74:** 1015–1023.

23. Wu, T. L., K. L. Insogna, L. Milstone & A. F. Stewart. 1987. Skin-derived fibroblasts respond to human PTH-like adenylate cyclase-stimulating proteins. J. Clin. Endocrinol. Metab. **65:**105–109.

24. Insogna, K., E. Weir, T. Wu, A. Stewart, A. Broadus, W. Burtis & M. Centrella. 1987. Co-purification of transforming growth factor beta-like activity with PTH-like and bone-resorbing activities from a tumor associated with humoral hypercalcemia of malignancy. Endocrinology **120:** 2183–2185.

25. Stewart, A., T. Wu, W. Burtis, E. Weir, A. Broadus & K. Insogna. 1987. The relative potency of a human tumor-derived PTH-like adenylate cyclase-stimulating preparation in three bioassays. J. Bone Miner. Res. **2:** 37–43.

26. Goldring, S. R., J. E. Mahaffey, S. M. Krane, J. T. Potts, J. M. Dayer & M. Rosenblatt. 1979. Parathyroid hormone inhibitors: Comparison of biological activity in bone- and skin-derived tissue. J. Clin. Endocrinol. Metab. **48:** 655–659.

27. Goldring, S. R., G. A. Tyler, S. M. Krane, J. T. Potts & M. Rosenblatt. 1984. Photoaffinity labeling of parathyroid hormone receptors: Comparison of receptors across species and target tissues and after desensitization. Biochemistry **23:** 498–502.

28. Silve, C., A. Santora & A. Spiegel. 1985. A factor produced by cultured rat Leydig tumor (Rice-500) cells associated with humoral hypercalcemia stimulates adenosine 3′,5′-monophosphate production via the parathyroid hormone receptor in human skin fibroblasts. J. Clin. Endocrinol. Metab. **60:** 1144–1147.

29. Fryer, M. J., S. R. Fritz & H. Heath. 1986. Accumulation of cyclic 3′,5′-adenosine monophosphate in cultured neonatal human dermal fibroblasts exposed to parathyroid hormone and prostaglandin E_2. Mayo Clin. Proc. **61:** 263–267.

30. Insogna, K., E. Weir, T. McCarthy, E. Burres, A. Broadus & M. Centrella. 1987. Highly purified adenylate cyclase-stimulating activity from tumors associated with humoral hypercalcemia of malignancy is mitogenic in primary bone cell culture. J. Bone Miner. Res. **2**(Suppl. 1): Abstr. 88.

31. Stewart, A., K. L. Insogna, W. J. Burtis, A. Aminiafshar, T. Wu, E. C. Weir & A. Broadus. 1986. Frequency and partial characterization of adenylate cyclase-stimulating

activity in tumors associated with humoral hypercalcemia of malignancy. J. Bone. Miner. Res. **1:** 267–276.

32. GREEN, H. 1978. Cyclic AMP in relation to proliferation of the epidermal cell: A new view. Cell **15:** 801–811.
33. OKADA, N., Y. KITANO & K. ICHIBARA. 1982. Effects of cholera toxin on proliferation of cultured human keratinocytes in relation to intracellular cyclic AMP levels. J. Invest. Dermatol. **79:** 42–47.
34. HENNINGS, H., D. MICHAEL, C. CHENG, P. STEINERT, K. HOLBROOK & S. H. YUSPA. 1980. Calcium regulation of growth and differentiation of mouse epidermal cells in culture. Cell **19:** 245–254.
35. MILSTONE, L. M. 1987. Calcium regulation of proliferation in confluent keratinocyte cultures. Epithelia **1:** 129–140.
36. SMITH, E. L., N. C. WALWORTH & M. F. HOLICK. 1986. Effect of 1-alpha,-25-dihydroxyvitamin D_3 on the morphologic and biochemical differentiation of cultured human epidermal keratinocytes grown in serum-free conditions. J. Soc. Invest. Dermatol. **86:** 709–714.
37. BIKLE, D. D., M. K. NEMANIC, E. GEE & P. ELIAS. 1986. 1,25-dihydroxy-vitamin D production by human keratinocytes. J. Clin. Invest. **78:** 557–566.
38. RAGHOW, R., A. E. POSTLETHWAITE, J. KESKI-OJA, H. L. MOSES & A. H. KANG. 1987. Transforming growth factor-β increases steady state levels of type I procollagen and fibronectin messenger RNAs posttranscriptionally in cultured human dermal fibroblasts. J. Clin. Invest. **79:**1285–1288.
39. WIKNER, N. E., K. A. PERISCHITTE & R. A. F. CLARK. 1988. Transforming growth factor-beta induces the synthesis of plasminogen activator inhibitor by human keratinocytes (abstract). Clin. Res. **36:** 254A.
40. SHIPLEY, G. D., M. R. PITTELKOW, J. J. WILLE, R. E. SCOTT & H. L. MOSES. 1986. Reversible inhibition of normal human prokeratinocyte proliferation by type B transforming growth factor—growth inhibitor in serum-free medium. Cancer Res. **46:** 2068–2071.

Epidermal Keratinocytes Secrete Apolipoprotein E[a]

ELIZABETH S. FENJVES,[b] DAVID A. GORDON,[b]
DAVID L. WILLIAMS,[c] AND LORNE B. TAICHMAN[b]

[b]Department of Oral Biology and Pathology
School of Dental Medicine

[c]Department of Pharmacological Sciences
School of Medicine
Health Sciences Center
State University of New York at Stony Brook
Stony Brook, New York 11794-8702

INTRODUCTION

Apolipoprotein E (apo E) is a glycoprotein found in plasma chylomicrons, very low density lipoproteins (VLDL) and high density lipoproteins (HDL). Apo E serves as a major structural component of plasma lipoproteins and also mediates lipoprotein uptake through interaction with apo E or apo B/E specific receptors on cellular membranes. Apo E thus plays an important role in the transport and removal of cholesterol-laden lipoproteins from the circulation.[1,2]

Most apolipoproteins are synthesized in the liver and small intestine. However, apo E is also synthesized in peripheral (nonhepatic) tissues.[3,4,5] Skin contains apo E mRNA, but it is not known which cells produce it, and if these cells synthesize and secrete the protein. We have examined apo E synthesis and secretion by cultured human keratinocytes. It was observed that these cells synthesize and secrete apo E protein at a rate of 890–950 pg/hr/10^6 cells. This was 16–20% of the rate seen with the human hepatocarcinoma cell line Hep G2. We conclude that keratinocytes in culture and probably those in skin, produce apo E. The significance of this is discussed.

MATERIALS AND METHODS

Culture

Human skin was obtained either from the foreskin of newborns (HFK denotes human foreskin keratinocytes) or from the abdomen of adults (HAK denotes human abdominal keratinocytes). The culture method of Rheinwald and Green[6] was used along with mitomycin C-inactivated 3T3 feeder layers.[7] The cultures were fed medium containing fetal calf serum.[8] Hep G2, a human hepatocellular carcinoma line[9] was grown in Dulbecco's Modified Eagle's Medium (DMEM) supplemented with 10% fetal calf serum and 5 μg/ml insulin.

[a]This research was supported by grants from the National Institutes of Health (DE 04511 to L.B.T., and HL 32868 and DK 18171 to D.L.W.) and from the Center for Biotechnology, SUNY Stony Brook.

160

Northern Analysis and Solution Hybridization

Total cellular RNA was isolated from confluent cultures by the guanidine isothiocyanate method.[10] After lysis, ethanol precipitation, and phenol extraction, RNA pellets were dried by lyophilization and dissolved in diethylpyrocarbonate (DEPC)-treated double distilled water. RNA concentrations were determined by UV absorbance at 260 nm.

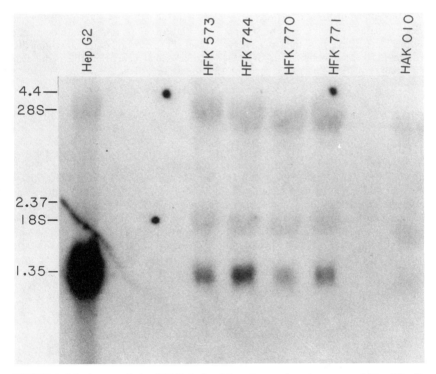

FIGURE 1. Northern analysis of RNA isolated from human keratinocytes and Hep G2 cells. Keratinocytes were cultured from four specimens of foreskin (HFK), one specimen of abdominal skin (HAK) and Hep G2 cells. Total RNA was isolated from each and 25 μg were electrophoresed on 1.2% formaldehyde agarose gels, transferred onto nylon filters and probed with ^{32}P labelled apo E cDNA. Apo E mRNA is detectable in all samples as a 1.2-kb band. The positions of standard molecular weight markers as well as 18S and 28S rRNA are shown on the *left.*

RNA samples treated with 6% formaldehyde were analyzed by electrophoresis in 1.2% agarose gels containing 6% formaldehyde. Transfer to nylon membranes (Genescreen/New England Nuclear, Boston, MA) was accomplished by electroblotting. Apo E mRNA was visualized by hybridization with an apo E cDNA probe obtained by nick translation of plasmid pE368 (kindly provided by J. Breslow, Rockefeller University, New York, NY).[11]

A DNA excess solution hybridization assay for human apo E mRNA was performed as described.[5] Total RNA was hybridized to completion with excess single stranded probe, and S1 nuclease-resistant hybrids were acid precipitated and counted by scintillation spectrometry as previously described.[5,12] Apo E mRNA values were determined by reference to a standard curve constructed with probe template DNA as the hybridization standard.

Metabolic Labelling

For labelling experiments, preconfluent cultures of keratinocytes or Hep G2 cells in 35 mm Petri dishes were rinsed with warm phosphate buffered saline (PBS) and then incubated for 30 minutes in Minimal Essential Medium lacking methionine and serum. The cultures were then incubated in 1.5 ml of the same medium containing 200 μCi/ml ^{35}S-methionine (1200 Ci/mM). After a 4-hour incubation, the media were collected, centrifuged at 10,000 rpm for 15 minutes, and dialyzed against PBS + 25 mM methinonine.

Immunoprecipitation and Electrophoresis

Aliquots of culture medium containing equal amounts of trichloroacetic acid (TCA) precipitable radioactivity were processed as by the previously described double antibody procedure[3] employing rabbit anti-apo E antiserum as the primary antibody and goat anti-rabbit gamma globulin (Cappell Biochem, Malvern, PA) as the second antibody. Primary antibody was used in excess to ensure quantitative immunoprecipitation. Control immunoprecipitation was carried out on all samples using preimmune rabbit serum as primary antibody. Routinely, the number of TCA precipitable counts used for precipitation from Hep G2 samples were about one third of those taken from keratinocyte samples (3×10^5 and 1×10^6, respectively). Immunoprecipitated samples were analyzed by sodium dodecyl sulfate (SDS)-10% polyacrylamide gel electrophoresis (PAGE) as described by Laemmli.[13] Gels were processed for fluorography[14] and exposed to Kodak X-OMAT AR film for 48–72 hours. Protein molecular weight standards (Biorad, Richmond, CA) and ^{14}C labeled human plasma apo E were routinely run for calibration.

Enzyme Linked Immunosorbent Assay (ELISA)

For the ELISA analyses, serum-free media were collected from subconfluent 100-mm culture dishes after a 12- and a 24-hour incubation, supplemented with 0.2 mg/ml phenylmethylsulfonyl fluoride, and spun at 6,000 rpm for 1 hour. ELISA analyses were carried out according to a modification of the procedure of Voller.[15] Briefly, 0.2 ml of a 1:5000 dilution of anti-apo E monoclonal antibody D3 (kindly supplied by E. Koren, Oklahoma Medical Research Foundation, Oklahoma City, OK) was covalently attached to microtiter plates in pH 9.6 carbonate buffer for 18 hours at room temperature. After three washes with PBS-TWEEN buffer, 0.2 ml dilutions of cell culture media and appropriate standards were added for 2 hours at room temperature. The wells were washed as above and 0.2 ml of the IgG fraction of rabbit anti-apo E antiserum diluted 1:500 were added for two hours. After washing, 0.2 ml of 1:500 goat anti-rabbit IgG conjugated to alkaline phosphatase (Sigma, St. Louis, MO) were added for 2 hours. Finally, 0.2 ml of 1mg/ml p-nitrophenylphosphate disodium salt (Sigma) were added for 45 minutes. The reactions were stopped by addition of 0.05 ml 1 N NaOH and absorbance was read at 405 nm.

RESULTS

To determine whether keratinocytes contain apo E mRNA, total cellular RNA was isolated from cultures of HAK and HFK and analyzed by Northern hybridization. FIGURE 1 is a Northern blot hybridized with ^{32}P labeled apo E cDNA probe. The results clearly show the presence of a band in all the keratinocyte culture media with the same molecular weight as apo E mRNA from Hep G2 cells. Hep G2, a human hepatocellular carcinoma cell line,[9] synthesizes apo E that is qualitatively comparable to that produced in liver. Using the highly accurate and sensitive DNA-excess solution hybridization assay, keratinocytes were shown to contain an average of 30 copies of apo

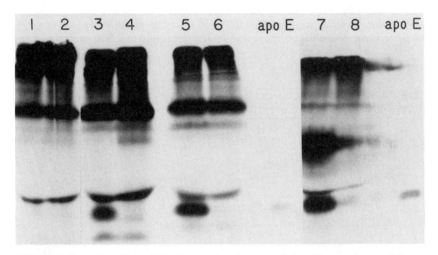

FIGURE 2. Secretion of apo E by human keratinocytes. Subconfluent cultures of human keratinocytes, Hep G2 cells and mitomycin C-treated 3T3 cells were labelled with ^{35}S methionine. The media were analyzed for the presence of apo E by immunoprecipitation with anti-human apo E raised in rabbits as a first antibody and goat anti-rabbit IgG as a second antibody (*odd numbered lanes*). As a control, the same samples were treated with preimmune serum and goat anti-rabbit IgG (*even numbered lanes*). HAK strain 010 (*lanes 3 and 4*), as well as HFK strain 771 (*lanes 5 and 6*), were compared to the hepatocarcinoma line Hep G2 (*lanes 7 and 8*). In all experimental lanes, immunoprecipitates show a band with an electrophoretic mobility equal to that of apo E. Immunoprecipitates from the mitomycin C-treated 3T3 cells (*lanes 1 and 2*) showed no apo E bands. Lanes marked "apo E" contain ^{14}C labelled plasma apo E.

E mRNA per cell. This value is 5–10% of the apo E mRNA content of Hep G2 cells (N. Dashti and D. L. Williams, unpublished data).

Apo E protein synthesis and secretion were examined by labelling keratinocytes and Hep G2 cells with ^{35}S-methionine and immunoprecipitating the culture media. The precipitates were analyzed by electrophoresis. The fluorogram in FIGURE 2 indicates that immunoprecipitates from keratinocyte media contain a band (lane 3 HAK 010 and lane 5 HFK 771) with the same electrophoretic mobility as apo E produced by Hep G2 cells (lane 7). The band immunoprecipitated from keratinocyte media also comigrated with ^{14}C labelled apo E from plasma. This band was not seen when the medium was precipitated with preimmune serum as primary antibody (even lanes) or when media from mitomycin C-treated 3T3 cells was precipitated (lanes 1 and 2) and was thus identified as human apo E. Further evidence for the identity of the apo E secreted from keratinocytes was obtained by demonstrating that this protein had the same isoprotein pattern as apo E from Hep G2 cells as judged by high resolution two-dimensional gel electrophoresis (data not shown). We conclude from these results that cultured epidermal keratinocytes synthesize and secrete apo E.

The absolute rate of apo E secretion by keratinocytes was determined by an ELISA analysis (TABLE 1). The presence of apo E in the media at concentrations of 1.5 ng/ml in 12 hours and 5 ng/ml in 24 hours corresponds to a production rate of 890–950 pg/hr/10^6 cells. This assumes complete recovery of the protein and the absence of protein degradation. The absolute rate of keratinocyte apo E production was estimated to be 16–20% of the rate seen with Hep G2 cells. The difference in apo E production rate between keratinocytes and Hep G2 cells is similar to the difference in apo E mRNA content. These data confirm that keratinocyte apo E mRNA is translated and show that the protein is secreted.

TABLE 1. Rate of Secretion of Apo E by Keratinocytes as Determined by Enzyme Linked Immunosorbent Assay (ELISA)

Cell Type	[Apo E] (ng/ml Media)	Rate of Apo E Secretion (ng/hr/10^6 Cells)
HAK 010		
12 hrs.	1.5	0.89
24 hrs.	5	0.95
HEP G2		
12 hrs.	60	5.6
24 hrs.	110	3.8

DISCUSSION

The primary function of keratinocytes has long been considered to be formation of an impermeable protective barrier. Recently, other metabolic functions have been identified and accorded special attention as reflected in this volume.

The present study establishes the synthesis and secretion of apo E by human keratinocytes in culture. Apo E made by cultured keratinocytes was identical to apo E secreted by Hep G2 cells. This is true in terms of electrophoretic mobility, immuno-reactivity, and isoform patterns on 2-D gels. Similarly, mRNA for apo E seems to be moderately abundant in human keratinocytes. ELISA studies indicate that, on a per cell basis, apo E is secreted at a rate that is approximately 16–20% the rate of Hep G2 cells. Synthesis and secretion of apo E in cultured keratinocytes indicate that keratinocytes within skin have the capacity to produce this protein.

The functional significance of peripheral apo E synthesis is not known, but potential functions of apo E made in other tissues have been proposed.[3–5,16] Since as much as 30–40% of total body apo E appears to arise from peripheral tissues, peripheral apo E may be of quantitative importance in reverse cholesterol transport, the process by which excess cholesterol is removed from peripheral tissues and delivered to the liver.[3–5,16] Apo E made in keratinocytes may contribute to this process although the amounts produced by epidermis are likely to be low in comparison to the amounts produced by the liver. Peripherally synthesized apo E may also serve as a shuttle protein for local redistribution of cholesterol within a tissue.[3,16] For example, apo E is made in the central nervous system.[4,5] Since there is little or no transfer of lipoproteins across the blood-brain barrier, synthesis of apo E within the brain[4,5] suggests a local role for apo E.[16] It is possible that keratinocyte apo E may play a similar role in local cholesterol transport and redistribution within the epidermis.

Apo E produced in skin may participate in aspects of lipid metabolism that are unique to skin. Epidermal lipids are believed to be an essential component of the permeability barrier of normal skin. In contrast to cells of the malpighian layer, cells of the stratum corneum are virtually depleted of intracellular lipids.[17] Lipids in the stratum corneum are located primarily in the intercellular spaces. It is these lipids that are thought to render the stratum corneum impermeable to aqueous solvent.[17]

It has also been suggested that changes in the composition of extracellular lipids during keratinocyte differentiation and migration are responsible for normal desquamation.[18] In recessive X-linked ichthyosis, an inherited absence of cholesterol sulfatase results in a failure to hydrolyze cholesterol sulfate to free cholesterol in the extracellular spaces. The scaling that is part of this syndrome and that represents a prolongation of the normal cohesiveness among stratum corneum cells is thought to be a result of this failure to metabolize cholesterol sulfate and of an abnormal lipid content.

Among the roles envisaged for epidermal lipids in maintaining both permeability and cohesiveness, transport of lipids from the intracellular to the extracellular compartment is an essential event. The formation of lamellar granules in the cytoplasm of spinous and granular cells and the extrusion of their membranous leaflets into the intercellular spaces in the upper granular layer is an example of this transport.[17] Since movement of lipids out of cells usually requires some apoprotein-containing acceptor, it is conceivable that apo E produced by keratinocytes in epidermis is utilized in the redistribution of lipids within this tissue.

In most cells, including skin fibroblasts, cholesterol synthesis is regulated by exogenous low density lipoprotein (LDL) cholesterol. Exogenous LDL are bound and internalized by cells in the well characterized receptor-mediated process.[19] The uptake of LDL suppresses de novo sterologenesis by inhibiting 3-hydroxy-3-methylglutaryl coenzyme A reductase. Epidermal cells, however, are not always sensitive to regulation by external LDL.[20] Young subconfluent cultures of keratinocytes bind LDL and downregulate sterologenesis. However, confluent cultures, which contain more suprabasal layers of differentiated cells, do not bind LDL and do not repress cholesterol biosynthesis. LDL receptors have been demonstrated in cultured squamous cell carcinoma[21] and in normal keratinocytes maintained in abnormally low calcium concentrations.[22] These results have led to the suggestion that LDL receptors are characteristic of undifferentiated keratinocytes and are lost upon differentiation. Changes in apo E synthesis may also accompany keratinocyte differentiation.

ACKNOWLEDGMENTS

We would like to thank Dr. Donald Cox (School of Dental Medicine, SUNY Stony Brook) for his invaluable assistance in the development of the ELISA. The assistance of Iris Kleinman is gratefully acknowledged. We also thank Jim Skillman for preparing the figures for the paper.

REFERENCES

1. MAHLEY, R. W. & T. L. INNERARITY. 1983. Biochim. Biophys. Acta **737**: 197–222.
2. SHERRIL, B. C., T. L. INNERARITY & R. W. MAHLEY. 1980. J. Biol. Chem. **255**: 1804–1807.

3. BLUE, M. L., D. L. WILLIAMS, S. ZUCKER, S. A. KHAN & C. B. BLUM. 1983. Proc. Natl. Acad. Sci. USA **80:** 283–287.
4. WILLIAMS, D. L., P. A. DAWSON, T. C. NEWMAN & L. L. RUDEL. 1985. J. Biol. Chem. **260:** 2444–2452.
5. NEWMAN, T. C., P. A. DAWSON, L. L. RUDEL & D. L. WILLIAMS. 1985. J. Biol. Chem. **260:** 2452–2457.
6. RHEINWALD, J. G. & H. GREEN. 1975. Cell **6:** 317–330.
7. TAICHMAN, L. B., S. REILLY & P. GARANT. 1979. Arch. Oral Biol. **24:** 335–341.
8. CONNELL, N. D. & J. G. RHEINWALD. 1983. Cell. **34:** 245–253.
9. ADEN, D. P., S. FOGEL, S. PLOTKIN, I. DAMJANOV & B. B. KNOWLES. 1979. Nature **282:** 615–616.
10. MANIATIS, T. E., E. F. FRITSCH & J. SAMBROOK. 1982. *In* Molecular Cloning: A Laboratory Manual. Cold Spring Harbor Laboratory. Cold Spring Harbor, NY.
11. ZANNIS, V. I., J. MCPHERSON, G. GOLDBERGER, S. K. KARATHANASIS & J. BRESLOW. 1984. J. Biol. Chem. **259:** 6498–6504.
12. WILLIAMS, D. L., T. C. NEWMAN, G. S. SHELNESS & D. A. GORDON. 1986. Methods Enzymol. **129:** 670–679.
13. LAEMMLI, U. K. 1970. Nature (London) **227:** 680–685.
14. BONNER, W. M. & R. A. LASKEY. 1974. Eur. J. Biochem. **46:** 83–88.
15. VOLLER, A., D. BIDWELL & A. BARTLETT. 1976. *In* Manual of Clinical Immunology. N. Rose & H. Freedman, Eds. American Society of Microbiology. Washington, DC
16. DAWSON, P. A., N. SCHECHTER & D. L. WILLIAMS. 1986. J. Biol. Chem. **261:** 5681–5684.
17. ELIAS, P. M. 1981. Int. J. Dermatol. **20:**1–19.
18. ELIAS, P. M., M. L. WILLIAMS, M. E. NALONEY, J. A. BONIFAS, B. E. BROWN, S. GRAYSON & E. H. EPSTEIN, JR. 1984. J. Clin. Invest. **74:** 1414–1421.
19. BROWN, M. S. & J. L. GOLDSTEIN. 1976. Science **191:** 150–154.
20. PONEC, M., L. HAVEKES, J. KEMPENAAR & B. J. VERMEER. 1983. J. Invest. Dermatol. **81:** 125–130.
21. PONEC, M., L. HAVEKES, J. KEMPENAAR & B. J. VERMEER. 1984. J. Invest Dermatol. **81:** 436–440.
22. PONEC, M., L. HAVEKES, J. KEMPENAAR, J. LAVRIJSEN *et al.* 1985. J. Cell. Physiol. **125:** 98–106.
23. WILLIAMS, M. L., A. M. MOMMAAS, S. L. RUTHERFORD, S. GRAYSON, B. J. VERMEER & P. ELIAS. 1987. J. Cell. Physiol. **132:** 428–440.

Do Keratinocytes Regulate Fibroblast Collagenase Activities During Morphogenesis?[a]

BARBARA JOHNSON-WINT[b]

Developmental Biology Laboratory
Medical Services
Massachusetts General Hospital
Harvard Medical School
Boston, Massachusetts 20114

INTRODUCTION

Epidermis and dermis, derived from embryonic ectoderm and mesoderm respectively, are the two tissue types that make up the skin. Epithelial-mesenchymal interactions involving an exchange of information between epidermal and connective tissue cells are thought to regulate skin morphogenesis during development,[1] and probably maintain integumentary structure in the adult. Collagenase to the interstitial collagens is a constitutive part of the papillary dermis at the dermal-epidermal interface in adult human skin.[2] In addition, epithelial cell cytokines from adult rabbit cornea,[3,4] or adult or fetal rabbit skin[5] can modulate collagenase production by connective tissue cells. These two conditions suggest that the epidermal-dermal interface may be a site of continuous cellular communication regulating collagen breakdown.

Under physiological conditions degradation of interstitial collagen (Types I, II and III) is specifically initiated by the enzyme collagenase, since, in its native form an interstitial collagen molecule is essentially resistant to digestion by other proteases.[6,7] In an increasing number of *in vitro* cases modulation of connective tissue cell collagenase production appears to involve cell-cell communication.

In general, communication between cells is required to regulate their development and organization into tissues, to control their growth and division, and to coordinate their diverse activities.[8] Cells appear to communicate in three ways: 1) by secreted chemicals that signal cells some distance away; 2) by plasma-membrane-bound signaling molecules that influence other cells through direct physical contact; and 3) by gap junction formation that directly joins interacting cell cytoplasms. Signaling molecules that are secreted can be subdivided into three groups according to delivery mechanism: 1) local cytokines; 2) hormones; and 3) neurotransmitters. All cells contain a distinctive set of receptor proteins that enable them to bind and respond to a complementary set of signaling molecules in a characteristic way.

Of these communication mechanisms, collagenase production has been shown to be

[a]This work was supported by National Institutes of Health Grants EY06233 and AM03564. This is publication no. 1043 of the Robert W. Lovett Group for the Study of Diseases Causing Deformities.

[b]Present address: Department of Biological Sciences, Northern Illinois University, De Kalb, Illinois 60115-2861.

regulated by both local chemical mediators and hormones. Lymphokines,[9] prostaglandins,[10] and bacterial endotoxins[11] stimulate collagenase production by macrophages. A partially purified mononuclear cell factor (MCF, an interleukin-1-like factor) from monocyte/macrophage cultures enhances collagenase secretion by passaged human rheumatoid synovial cells.[12] Rabbit peritoneal macrophages secrete a cytokine, which has been partially purified, for collagenase production by isolated rabbit cartilage cells.[13] Medium conditioned by rabbit blood mononuclear cells, stimulates collagenase production by rabbit corneal cells from alkali-burned eyes.[14] A purified interleukin-1-like factor, catabolin, from cultured synovial tissue, stimulates collagenase production by isolated chondrocytes.[15] Partially purified stimulatory or inhibitory cytokines from normal corneal epithelial cells *in vitro,* as well as a similar stimulatory cytokine from epidermal cell cultures, regulate collagenase production by isolated primary corneal stromal cells[3,4] and passaged human skin fibroblasts.[16] Collagenase release by post partum uterus is inhibited by progesterone[17] and cAMP. Similarly, estradiol and progesterone inhibit enzyme production by endotoxin-stimulated macrophages.[18] Hydrocortisone inhibits collagenase release by skin organ cultures.[19] Medroxyprogesterone and dexamethasone prevent collagenase production by corneal tissue.[20]

To date several unique purified proteins have been shown to induce collagenase production by connective tissue cells. These include recombinant Interleukin-1β[21] (IL-1β), tumor necrosis factor[22] (TNF), platelet-derived growth factor[23] (PDGF), epidermal growth factor[23] (EGF), eye-derived growth factor[24] (EDGF) and substance P.[25]

We have previously shown that primary cultures of adult rabbit corneal epithelial cells,[4] and adult and fetal skin epidermal cells[5] release cytokine(s) into culture medium which stimulate connective tissue cells *in vitro* to produce and secrete latent collagenase, thus, demonstrating that these cell types participate in a paracrine cell-cell interaction. The shapes of the dose-response curves for both adult and fetal epidermal cell-conditioned medium when used to induce collagenase production by connective tissue cells were biphasic[5] indicating the presence of more than one cytokine in the epidermal cell preparations. Upon AcA54 molecular sieve chromatography conditioned medium from rabbit skin epidermis and corneal epithelium yields a 21,000-dalton (21-kDa)[5] and a 19,000-dalton (19-kDa)[4] cytokine peak respectively. To better characterize and more accurately evaluate molecular weight and heterogeneity of epithelial cytokines from skin and cornea, the 21-kDa and 19-kDa cytokine peaks from these epithelial sources have been subject to further purification and dose-response analysis.

MATERIALS AND METHODS

Preparation of Epithelial Cells

Primary epithelial cells were obtained from New Zealand White rabbits by splitting excised corneas or ear skin with trypsin.[4,5] Briefly, corneas or skin were incubated in Ca^{2+}-Mg^{2+} free Hanks' balanced salt solution plus 0.25% trypsin (Grand Island Biological Company, Grand Island, NY) for 12–16 hours at 4°C. After incubation the epithelium was gently pushed from the corneal stromal surface with the edge of a scalpel blade or peeled as a sheet from the dermis with tweezers. The isolated epithelial cells, cell clumps or sheets were collected, centrifuged, and washed several times in Dulbecco's modified Eagle's medium (DME) (high glucose) plus 5% fetal calf serum prior to plating in the same medium.

Primary epidermal cells were also obtained from the skin of 4-week-old fetal

FIGURE 1. Biphasic dose-response curves of epithelial cytokines on collagenase production by corneal stromal cells. Stromal cells were exposed to different concentrations of each cytokine for six days. Cytokines were purified from conditioned medium by ultrogel AcA54 gel filtration chromatography. *Open squares:* 19-kDa cytokine from adult corneal epithelial cells (n = 8); *closed squares:* 21-kDa cytokine from 4-week-old fetal skin (n = 8).

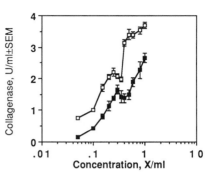

rabbits.[5] The body skin from the whole torso of such fetuses was removed and processed as mentioned above.

Determination of Cell Number

The concentration of freshly isolated stromal cells was determined with a hemocytometer. Epithelial cell number was determined by DNA assay[26] using appropriate standards.

Culture of Epithelial Cells for Production of Stimulatory Cytokines

Cell culture was carried out as previously described.[4,5] Medium conditioned with stimulatory cytokines was obtained from high-density epidermal cell cultures in serum-free medium[16] and low-density corneal epithelial cell cultures maintained with cytochalasin B (CB).[4]

Purification of Cytokines

Stimulatory cytokines from epithelial cell-conditioned medium were purified by AcA54 molecular seive chromatography.[5,16]

The "21-kDa" cytokine peak from AcA54 chromatography was concentrated 100-fold on a YM5 Amicon filter, and run unreduced on a 12.5% polyacrylamide gel

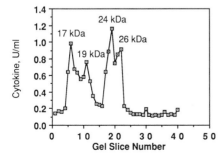

FIGURE 2. Purification of 21-kDa cytokine from adult skin epidermal cells by electrophoresis in a 12.5% polyacrylamide SDS gel. Consecutive 1-mm slices of the gel from the front to the level of the 43-kDa standard were eluted and bioassayed for stimulatory cytokine (n = 4). Each cytokine peak is labelled with its calculated molecular weight.

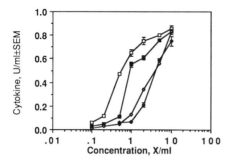

FIGURE 3. Dose-response curves of individual polyacrylamide gel purified epithelial cytokines. Each cytokine was bioassayed on corneal stromal cells for six days (n = 4). *Open squares:* 17-kDa cytokine; *closed diamonds:* 19-kDa cytokine; *closed squares:* 24-kDa cytokine; *open diamonds:* 26-kDa cytokine.

containing sodium dodecyl sulphate (SDS).[27] The gel was cut into consecutive 1-mm-wide slices from the dye front to the level of the 43,000 dalton protein standard and was eluted with gel extraction buffer (100 mM Tris-HCl, 0.1% SDS, 1 mM tetrasodium ethylenediamine tetraacetic acid, 0.04% sodium azide, pH = 8.0) at room temperature with shaking overnight. Eluates were dialyzed against DME salts plus antibiotics containing 0.1% ethanol at 4°C, supplemented to complete culture medium[16] and cytokine activity bioassayed. Protein molecular weight standards (43,000 daltons, 30,000 daltons, 21,000 daltons and 14,000 daltons) were run on a portion of the gel and used to calculate the molecular weights of the cytokines.

Bioassays for Cytokines

The ability of cytokine preparations to stimulate collagenase production by a connective tissue was tested *in vitro* on primary corneal stromal cells in the presence of CB.[3–5]

One unit (U) of stimulatory cytokine was defined as the amount of cytokine that stimulates 3×10^5 primary stromal cells in one ml of culture medium at 37°C to produce 1 U of collagenase in 6 days.

Collagenase Assay

Culture medium from target stromal cells in the bioassay was assayed directly for collagenase activity by the ^{14}C-collagen fibril film method[28] after trypsin activation of the latent form of the enzyme. Collagen degradation was calculated by subtracting

FIGURE 4. Dose-response curves of individual and mixed 17-kDa and 19-kDa polyacrylamide gel purified epithelial cytokines. Each cytokine and mixture was bioassayed on corneal stromal cells for six days (n = 4). *Open squares:* 17-kDa cytokine; *closed diamonds:* 19-kDa cytokine; *closed squares:* 1:1 mixture of 17-kDa and 19-kDa cytokine.

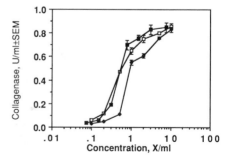

buffer blank values from experimental values. One U of collagenase was defined as the amount of enzyme that degrades 1 μg of collagen fibrils per minute at 37°C.

RESULTS AND DISCUSSION

Collagenase production by stromal cells in response to different concentrations of either the 19-kDa adult corneal or 21-kDa fetal skin stimulatory epithelial cytokine was biphasic (FIG. 1). Both dose-response curves plateau at an intermediate concentration over the range examined. This pattern of response results in two stable levels of enzyme production, and may reflect the way in which normal collagen turnover and massive collagen remodelling or pathology are stabilized in cornea or skin *in situ*. This type of pattern may be caused by the presence of two or more cytokines with different receptors which have an additive effect on connective tissue collagenase induction.

To study epithelial cytokine complexity, the 21-kDa cytokine peak from AcA54 chromatography of adult epidermal cell-conditioned medium was concentrated on a YM5 Amicon filter, and run unreduced on a 12.5% SDS polyacrylamide gel. The 21-kDa cytokine peak eluted from the polyacrylamide gel as four discrete active peaks

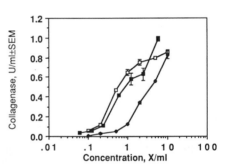

FIGURE 5. Dose-response curves of individual and mixed 17-kDa and 23-kDa polyacrylamide gel purified epithelial cytokines. Each cytokine and mixture was bioassayed on corneal stromal cells for six days (n = 4). *Open squares:* 17-kDa cytokine; *closed diamonds:* 23-kDa cytokine; *closed squares:* 1:1 mixture of 17-kDa and 23-kDa cytokine.

(FIG. 2) with molecular weights of 17,000 daltons (17 kDa), 19,000 daltons (19 kDa), 24,000 daltons (24 kDa) and 26,000 daltons (26 kDa). The 17-kDa and 24-kDa peaks were the major species and each contained 2–3 times as much activity as the 19-kDa and 26-kDa peaks.

To determine the relationship between these four cytokine peaks, the dose-response characteristics of each peak alone and in combination with others was determined. All were found to be individually semi-log and monophasic on target stromal cells (FIG. 3), as was a mixture of 17-kDa and 19-kDa cytokine (FIG. 4). However, mixtures of 17-kDa and 24-kDa cytokine (FIG. 5) or 19-kDa and 24-kDa cytokine (not shown) yielded biphasic, semi-log dose-response curves. The latter two combinations recreated the biphasic curve produced by the original 21-kDa peak from AcA54 chromatography. These results indicate that 17-kDa and 19-kDa cytokine are using the same receptor on the stromal cell, while the 24-kDa cytokine is using a different receptor than these two. The 26-kDa cytokine has not been evaluated in combination with the others yet.

Hence, at minimum there are at least two classes of epidermal cytokine: the 17-kDa/19-kDa cytokines in one class and the 24-kDa cytokine in another. At maximum all four cytokine peaks are different proteins. In addition, these results show that target stromal cells contain at least two classes of receptor which stimulate collagenase production and whose effects are additive.

Several different cytokines are known to induce collagenase production by connective tissue cells. These include II-1[21], TNF[22] and PDGF[23] all of which are produced by the monocyte/macrophage.[29] The present study indicates that epidermis also is capable of producing several collagenase-inducing cytokines. Since, epidermis is made up of keratinocytes and Langerhans cells, the precise cellular source of cytokine heterogeneity, however, remains to be elucidated. One of the epidermal cytokines is probably II-1 which both cell types produce.[30,31]

Corneal epithelium is made up exclusively of keratinocytes, and its 19-kDa cytokine induces a biphasic dose-response from connective tissue cells. In this case, it appears that keratinocytes make at least two distinct cytokines. Similarly, the biphasic dose-response kinetics of 21-kDa cytokine from fetal skin shows that multiple cytokines from epidermal cells which regulate fibroblast collagen degradation appear early in integument development.

The local use by epithelia of multiple cytokines whose effects can be additive to induce connective tissue collagenase production, may be part of the mechanism underlying normal skin and corneal morphogenesis, maintenance and wound healing, or pathological ulceration.

ACKNOWLEDGMENTS

The author wishes to thank Dr. Jerome Gross for partial support (NIH Grant AM03564) of and many fruitful discussions about this work.

REFERENCES

1. COHEN, J. 1969. *In* Advances in Biology of Skin. W. Montagna & R. L. Dodson, Eds. Vol. 9: 1–18. Pergamon Press. New York, NY.
2. REDDICK, M. E., E. A. BAUER & A. Z. EISEN. 1974. J. Invest. Dermatol. **62:** 361–366.
3. JOHNSON-MULLER, B. & J. GROSS. 1978. Proc. Natl. Acad. Sci. USA **75:** 4417–4421.
4. JOHNSON-WINT, B. 1980. Proc. Natl. Acad. Sci. USA **77:** 5331–5335.
5. JOHNSON-WINT, B. & J. GROSS. 1984. J. Cell Biol. **98:** 90–96.
6. GROSS, J. 1976. *In* Biochemistry of Collagen. G. N. Ramachandran & A. M. Reddi, Eds. 275–317. Plenum. New York, NY.
7. MURPHY, G. & A. SELLERS. 1980. *In* Collagenase in Normal and Pathological Connective Tissues. D. E. Woolley & J. M. Evanson, Eds. 65–81. John Wiley and Sons, Ltd. New York, NY.
8. ALBERTS, B., D. BRAY, J. LEWIS, M. RAFF, K. ROBERTS & J. D. WATSON, Eds. 1983. Molecular Biology of the Cell. Ch. **13:** 717–765. Garland Publ., Inc. New York, NY.
9. WAHL, L. M., S. M. WAHL, S. E. MERGENHAGEN & G. R. MARTIN. 1975. Science **187:** 261–263.
10. WAHL, L. M., C. E. OLSEN, A. L. SANDBERG & S. E. MERGENHAGEN. 1977. Proc. Natl. Acad. Sci. USA **74:** 4955–4958.
11. WAHL, L. M., S. M. WAHL, S. E. MERGENHAGEN & G. R. MARTIN. 1974. Proc. Natl. Acad. Sci. USA **71:** 3598–3601.
12. DAYER, J.-M., J. BREARD, L. CHESS & S. M. KRANE. 1979. J. Clin. Invest. **64:** 1386–1392.
13. DESHMUKH-PHADKE, K., M. LAWRENCE & S. NANDA. 1978. Biochem. Biophys. Res. Commun. **85:** 490–496.
14. NEWSOME, D. A. & J. GROSS. 1979. Cell **16:** 895–900.
15. SAKLATVALA, J., L. M. C. PILSWORTH, S. J. SARSFIELD, J. GAVRILOVIC & J. K. HEATH. 1984. Biochem. J. **224:** 461–466.

16. JOHNSON-WINT, B. & E. A. BAUER. 1985. J. Biol. Chem. **260**: 2080–2085.
17. KOOB, T. J. & J. J. JEFFREY. 1974. Biochim. Biophys. Acta **354**: 61–70.
18. WAHL, L. M. 1977. Biochem. Biophys. Res. Commun. **74**: 838–845.
19. KOOB, T. J., J. J. JEFFREY & A. Z. EISEN. 1974. Biochem. Biophys. Res. Commun. **61**: 1083–1088.
20. NEWSOME, D. A. & J. GROSS. 1977. Invest. Ophthalmol. **16**: 21–31.
21. DAYER, J. M., B. ROCHEMONTEIX, B. BURRUS, S. DEMCZUK & C. A. DINARELLO. 1986. J. Clin. Invest. **77**: 645–648.
22. DAYER, J. M., B. BEUTLER & A. CERAMI. 1985. J. Exp. Med. **162**: 2163–2168.
23. CHUA, C. C., D. E. GEIMAN, G. H. KELLER & R. L. LADDA. 1985. J. Biol. Chem. **260**: 5213–5216.
24. CHUA, C. C., D. BARRITAULT, D. E. GEIMAN & R. L. LADDA. 1985. Collagen Rel. Res. **7**: 277–284.
25. LOTZ, M., D. A. CARSON & J. H. VAUGHAN. 1987. Science **235**: 893–895.
26. JOHNSON-WINT, B. & S. HOLLIS. 1982. Anal. Biochem. **122**: 338–344.
27. LAEMMLI, U. K. 1970. Nature **227**: 680–685.
28. JOHNSON-WINT, B. 1980. Anal. Biochem. **104**: 175–181.
29. NATHAN, C. F. 1987. J. Clin. Invest. **79**: 319–326.
30. HAUSER, C., J.-H. SAURAT, F. JAUNIN, S. SIZONENKO & J. M. DAYER. 1985. Biochim. Biophys. Acta **840**: 350–355.
31. SCHMITT, A., C. HAUSER, F. JAUNIN, J. M. DAYER & J. H. SAURAT. 1985. Lymphokine Res **5**: 105–118.

Keratinocyte- and Tumor-Derived Inducers of Collagenase[a]

E. A. BAUER,[b] A. P. PENTLAND, A. KRONBERGER,[b]
S. M. WILHELM, G. I. GOLDBERG, H. G. WELGUS,
AND A. Z. EISEN

Division of Dermatology
Campus Box 8123
Washington University School of Medicine
660 South Euclid Avenue
St. Louis, Missouri 63110

INTRODUCTION

The basement membrane zone (BMZ) is a complex region in which the interstitial collagens of the dermis and the proteins of the dermal-epidermal interface form an elaborate network of structural macromolecules. Remodeling of the BMZ is required in morphogenesis, wound healing and tumor invasion. Because of the structural importance of the interstitial collagens (types I and III) and their close association in the papillary dermis with basement membrane collagens, especially type VII collagen-containing anchoring fibrils and anchoring plaques,[1] cleavage of the interstitial collagens in that location may destabilize the entire zone, as in tumor invasion[2-4] and some forms of epidermolysis bullosa.[5]

Epithelial modulation of stromal elements may play an important role in this process. One possible explanation for the events which occur during morphogenesis and wound repair is that factors released by the epithelium modulate collagenase and/or tissue inhibitor of metalloproteases (TIMP) expression by subadjacent fibroblasts. For example, cytokines found in rabbit corneal epithelium stimulate collagenase synthesis by stromal cells[6] through mechanisms involving the presence of increased translatable collagenase mRNA.[7] Furthermore, in human basal cell carcinomas (BCC), collagenase is localized to the stroma rather than to the tumor islands,[4] suggesting that the tumor epithelium elaborates one or more cytokines which stimulate collagenase expression in surrounding skin fibroblasts.[8,9] To explore this putative mechanism we employed both keratinocytes cultured from normal human epidermis and extracted basal cell carcinoma epithelium.

[a]This work was supported in part by United States Public Health Service Grants AR 19537, AR 12129, and TO-AR 07284, by National Institutes of Health Grant RR 00036, by the Washington University-Monsanto Biomedical Research Agreement, and by the Washington University/DEBRA Center for Research and Therapy of Epidermolysis Bullosa.

[b]Current address: Department of Dermatology, Stanford University School of Medicine, Stanford, CA 94305.

METHODS

Cell Cultures

Human skin keratinocytes were cultured as previously described.[10] Briefly, normal human skin obtained at the time of cosmetic surgery was rinsed in a neutral salt solution containing 2 μg/ml of amphotericin B after which the deep dermis and subcutaneous fat were removed with a scalpel. Following trypsinization, the epidermis was removed and placed in Dulbecco's Modified Eagle's Medium (DMEM) containing 10% fetal calf serum (FCS) and plated on collagen-coated dishes in a 5% CO_2 atmosphere at 37°C.

For experiments employing target cells, human skin fibroblasts having low constitutive levels of collagenase expression (AG 4431, Human Genetic Mutant Cell Repository, Camden, NJ) were cultured in DMEM with 10% FCS until confluency was reached.[11] This medium was then removed and replaced with DMEM containing either 1% FCS alone (control) or DMEM/1% FCS with varying dilutions of the conditioned media or tissue extracts containing the putative cytokines.

Tissue Extracts of Basal Cell Carcinomas

For preparation of the BCC extracts, tumors which had been therapeutically excised were sliced under sterile conditions. Under a dissecting microscope the tumor nodules were removed using blunt-dissection and homogenized. Following homogenization the suspension was subjected to sonication at an intensity of 60 for 30 sec and centrifuged at 10,000 g for 10 min at 4°C, and the insoluble pellet was discarded. The tissue extracts were then diluted as needed with DMEM prior to incubation with normal human skin fibroblast target cells.[8]

Collagenase and TIMP Assays

Human skin procollagenase in the culture medium was activated proteolytically with trypsin as described previously.[12] Each mixture was assayed for collagenase activity at 37°C in 50 mM Tris-HCl (pH 7.5) in the presence of 10 mM $CaCl_2$ using native reconstituted ^{14}C-glycine-labeled collagen fibrils containing approximately 5000 cpm per substrate gel.[13] Immunoreactive collagenase protein and TIMP immunoreactive protein were quantitated using the enzyme-linked immunosorbent assay (ELISA) reported earlier.[14,15]

Thymocyte Proliferation Assay

Samples were assayed for augmentation of lectin-induced thymocyte proliferation using the assay of Mizel, Oppenheim and Rosenstreich.[16] Briefly, 1.5×10^6 C3H/HeJ thymocytes were cultured for 72 hr in RPMI 1540 medium containing 5% FCS, 1 μg/ml phytohemagglutinin and a 1/4 dilution of the samples to be tested. Cultures were pulsed with 0.5 μCi of ^3H-TdR (1.9 Ci/mmol, New England Nuclear, Boston, MA) for the final 6 hr. Cells were then collected on filter paper and the results were expressed as the cpm of ^3H-TdR incorporated (mean $\pm$ SE of triplicate cultures).

TABLE 1. Effect of Keratinocyte Growth on Fibroblast Collagenase Stimulation by Conditioned Media[a]

Keratinocyte-Conditioned Medium	Number	Immunoreactive Fibroblast Collagenase (μg/ml)
None[b]	7	0.30 ± 0.07
Subconfluent[c]	13	2.07 ± 0.58
Confluent[c]	25	1.60 ± 0.23

[a]Keratinocyte-conditioned media were diluted 1/2 in DMEM/1% FCS and incubated for 24 hr on normal skin fibroblast target cells.
[b]Baseline collagenase expression by target cells alone.
[c]Collagenase concentrations in the keratinocyte-conditioned media before incubation with the fibroblast target cells were 0.69 ± 0.15 μg/ml (subconfluent) and 0.18 ± 0.03 μg/ml (confluent).

RESULTS

We sought evidence for epithelial-stromal interactions by probing the effects of keratinocyte-conditioned medium on normal human skin fibroblasts used as target cells. As shown in TABLE 1, in medium containing DMEM/1% FCS alone the control target fibroblasts accumulated 0.30 ± 0.07 μg/ml of collagenase per 24 hr. The addition of conditioned medium from subconfluent keratinocyte cultures (diluted ½) stimulated collagenase expression about 7-fold. Conditioned media harvested from confluent keratinocyte cultures also stimulated collagenase expression in the target cells, although consistently to a lower degree (~5-fold).

In a similar fashion the conditioned medium from subconfluent keratinocyte cultures elicited a 5-fold increase in TIMP expression by the target cells (TABLE 2). However, the conditioned media harvested from confluent keratinocyte cultures repeatedly contained only about one-half the TIMP-stimulatory capacity of that of the subconfluent cultures.

The effects of the keratinocyte-conditioned media on fibroblast collagenase expression were remarkably similar to those seen with tissue extracts of tumor islands from human BCC.[8] As shown in TABLE 3, incubation of pooled BCC tumor extracts resulted in a 7- to 10-fold increase in collagenase activity after 24-hr incubation of skin fibroblast target cells, whereas incubation of the same dilution of BCC extract on

TABLE 2. Effect of Keratinocyte Growth on Fibroblast TIMP Stimulation by Conditioned Media[a]

Keratinocyte-Conditioned Medium	Number	Immunoreactive Fibroblast TIMP (μg/ml)
None[b]	3	0.20 ± 0.02
Subconfluent[c]	4	1.05 ± 0.16
Confluent[c]	6	0.47 ± 0.09

[a]Keratinocyte-conditioned media were diluted 1/2 in DMEM/1% FCS and incubated for 24 hr on normal skin fibroblast target cells.
[b]Baseline TIMP expression by target cells alone.
[c]TIMP concentrations in the keratinocyte-conditioned media before incubation with the fibroblast target cells were 0.01 ± 0.01 μg/ml (subconfluent) and 0 μg/ml (confluent).

TABLE 3. Comparison of Collagenase- and Lymphocyte-Stimulatory Activities in BCC Extracts[a]

Extract Dilution	Collagenase Activity[b]		Lymphocyte Activation[c]	
	Control	BCC	Control	BCC
1/4	182 ± 3	1,345 ± 128	nd	nd
1/2	116 ± 9	1,232 ± 128	378 ± 23	731 ± 180

[a]Data are adapted from Goslen *et al.*[8]

[b]Normal skin fibroblast target cells were cultured either alone (control) or in the presence of a 1/2 or 1/4 dilution of a BCC tumor nodule extract (BCC). Collagenase activity was then measured in the fibroblast culture medium. Data are expressed as cpm (mean ± SE) ^{14}C-collagen solubilized.[13]

[c]Data are expressed as cpm (mean ± SE) ^{3}H-Tdr incorporated.[16]

thymocyte target cells followed by a pulse with ^{3}H-thymidine revealed only a 2-fold increase in lymphocyte activation. Serial dilution of the extract from 1/16 to 1/64 resulted in no more than an ~15% increase in lymphocyte activation, suggesting that the extract did not contain an inhibitor of interleukin 1-like activity.[8]

It is of interest that experiments employing nontumor-derived material also suggested an epithelial source for the cytokine(s). Only the tumor extracts and normal adjacent epidermis exhibited any stimulatory effect on the target cells, whereas the stromal extracts displayed no stimulation (TABLE 4).

DISCUSSION

It is likely that epithelial-stromal interactions operate in the remodeling process during both differentiation and repair. Indeed, keratinocyte-conditioned media contained collagenase- and TIMP-stimulatory factor(s) which induced stromal cells to produce greater amounts of these gene products (TABLES 1 and 2). Stimulation was maximally expressed in the subconfluent cultures, *i.e.*, during *in vitro* proliferation of the cells that might correspond most closely to the *in vivo* situations of developmental proliferation or wound repair. However, these and other[9] studies also demonstrate that one or more factors with a significant capacity to stimulate collagenase production by human skin fibroblasts are present in BCC-derived tissue extracts and may play a role in tumorigenesis.

TABLE 4. Effect of Various Tissue Extracts on Collagenase Expression in Normal Fibroblast Target Cells[a]

Extract	Number	Collagenase Activity[b] (cpm)
None	4	744 ± 57
Stroma	6	622 ± 48
Normal Epidermis	5	812 ± 51
BCC Tumor	4	1014 ± 64

[a]Normal skin fibroblast target cells were cultured either alone or in the presence of a 1/4 dilution of an extract of BCC tumor nodules, keratomed epidermis, or stoma obtained at surgery from connective tissue adjacent to the tumor. Data are adapted from Goslen *et al.*[8]

[b]Data are expressed as cpm (mean ± SE) ^{14}C-collagen solubilized.[13]

The change in the stimulatory capacity of keratinocytes as a function of growth can be related to observations of Johnson-Wint and Gross.[17] These investigators found stimulators of collagenase production in medium from cultured corneal and cutaneous epithelium of rabbits and observed approximately ten times more stimulatory activity from cultured fetal epidermal cells than from adult epidermal cells of the rabbit.[6,17] Although the cytokines remained unidentified, these investigators detected both high-M_r (~55 kDa) and low-M_r (~19 kDa) stimulatory factors.[17] These cytokines have been shown to act at a pretranslational level to stimulate collagenase synthesis, probably through enhanced transcription of collagenase messenger RNA.[7]

Furthermore, the identity of the epidermally derived collagenase- and the TIMP-stimulatory factor(s) described here is as yet unknown. However, epidermal cells produce a T cell stimulatory factor (ETAF, epidermal cell thymocyte activating factor)[18] which has been shown to be identical to interleukin-1 (IL-1).[19] IL-1 itself stimulates collagenase production in fibroblast cultures, as shown both by the use of purified preparations from monocyte-macrophage cells[20,21] and by employing recombinant IL-1.[22] Our experiments support the idea that an IL-1-like cytokine(s) may account for at least some of the stimulation, since IL-1 activity was detected both in BCC extracts (TABLE 3) and in culture medium of the keratinocyte cultures (not shown).

The observations in this report support the notion that indirect (cytokine) mechanisms exist for mediating turnover of the collagens of the BMZ. From this we speculate that when profound changes in connective tissue architecture are required, it may be advantageous for the organism to engage the stromal cells to synthesize collagenase maximally. Epithelial-stromal interactions, mediated by epidermally derived cytokines, would represent an ideal mechanism for regulation in such circumstances. This scheme would also support the observation of a coordinate increase in TIMP expression by the target fibroblasts, since such an increase would presumably offer the greatest possibility for tight local control of proteolytic activity.

REFERENCES

1. KEENE, D. R., L. Y. SAKAI, G. P. LUNSTRUM, N. P. MORRIS & R. E. BURGESON. 1987. Type VII collagen forms an extended network of anchoring fibrils. J. Cell Biol. **104:** 611–621.

2. DRESDEN, M. H., S. A. HEILMAN & J. D. SCHMIDT. 1972. Collagenolytic enzymes in human neoplasms. Cancer Res. **32:** 993–996.

3. YAMANISHI, Y., M. K. DABBOUS & K. HASHIMOTO. 1972. Effect of collagenolytic activity in basal cell epithelioma of the skin on reconstituted collagen and physical properties and kinetics of the crude enzyme. Cancer Res. **32:** 2551–2560.

4. BAUER, E. A., J. M. GORDON, M. E. REDDICK & A. Z. EISEN. 1977. Quantitation and immunocytochemical localization of human skin collagenase in basal cell carcinoma. J. Invest. Dermatol. **69:** 363–367.

5. BAUER, E. A. 1986. Collagenase in recessive dystrophic epidermolysis bullosa. Ann. N.Y. Acad. Sci. **460:** 311–320.

6. JOHNSON-MULLER, B. & J. GROSS. 1978. Regulation of corneal collagenase production: Epithelial-stromal cell interactions. Proc. Natl. Acad. Sci. USA **75:** 4417–4421.

7. JOHNSON-WINT, B. & E. A. BAUER. 1985. Stimulation of collagenase synthesis by a 20,000 dalton epithelial cytokine: Evidence for pre-translational regulation. J. Biol. Chem. **260:** 2080–2085.

8. GOSLEN, J. B., A. Z. EISEN & E. A. BAUER. 1985. Stimulation of skin fibroblast collagenase production by a cytokine derived from basal cell carcinomas. J. Invest. Dermatol. **85:** 161–164.

9. HERNANDEZ, A. D., M. S. HIBBS & A. E. POSTLETHWAITE. 1985. Establishment of basal cell carcinoma in culture: Evidence for a basal cell carcinoma-derived factor(s) which stimulates fibroblasts to proliferate and release collagenase. J. Invest. Dermatol. **85**: 470–475.

10. PENTLAND, A. P. & P. NEEDLEMAN. 1986. Modulation of keratinocyte proliferation *in vitro* by endogenous prostaglandin synthesis. J. Clin. Invest. **77**: 236–251.

11. BAUER, E. A. 1977. Cell culture density as a modulator of collagenase expression in normal human fibroblast cultures. Exp. Cell Res. **107**: 269–276.

12. BAUER, E. A., G. P. STRICKLIN, J. J. JEFFREY & A. Z. EISEN. 1975. Collagenase production by human skin fibroblasts. Biochem. Biophys. Res. Commun. **64**: 232–240.

13. NAGAI, Y., C. M. LAPIERE & J. GROSS. 1966. Tadpole collagenase: Preparation and purification. Biochemistry **5**: 3123–3130.

14. COOPER, T. W., E. A. BAUER & A. Z. EISEN. 1983. Enzyme-linked immunoadsorbent assay for human skin collagenase. Collagen Relat. Res. **3**: 205–216.

15. WELGUS, H. G. & G. P. STRICKLIN. 1983. Human skin fibroblast inhibitor. Comparative studies in human connective tissues, serum, and amniotic fluid. J. Biol. Chem. **258**: 12259–12264.

16. MIZEL, S. B., J. J. OPPENHEIM & D. L. ROSENSTREICH. 1978. Characterization of lymphocyte-activating factor (LAF) produced by the macrophage cell line, P388D1. J. Immunol. **120**: 1504–1508.

17. JOHNSON-WINT, B. & J. GROSS. 1984. Regulation of connective tissue collagenase production: Stimulators from adult and fetal epidermal cells. J. Cell Biol. **98**: 90–96.

18. SAUDER, D. N., C. S. CARTER, S. I. KATZ & J. J. OPPENHEIM. 1982. Epidermal cell production of thymocyte activating factor (ETAF). J. Invest. Dermatol. **79**: 34–39.

19. KUPPER, T. S., D. W. BALLARD, A. O. CHUA, J. S. McGUIRE, P. M. FLOOD, M. C. HOROWITZ, R. LANGDON, L. LIGHTFOOT & U. GRUBER. 1986. Human keratinocytes contain mRNA indistinguishable from monocyte interleukin 1α and β mRNA. Keratinocyte epidermal cell-derived thymocyte-activating factor is identical to interleukin 1. J. Exp. Med. **164**: 2095–2100.

20. DAYER, J.-M., J. BREARD, L. CHESS & S. M. KRANE. 1979. Participation of monocyte-macrophages and lymphocytes in the production of a factor that stimulates collagenase and prostaglandin release by rheumatoid synovial cells. J. Clin. Invest. **64**: 1386–1392.

21. POSTLETHWAITE, A. E., L. B. LACHMAN, C. MAINARDI & A. H. KANG. 1983. Interleukin 1 stimulation of collagenase production by cultured fibroblasts. J. Exp. Med. **157**: 801–806.

22. DAYER, J.-M., B. DE ROCHEMONTEIX, B. BURRUS, S. DEMCZUK & C. A. DINARELLO. 1986. Human recombinant interleukin 1 stimulates collagenase and prostaglandin E_2 production by synovial cells. J. Clin. Invest. **77**: 645–648.

Paracrine Stimulation of Melanocytes by Keratinocytes through Basic Fibroblast Growth Factor[a]

R. HALABAN,[b] R. LANGDON,[b] N. BIRCHALL,[b]
C. CUONO,[c] A. BAIRD,[d] G. SCOTT,[b] G. MOELLMANN,[b]
AND J. MCGUIRE[b]

*Departments of [b]Dermatology and [c]Plastic
and Reconstructive Surgery
Yale University School of Medicine
333 Cedar Street
P.O. Box 3333
New Haven, Connecticut 06510
and
[d]Department of Neuroendocrinology
Salk Institute for Biological Studies
La Jolla, California 92037*

INTRODUCTION

Epidermal melanocytes survive and may even proliferate for several weeks when cultured in the presence of keratinocytes.[1] In contrast, melanocytes die within a week when cultured alone in routine media[2] or medium optimal for keratinocyte proliferation (our unpublished results), because melanocytes require two classes of mitogen. One class is represented by TPA (12-O-tetradecanoyl-phorbol-13-acetate), the other by substances such as cholera toxin, isobutylmethyl-xanthine (IBMX) or dibutyryl-cyclic-adenosine-monophosphate (dbcAMP), which increase intracellular cyclic-adenosine-monophosphate (cAMP).[3-4] Recently, we have shown that basic fibroblast growth factor (bFGF) can substitute for TPA and is the melanocyte mitogen in extracts of human melanomas and placenta, bovine pituitary gland, and other bovine organs.[5] Since, *in situ,* normal human melanocytes are surrounded by keratinocytes and in culture can be sustained by contact with keratinocytes, we asked whether keratinocytes might stimulate melanocytes via bFGF. The studies described here demonstrate that proliferating human keratinocytes in culture contain a mitogen for human melanocytes, which is bFGF. These findings suggest that the viability of the epidermal melanocyte population *in vivo* is regulated by bFGF from basal, proliferative keratinocytes.

[a]This work was supported by National Institutes of Health Grants 5 RO1 CA04679-26, 5 PO1 AR25252-09, 5 RO1 AR13929-18, HB 09690, and DK 18811.

180

MATERIALS AND METHODS

Cell Culture

Melanocyte cultures were initiated from neonatal foreskins in *TIP* medium, consisting of 85 nm *T*PA (LC Services Corporation, Woburn, MA), 0.1 mM *I*BMX (Sigma, St. Louis, MO) and 10–20 μg protein/ml *p*lacental extract in Ham's F-10 medium (American Biorganics, Inc., N. Tonawanda, NY) supplemented with 10% newborn calf serum (GIBCO Laboratories, Grand Island, NY), 200 units/ml penicillin, 100 μg/ml streptomycin.[4] Contaminating fibroblasts were eliminated by incubating the cultures for 3–4 days in TIP medium supplemented with 100 μg/ml geneticin (G418 sulfate, GIBCO Laboratories, Grand Island, NY).[6] Normal human melanocytes used in the experiments had been in culture for no longer than four months and had been passed no more than five times, at a ratio of 1:3.

Keratinocyte cultures were initiated from neonatal foreskins and adult skin on collagen-coated Petri dishes in modified MCDB-153 (Irvine Scientific, Santa Ana, CA; low calcium medium containing 0.03 mM $CaCl_2$), supplemented with antibiotics as described above, 0.63 μg/ml fungizone (Flow Laboratories, Inc., McLean, VA), 70 μg protein/ml bovine pituitary extract (Pel Freeze Biologicals, Rogers, AR, prepared as described in Halaban et al., 1987),[5] 1 ng/ml epidermal growth factor, 50 μM hydrocortisone (both from GIBCO Laboratories, Grand Island, NY), 0.1 mM ethanolamine and 0.1 mM phosphoethanolamine (Sigma, St. Louis, MO). Subsequently, keratinocytes were passed in the same medium in uncoated culture flasks. Adult keratinocytes were derived from unburned skin of burn patients or from cadavers. Where indicated, the cells were maintained in DMEM (GIBCO, containing 3.0 mM $CaCl_2$) supplemented with 20% fetal calf serum (Hyclone Laboratories, Inc., Logan, UT), in which keratinocytes stratify. Keratinocytes were used after two passages in culture. To test the effect of interleukin 1 (IL-1), keratinocytes were grown in MCDB-153 medium and incubated for 24 hr with 0.02 μg/ml human recombinant IL-1 (Hoffman LaRoche, Nutley, NJ).

Fibroblasts from neonatal foreskins were grown in DMEM supplemented with 10% calf serum and were used after the 2nd or 3rd passage. SK-HEP-1, a human hepatoma cell line, was grown in the same medium.

In order to compare the viability of melanocytes with and without keratinocytes, melanocytes obtained from the foreskin of a black infant were seeded onto (a) confluent cultures of allogeneic neonatal keratinocytes; (b) into wells in which approximately half the surface was occupied by keratinocytes; and (c) into wells without keratinocytes. Similar cocultures were set up with melanocytes and dermal fibroblasts. The cultures received MCDB-153, DMEM, or TIP medium; these media support, respectively, keratinocyte proliferation, keratinocyte stratification, or melanocyte proliferation.

UVB Irradiation

Cells were irradiated with 50 mJ/cm^2 of UVB light (290–310 nm wavelength) from a panel of 4 lamps (FS20 Sun Lamp, Westinghouse Electric Corp., Pittsburgh, PA) at a dose of 1.5 mW/cm^2. Incident dose at the cell surface was measured through one layer each of tissue culture plastic and medium by means of a UVX digital radiometer (Ultraviolet Products, Inc., San Gabriel, CA).[7] Keratinocytes and fibroblasts, grown in 150 cm² flasks, were collected 7–9 hr and 24 hr after UVB irradiation. Extracts of these cells were used to test for mitogenic activity toward melanocytes.

Preparation of Cell Extracts and Assay for Mitogenic Activity Toward Melanocytes

To prepare extracts, cells were scraped off the culture surface, suspended in phosphate-buffered saline (PBS), centrifuged and washed twice with PBS. The cell pellets were resuspended in 0.2–0.5 ml PBS or double-distilled water and sonicated on ice. Alternatively, cells were disrupted by 3 cycles of freeze-thawing. Mitogenic activity of the extracts was similar, regardless of the procedure used. The disrupted cells were centrifuged at 13,500 g for 10 min at 4°C, and 5 μl aliquots of the supernatants were taken for protein determination by the Bio Rad assay. Bovine serum albumin served as a control. To test the mitogenic activity of cell extracts toward melanocytes, melanocyte cultures, 24,000–80,000 cells/4 cm² well (Costar, Cambridge, MA), were incubated overnight in PC-1 defined medium (Ventrex Laboratories, Inc., Portland, ME) without serum and without added growth factors.[5] This medium was then removed and experimental media (1 ml/well) were added.

Experimental media were prepared by adding known amounts of cell extract to PC-1 medium which in some experiments was then passed through a 0.22-μm filter (Millex-GV, Millipore Corporation, Bedford, MA). The mitogenic activity was not affected by filtration. Stimulation of melanocyte growth by the extracts was assayed at the end of 24 or 48 hr by determining ³H-thymidine incorporation as a measure of DNA synthesis. For this purpose, the experimental media were exchanged with assay medium consisting of minimal essential medium without calcium and magnesium (MEMS, GIBCO, Grand Island, NY), and containing 5 μCi/ml ³H-thymidine (90 Ci/mmole, Amersham, 0.5 ml/well). At the end of 1–3 hr incubation, 0.3 ml of trypsin-EDTA solution in MEMS was added to detach the cells from the culture dishes. The detached cells were trapped onto #30 glass filters in the Minifold apparatus of Schleicher and Schuell (Keen, NH; American Bioanalytical, Natick, MA). The filters were washed four times with distilled water, dried, and placed in scintillation fluid. Radioactivity was determined in a scintillation counter.

Northern Blot Analysis

RNA was extracted by the procedure of Chomczynski and Sacchi.[8] Total or polyadenylated RNAs were fractionated on 1.5% agarose denaturing gels, containing formaldehyde, and transferred to nitrocellulose filters. The blots were pretreated with BLOTTO buffer, containing 50% formamide, and hybridized in the same buffer, supplemented with 10% dextran sulfate[9] and ³²P-labeled 1.4 kb EcoRI fragment of cDNA for basic fibroblast growth factor (bFGF). The cDNA fragment (for bovine bFGF in pBR322 plasmid, known also as pJJ11-1[10]) was obtained from Drs. J. A. Abraham and J. C. Fiddes, California Biotechnology, Inc., Mountain View, CA.

RESULTS

Evidence That Keratinocytes Synthesize a Mitogen toward Melanocytes

The light microscopic appearance of melanocytes in MCDB-153 medium with and without keratinocytes is shown in FIGURE 1. In mixed culture, those melanocytes that were in direct contact with a keratinocyte survived for over 2 weeks and sprouted dendrites toward neighboring keratinocytes. In pure culture, and in mixed cultures in which keratinocytes were sparse, the melanocytes became spindle-shaped, then rounded up, detached, and lost their viability after one week. Loss of viability was defined as an inability to incorporate ^{3}H-thymidine one day after restimulation with TPA and IBMX.[2] Conditioned medium from keratinocyte cultures did not support proliferation or survival of melanocytes.[10a] These results indicated that direct contact

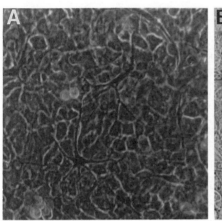

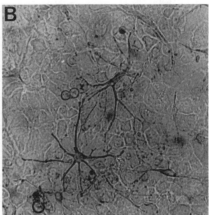

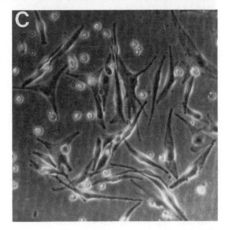

FIGURE 1. Morphology of melanocytes cultured with and without keratinocytes. (**A**) Phase-contrast and (**B**) bright field photomicrographs, respectively, of human keratinocytes and melanocytes cultivated together in MCDB-153 medium for 2 weeks. Melanocytes are viable and remain highly dendritic. (**C**) Phase-contrast photomicrograph of a pure culture of melanocytes grown in MCDB-153 medium as above for 5 days. The rounded cells are dying melanocytes. All magnifications 133x. (From Halaban *et al.*[10a] Reprinted by permission from the *Journal of Cell Biology*.)

with keratinocytes supported the viability of melanocytes and suggested that the growth factor for melanocytes, produced by the keratinocytes, was not secreted freely into the medium. Keratinocytes were more effective than fibroblasts in supporting the survival of melanocytes because fibroblasts outgrew and hence overcrowded and displaced the melanocytes (data not shown).

Mitogenic activity toward normal human melanocytes in extracts of keratinocytes is shown in TABLE 1. The data demonstrate that keratinocyte extract without addition of cAMP stimulated DNA synthesis in melanocytes. That stimulation was enhanced

TABLE 1. Mitogenic Activity Toward Human Melanocytes in Keratinocyte Extracts[a]

	Additions	^{3}H-thymidine Incorporation cpm/Well/Hr
Experiment 1	none	not detectable
	dbcAMP	not detectable
	KE	1,400
	KE + dbcAMP	20,200
Experiment 2	none	not detectable
	dbcAMP	not detectable
	KE	300
	KE + dbcAMP	1,240

[a]Normal human melanocytes were seeded in high-calcium (3.0 mM)PC-1 medium in 12-well cluster plates (80,000 cells/4 cm^2 well), and experimental media were added the following day. In experiment 1, extract (KE) from keratinocytes grown in MCDB-153 medium was added at 25 μg protein/ml, and in experiment 2, extract from keratinocytes grown in DMEM medium for 10 days was added at 20 μg protein/ml. The concentration of dbcAMP (dibutyryl-cyclic-adenosine-monophosphate) was 1 mM. ^{3}H-thymidine incorporation was carried out during the last 2–3 hours of a 24-hr incubation with experimental media. Values are averages of cpm from duplicate wells.

approximately 15-fold by 1 mM dbcAMP. The medium in which the keratinocytes were maintained affected the level of melanocyte mitogen in the extracts. Extract from keratinocytes grown in serum-free, low-calcium medium (MCDB-153), optimal for keratinocyte proliferation, had about 16-fold higher mitogenic activity toward melanocytes than a similar amount of extract from keratinocytes grown in DMEM medium, containing serum and high calcium, optimal for keratinocyte stratification. Differences in the levels of melanocyte mitogen between rapidly proliferating versus stratifying keratinocytes were observed with and without dbcAMP. There was a sharp drop in the level of mitogenic activity on the third day of culture in DMEM. This decline continued over the following days.[10a]

High levels of mitogenic activity toward melanocytes were found also in extracts of dermal fibroblasts, with 5 μg protein/ml exerting optimal stimulation of growth.[10a] Stimulation by fibroblast extract required dbcAMP. Conditioned medium from fibroblast cultures had no mitogenic activity toward melanocytes (data not shown).

UVB Light Induces the Melanocyte Mitogen in Keratinocytes

UVB irradiation is known to increase human skin pigmentation by increasing the synthesis of melanin and the transfer of melanin to keratinocytes. UVB light stimulates the proliferation of melanocytes and keratinocytes in murine skin.[11,12] We, therefore, tested whether UVB light regulates melanocyte proliferation indirectly by modulating the level of the melanocyte mitogen in keratinocytes. The data presented in FIGURE 2 show that UVB light increases the levels of the melanocyte mitogen in keratinocytes. At suboptimal doses of 2.5–5 μg protein/ml the level of mitogenic activity in rapidly proliferating keratinocytes harvested 7 hr after UVB irradiation was approximately 6-fold higher than that of nonirradiated keratinocytes. By 24 hr, the mitogenic activity had returned to control levels (data not shown). A response to UVB light was observed only in keratinocytes grown in MCDB-153 but not in DMEM.

IL-1, induced in keratinocytes in response to UVB irradiation,[7] did not stimulate the proliferation of melanocytes nor did it stimulate keratinocytes to produce higher levels of mitogenic activity toward melanocytes (data not shown).

The Melanocyte Mitogen in Keratinocytes Is Basic Fibroblast Growth Factor

Because we had shown previously that bFGF was a natural growth factor for melanocytes, able to substitute for TPA,[5] we used neutralizing anti-bFGF antibodies,[5] a synthetic peptide fragment of bFGF that inhibits bFGF activity,[13,14] and a bFGF cDNA to probe the nature of the mitogen in keratinocytes.

As demonstrated in TABLE 2, antibodies raised in rabbits against a synthetic peptide corresponding to a segment of the amino-terminal domain of bFGF (anti-bFGF-(1–24)) and known to neutralize bFGF activity toward melanocytes,[5] neutralized at least 90% of the mitogenic activity in extracts of stratifying keratinocytes and 70% in proliferating keratinocytes. That the mitogenic activity in proliferating keratinocytes was neutralized at a lower percentage is probably due to the higher specific mitogenic activity in these cultures. With less extract, over 90% of the mitogenic activity was inhibited.[10a] The mitogenic activity in dermal fibroblasts was completely abolished by the inhibiting anti-bFGF antibodies (data not shown).

A synthetic peptide that blocks the binding of ^{125}I-bFGF to PC12 pheochromocytoma cells,[13] baby hamster kidney (BHK) cells, 3T3 fibroblasts and vascular endothelial cells,[14] and normal human melanocytes (Baird and Halaban, unpublished) also blocked the mitogenic activity of bFGF and keratinocyte extract toward human melanocytes (TABLE 2).

Northern blot analysis using a bovine cDNA probe for bFGF, revealed that keratinocytes grown in MCDB-153 medium, but not in DMEM, produced bFGF gene transcripts (FIGURE 3). The amount of bFGF gene transcript varied from one culture to another (*e.g.*, lanes 4 and 5) in concordance with the observed variability in the levels of the melanocyte mitogen in keratinocytes (data not shown). Two keratinocyte mRNA species, of 7.0 and 3.7 kb hybridized to the bFGF-cDNA probe. The two cell types used as positive controls, human hepatoma (SK-HEP-1)[10] and normal dermal fibroblasts,[15] contained bFGF gene transcripts of sizes identical to those found in keratinocytes (lanes 6 and 7).

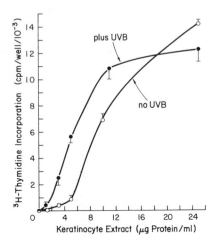

FIGURE 2. Increased melanocyte mitogen in UVB irradiated keratinocytes. Keratinocytes, derived from a newborn foreskin, were grown in MCDB-153 medium for a month and were passaged twice. Half of the keratinocyte cultures were exposed to 50 mJ/cm² UVB light and harvested 7 hr later (●). The control half was not exposed to UVB light (O) and was harvested at the same time as the experimental cultures. Extracts from these cultures at different concentrations were added to melanocyte cultures (24,000 cells/well) in PC-1 medium supplemented with 1 mM dbcAMP. ³H-thymidine incorporation into melanocytes was tested during the last 3 hr of 24-hr incubation with experimental or control media. Values are averages of cpm from 2 wells/3 hr. *Vertical bars* indicate standard errors. (From Halaban *et al.*[10a] Reprinted by permission from the *Journal of Cell Biology.*)

TABLE 2. The Melanocyte Mitogen in Keratinocytes is Related to bFGF[a]

Additions	^{3}H-Thymidine Incorporation cpm/Well/Hr
A. Inhibition of mitogenic activity by anti-bFGF antibodies	
Experiment 1: Keratinocytes grown in DMEM	
none	not detectable
KE (80 μg protein/ml)	2,100
KE (80 μg protein/ml) + anti-bFGF-(1–24) serum (10 μl/ml)	150
KE (80 μg protein/ml) + nonimmune serum (10 μl/ml)	2,200
Experiment 2: Keratinocytes grown in MCDB-153	
none	not detectable
KE (50 μg protein/ml) + anti-bFGF-(1–24) serum (20 μl/ml)	6,000
KE (50 μg protein/ml) + nonimmune serum (20 μl/ml)	19,400
B. Inhibition of mitogenic activity by a synthetic peptide fragment of bFGF	
none	not detectable
bFGF	5,000
bFGF + (1–10)OH	5,900
bFGF + (103–146)NH$_2$	400
KE (20 μg protein/ml) + (1–10)OH	3,700
KE (20 μg protein/ml) + (103–146)NH$_2$	190

[a]All additions were made to PC-1 defined medium supplemented with 1 mM dbcAMP. In experiment A1, extract was prepared from keratinocytes (KE) derived from adult skin grown in DMEM, and in experiments A2 and B, keratinocytes were derived from newborn foreskins grown in MCDB-153. Anti-bFGF-(1–24) serum was raised in rabbits as described before.[32] Basic fibroblast growth factor (bFGF, 95% pure, Collaborative Research, In., Bedford, MA) was added at 1 ng/ml. Synthetic peptides (1–10)OH and (103–146)NH$_2$ were prepared as described in REFERENCE 13 and were added at 300 μg/ml. ^{3}H-Thymidine incorporation was carried out during the last 2–3 hr of a 24-hr incubation with experimental media. Data are averages of cpm from 2 wells corrected for 1 hr of ^{3}H-thymidine incorporation. (From Halaban *et al.*[10a] Reprinted by permission from the *Journal of Cell Biology*.)

DISCUSSION

Human melanocytes differentiate not only in regard to pigment formation and cell shape but also in regard to a strict dependency on specific growth factors in order to be able to survive and proliferate in culture. The specific agents are bFGF (or TPA) plus substances that increase intracellular levels of cAMP. Unlike endothelial cells and fibroblasts, which are stimulated by bFGF as well as producing this polypeptide growth factor on their own,[15–18] bFGF was undetectable in melanocytes either as gene transcript or as immunoprecipitable protein.[18a] The absence of bFGF by these biochemical criteria was supported by biological assay. Melanocyte stimulating activity could not be detected in extracts of highly proliferative normal human melanocytes grown in TIP, indicating that bFGF was not induced in detectable amounts in response to mitogenic stimulation by TPA. We had shown before that human metastatic melanoma cells contained bFGF and depended on their intrinsic bFGF activity for continued proliferation, suggesting that bFGF acts as an oncogene for human melanomas.[18a] Our conclusion has recently been strengthened by experiments *in vitro,* demonstrating that the aberrant expression of bFGF by way of a transfected cDNA encoding for bFGF fused with sequences specifying a signal peptide, conferred the tumorigenic phenotype to NIH 3T3 cells.[19]

The melanocyte mitogen in keratinocytes is probably bFGF because, as in melanoma cells, it is inhibited by two agents that inhibit the activity of bFGF. Those are antibodies to a synthetic peptide of bFGF[5] and a synthetic fragment of bFGF that blocks the binding of bFGF.[13,14] These two agents also inhibit the mitogenic activity of purified bFGF toward melanocytes as demonstrated here and before.[5] The presence of mRNA species that hybridize with a bFGF-cDNA probe is direct evidence that keratinocytes produce bFGF. The levels of bFGF gene transcripts in keratinocytes, like the levels of the melanocyte mitogen, are not constant. Such fluctuations may explain the failure of other investigators to detect bFGF gene transcripts in keratinocytes.[15] The two bFGF-mRNA species of 7.0 and 3.7 kb, known to be present in other tissues and cells that produce bFGF, are detected easily in keratinocytes grown in MCDB-153, the medium that promotes keratinocyte proliferation and production of the melanocyte mitogen. The bFGF-mRNAs were not detected in keratinocytes grown in DMEM, a medium that suppresses the levels of melanocyte mitogen in keratinocytes. Dermal fibroblasts, rich in melanocyte mitogen, also express high levels of the two gene transcripts.

Melanocytes *in vivo,* at least in adults, can be triggered to divide in response to UVB.[11,12] The studies with human epidermal cells *in vitro,* described here, indicate that keratinocytes regulate the proliferation of melanocytes through bFGF, whose production may be increased directly in response to UVB light or, indirectly, as a consequence of the induction of proliferation by UVB. That UV irradiation induces DNA replication and can enhance the synthesis of selected proteins has been demonstrated for human fibroblasts.[20,21] These responses were suggested to have been generated through DNA damage because (a) other DNA damaging agents such as N-methyl-N-nitrosourea and N-acetoxy-2-acetylaminofluorene also induced DNA synthesis,[20] and (b) lower doses of UV light were sufficient to enhance the synthesis of protein in cells defective in DNA repair such as those from patients with Cockayne's syndrome or xeroderma pigmentosum.[21] In fibroblasts, UV light induces an increase in abundance of the same proteins that are induced by TPA.

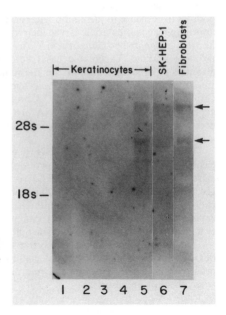

FIGURE 3. Expression of mRNA for bFGF in highly proliferative normal human kerotinocytes. RNA samples from keratinocytes (*lanes 1–5*), hepatoma SK-HEP-1 (*lane 6*) and dermal fibroblasts (*lane 7*) were subjected to Northern blot hybridization with a bovine cDNA probe for bFGF (a 1.4-kb EcoRI fragment of pJJ11-1). Lanes 1–2, neonatal keratinocytes and lane 3 adult keratinocytes, all grown in DMEM; lanes 4–5, neonatal keratinocytes grown in MCDB-153 medium; lanes 6–7, SK-HEP-1 and fibroblasts, respectively, harvested after 4 hr stimulation with serum. RNA quantities loaded onto the gel were as follows: lanes 1, 4–6: 20 μg total RNA; lanes 2–3: 1 μg of poly(A)$^+$ RNA; lane 7: 10 μg total RNA. *Arrows* indicate 7.0 and 3.7 kilobase gene transcripts. (From Halaban *et al.*[10a] Reprinted by permission from the *Journal of Cell Biology.*)

The studies described here demonstrate that, *in vitro,* rapidly proliferating keratinocytes produce higher levels of bFGF than do stratifying keratinocytes. This is a provocative finding that is in agreement with the preferred location and/or activity of melanocytes in intact skin. Studies of ^{3}H-thymidine uptake by normal palmar epidermis of humans and monkeys have shown that 80% of the labeled nuclei are in the tips of the deep rete ridges, indicating that the keratinocytes in these areas are highly proliferative.[22] These ridges are also more heavily pigmented than the shallow ridges or the interridge epidermis. It is thus possible that actively dividing keratinocytes stimulate neighboring melanocytes to divide and/or to produce more melanin, although an increase in the number of pigment cells in these areas could not be confirmed.[22] Another well-known site of growth-associated, cyclic melanocyte activity is in anagen hair follicles, where melanocytes come to lie in close proximity to the rapidly proliferating keratinocytes that constitute the cellular bulb matrix. Recently, it was demonstrated that anagen hair bulbs from rats, like bFGF,[23] are highly angiogenic, though the nature of the angiogenic substance was not yet determined.[24] In mice, UVB irradiation causes an increase in the mitotic frequency in melanocytes and basal keratinocytes, and the correlation between the number of mitotic figures in basal keratinocytes and melanocytes of irradiated versus nonirradiated mouse ear is positive.[12]

Clinically, some pigmented lesions, accentuated by sunlight, could be explained on the basis of melanocyte stimulation by proliferating keratinocytes. In solar lentigo, highly pigmented buds of proliferating kertinocytes project into an actinically damaged papillary dermis. Characteristically the bases of these projections are more heavily pigmented than are uninvolved rete ridges. Other candidate lesions include spreading pigmented actinic keratoses, pigmented seborrheic keratoses, pigmented squamous cell carcinoma, pigmented basal cell carcinomas, pigmented epidermal nevi, and melanoacanthoma. In each case, the proliferating epithelial component is associated with an increase in the number of melanocytes and/or level of pigmentation.

Conversely, there are clinical conditions in which epidermal melanocytes die, perhaps in response to ill-functioning keratinocytes. In vitiligo, a morphologic sign of impending depigmentation is vacuolar damage of basal keratinocytes in clinically healthy, pigmented portions of skin.[25] The depigmented epidermis in vitiligo appears to be more shallow. Vacuolated basal keratinocytes have also been observed in two disorders of hypopigmentation: the Hermansky-Pudlak syndrome[26] and piebaldism.[27] Interestingly, the piebald trait was suggested to be associated with deletion in chromosome band 4q12 to q21.1.[28] Chromosome 4 is also the site for the structural gene for bFGF.[29] Probing the DNA of piebald patients with cDNA for bFGF will determine whether the dominantly inherited piebald trait is associated with a deletion in the gene for bFGF.

The mechanism by which bFGF gets from keratinocytes to melanocytes is not clear. Basic FGF lacks a signal peptide[10,30] that would enable it to be secreted by classical exocytosis. However, bFGF is a heparin-binding protein[31] and binds to glycosaminoglycans of extracellular matrices such as are produced by vascular endothelial cells[17,32] and corneal epithelium and endothelium.[33] Melanocytes could be exposed to bFGF through direct contact with keratinocytes and by way of the extracellular matrix layed down by neighboring keratinocytes.

The finding that highly proliferative keratinocytes produce the angiogenic factor bFGF[23] may also explain clinical conditions that involve a combination of rapid proliferation of keratinocytes and microvascular endothelial cells, such as occur in wound healing and psoriasis.

SUMMARY

Melanocytes cultured in the presence of keratinocytes survive for weeks without added basic fibroblast growth factor (bFGF) and cyclic-adenosine-monophosphate (cAMP), the two factors needed for their proliferation *in vitro*. We show here that the growth factor for melanocytes produced by human keratinocytes is bFGF because its activity can be abolished by neutralizing antibodies to bFGF and by a bFGF synthetic peptide that inhibits the binding of the growth factor to its receptor. The melanocyte mitogen in keratinocytes is cell-associated and increases after irradiation with ultraviolet B (UVB). Northern blots reveal bFGF gene transcripts in keratinocytes but not melanocytes. These studies demonstrate that bFGF elaborated by keratinocytes *in vitro* sustains melanocyte growth and survival, and they suggest that keratinocyte-derived bFGF is the natural growth factor for normal human melanocytes *in vivo*.

ACKNOWLEDGMENTS

We thank Dr. A. B. Lerner for support, Drs. J. A. Abraham and J. C. Fiddes for the bovine bFGF cDNA, S. Straight and L. Kim for growing the cells, and J. Schreiber for secretarial assistance.

REFERENCES

1. PRUNIERAS, M., T. K. LEUNG & P. COLSON. 1964. Dissociation et recombinaison *in vitro* de l'epiderme de cobaye adulte. Ann. Dermatol. Syphiligr. **91:** 23–27.
2. HALABAN, R. 1988. Responses of cultured melanocytes to defined growth factors. Pigm. Cell Res. 1(Suppl. 1):18–26.
3. EISINGER, M. & O. MARKO. 1982. Selective proliferation of normal human melanocytes *in vitro* in the presence of phorbol ester and cholera toxin. Proc. Natl. Acad. Sci. USA **79:** 2018–2022.
4. HALABAN, R., S. GHOSH, P. DURAY, J. M. KIRKWOOD & A. B. LERNER. 1986. Human melanocytes cultured from nevi and melanomas. J. Invest. Dermatol. **87:** 95–101.
5. HALABAN, R., S. GHOSH & A. BAIRD. 1987. bFGF is the putative natural growth factor for human melanocytes. In Vitro Cell Dev. Biol. **23:** 47–52.
6. HALABAN, R. & F. D. ALFANO. 1984. Selective elimination of fibroblasts from cultures of normal human melanocytes. In Vitro **20:** 447–450.
7. KUPPER, T. S., A. O. CHUA, P. FLOOD, J. McGUIRE & U. GUBLER. 1987. Interleukin 1 gene expression in cultured human keratinocytes is augmented by ultraviolet irradiation. J. Clin. Invest. **80:** 430–436.
8. CHOMCZYNSKI, P. & N. SACCHI. 1987. Single-step method of RNA isolation by acid guanidinium thiocyanate-phenol-chloroform extraction. Anal. Biochem. **162:** 156–159.
9. SIEGEL, L. I. & E. BRESNICK. 1986. Northern hybridization analysis of RNA using diethylpyrocarbonate-treated nonfat milk. Anal. Biochem. **159:** 82–87.
10. ABRAHAM, J. A., A. MERGIA, J. L. WHANG, A. TUMOLO, J. FRIEDMAN, K. A. HJERRILD, D. GOSPODAROWICZ & J. C. FIDDES. 1986. Nucleotide sequence of a bovine clone encoding the angiogenic protein, basic fibroblast growth factor. Science **233:** 545–548.
10a. HALABAN, R., R. LANGDON, N. BIRCHALL, C. CUONO, A. BAIRD, G. SCOTT, G. MOELLMANN & J. McGUIRE. 1988. Basic fibroblast growth factor from human keratinocytes is a natural mitogen for melanocytes. J. Cell Biol. **107:** 1611–1619.
11. ROSDAHL, I. & G. SZABO. 1978. Mitotic activity of epidermal melanocytes in UV-irradiated mouse skin. J. Invest. Dermatol. **70:** 143–148.
12. ROSDAHL, I. K. 1978. Melanocyte mitosis in UVB-irradiated mouse skin. Acta Dermatovener. (Stockholm) **58:** 217–221.

13. SCHUBERT, D., N. LING & A. BAIRD. 1987. Multiple influences of a heparin-binding growth factor on neuronal development. J. Cell Biol. **104:** 635–643.
14. BAIRD, A., D. SCHUBERT, N. LING & R. GUILLEMIN. 1988. Receptor and heparin binding domains of basic fibroblast growth factor. Proc. Natl. Acad. Sci. USA. **85:** 2324–2328.
15. SHIPLEY, G. D., M. D. STERNFELD, R. J. COFFEY & M. R. PITTELKOW. 1988. Differentiation expression of type-2 heparin-binding growth factor mRNA in normal and transformed human cells. J Cell. Biochem., 12A(Suppl.): 125.
16. GOSPODAROWICZ, D., S. MASSOGLIA, J. CHENG & J. FUJI. 1986. Effect of fibroblast growth factor and lipoproteins on the proliferation of endothelial cells derived from bovine adrenal cortex, brain cortex, and corpus luteum capillaries. J. Cell. Physiol. **127:** 121–136.
17. VLODAVSKY, I., J. FOLKMAN, R. SULLIVAN, R. FRIDMAN, R. ISHAI-MICHAELI, J. SASSE & M. KLAGSBRUN. 1987. Endothelial cell-derived basic fibroblast growth factor: Synthesis and deposition into subendothelial extracellular matrix. Proc. Natl. Acad. Sci. USA **84:** 2292–2296.
18. SCHWEIGERER, L., G. NEUFELD, J. FRIEDMAN, J. A. ABRAHAM, J. C. FIDDES & D. GOSPODAROWICZ. 1987. Capillary endothelial cells express fibroblast growth factor, a mitogen that promotes their own growth. Nature **325:** 257–259.
18a. HALABAN, R., B. S. KWON, S. GHOSH, P. DELLI BOVI & A. BAIRD. 1988. bFGF as an autocrine growth factor for human melanomas. Oncogene Res. **3:** 177–186.
19. ROGELJ, S., R. A. WEINBERG, P. FANNING & M. KLAGSBRUN. 1988. Basic fibroblast growth factor fused to a signal peptide transforms cells. Nature **331:** 173–175.
20. COHEN, S. M., B. R. KRAWICZ, S. L. DRESLER & M. W. LIBERMAN. 1984. Induction of replicative DNA synthesis in quiescent human fibroblasts by DNA damaging agents. Proc. Natl. Acad. Sci. USA **81:** 4828–4832.
21. SCHORPP, M., U. MAILIC, H. J. RAHMSDORF & P. HERRLICH. 1984. UV-induced extracellular factor from human fibroblasts communicates the UV response to nonirradiated cells. Cell **31:** 861–868.
22. LAVKER, R. M. & T.-T. SUN. 1983. Epidermal stem cells. J. Invest. Dermatol. **81**(Suppl.): 121s–127s.
23. FOLKMAN, J. & M. KLAGSBRUN. 1987. Angiogenic factors. Science **235:** 442–447.
24. STENN, K. S., L. A. FERNANDEZ & S. J. TIRRELL. 1988. The angiogenic properties of the rat vibrissa hair follicle associated with the bulb. J. Invest. Dermatol. **90:** 409–411.
25. MOELLMANN, G., S. KLEIN-ANGERER, D. A. SCOLLAY, J. J. NORDLUND & A. B. LERNER. 1982. Extracellular granular material and degeneration of keratinocytes in the normally pigmented epidermis of patients with vitiligo. J. Invest. Dermatol. **79:** 321–330.
26. MOELLMANN, G. E., R. C. LANGDON, E. KUKLINSKA & J. J. NORDLUND. 1986. Langerhans-cell associated damage of keratinocytes in the Hermansky-Pudlak Syndrome (abstract) J. Invest. Dermatol. **87:** 156.
27. LERNER, A. B., R. HALABAN, S. N. KLAUS & G. E. MOELLMANN. 1987. Transplantation of human melanocytes. J. Invest. Dermatol. **89:** 219–224.
28. HOO, J. J., R. H. A. HASLAM & C. VAN ORMAN. 1986. Tentative assignment of piebald trait gene to chromosome band 4q12. Hum. Genet. **73:** 230–231.
29. MERGIA, A., R. EDDY, J. A. ABRAHAM, J. C. FIDDES & T. B. SHOWS. 1986. The genes for basic fibroblast growth factors are on different human chromosomes. Biochem. Biophys. Res. Commun. **138:** 644–651.
30. ABRAHAM, J. A., J. L. WHANG, A. TUMOLO, A. MERGIA, J. FRIEDMAN, D. GOSPODAROWICZ & J. C. FIDDES. 1986. Human basic fibroblast growth factor: Nucleotide sequence and genomic organization. EMBO J. **5:** 2523–2529.
31. GOSPODAROWICZ, D., J. CHENG, G.-M. LUI, A. BAIRD & P. BOHLEN. 1984. Isolation of brain fibroblast growth factor by heparin-Sepharose affinity chromatography: Identity with pituitary fibroblast growth factor. Proc. Natl. Sci. USA **81:** 6963–6967.
32. BAIRD, A. & N. LING. 1987. Fibroblast growth factors are present in the extracellular matrix produced by endothelial cells *in vitro:* Implications for a role of heparinase-like enzymes in the neovascular response. Biochem. Biophys. Res. Commun. **142:** 428–435.
33. JEANNY, J.-C., N. FAYEIN, M. MOENNER, B. CHAVALIER, D. BARRITAULT & Y. COURTOIS. 1987. Specific fixation of bovine brain and retinal acidic and basic fibroblast growth factors to mouse embryonic eye basement membranes. Exp. Cell Res. **171:** 63–75.

Signal Transduction for Proliferation and Differentiation in Keratinocytes

STUART H. YUSPA,[a] HENRY HENNINGS,[a]
ROBERT W. TUCKER,[b] SUSAN JAKEN,[c]
ANNE E. KILKENNY,[a] AND DENNIS R. ROOP[a]

[a]*Laboratory of Cellular Carcinogenesis and Tumor Promotion*
Building 37, Room 3B25
National Cancer Institute
Bethesda, Maryland 20892

[b]*Johns Hopkins Oncology Center*
Baltimore, Maryland 21205

[c]*W. Alton Jones Cell Science Center*
Lake Placid, New York 12946

INTRODUCTION

The analysis of epidermal growth and differentiation *in situ* and by the use of *in vitro* model analogues has been extraordinarily informative to cell and molecular biologists. These studies have revealed the relationship of growth factors to growth control,[1] the regulation of expression of keratins and other differentiation-specific genes[2-4] and aberrations in function associated with pathological states, especially neoplasia,[5] among other findings. These studies have also suggested a previously unappreciated function of epidermis as a secretory tissue. Secretory activity of keratinocytes has the potential to influence the behavior of the epidermis itself and to function in the maintenance of physiological homeostasis of the host or to modify the host response to pathological conditions.

Sufficient experience has accumulated to indicate that epidermal secretory activity for a particular effector is regulated by the differentiation state of the keratinocyte. Furthermore, the response of keratinocytes to exogenous effectors is often determined by the state of maturation. Grafted epidermis is being considered as a source for the expression of products for gene therapy, and expression of such products is likely to be modified by the keratinocyte differentiation state. For these reasons, an understanding of the signals which regulate epidermal differentiation is central to a full understanding of epidermal secretory activity.

The discovery that extracellular Ca^{2+} (Ca_o) regulates many aspects of epidermal differentiation *in vitro*[6] provided a model in which the major differentiation specific functions of epidermis could be modified. In this model, cells cultured in medium with Ca_o between 0.02 and 0.1 mM are phenotypically similar to basal epidermal cells[7] (FIG. 1). Elevation of Ca_o to >0.1 mM induced a rapid change in biochemistry and morphology in which the cells have many properties of the suprabasal phenotype *in vivo* (FIG. 1). Such cells become irreversibly committed to terminal differentiation after 48–72 hr of increased Ca_o and slough from the culture plate as mature squames.

Many aspects of this Ca_o-activated process of epidermal differentiation appear to be physiological. For example, at 1 mM Ca_o most maturing keratinocytes *in vitro* express pemphigus antigen,[8] form an extensive desmosomal network,[9] and produce

191

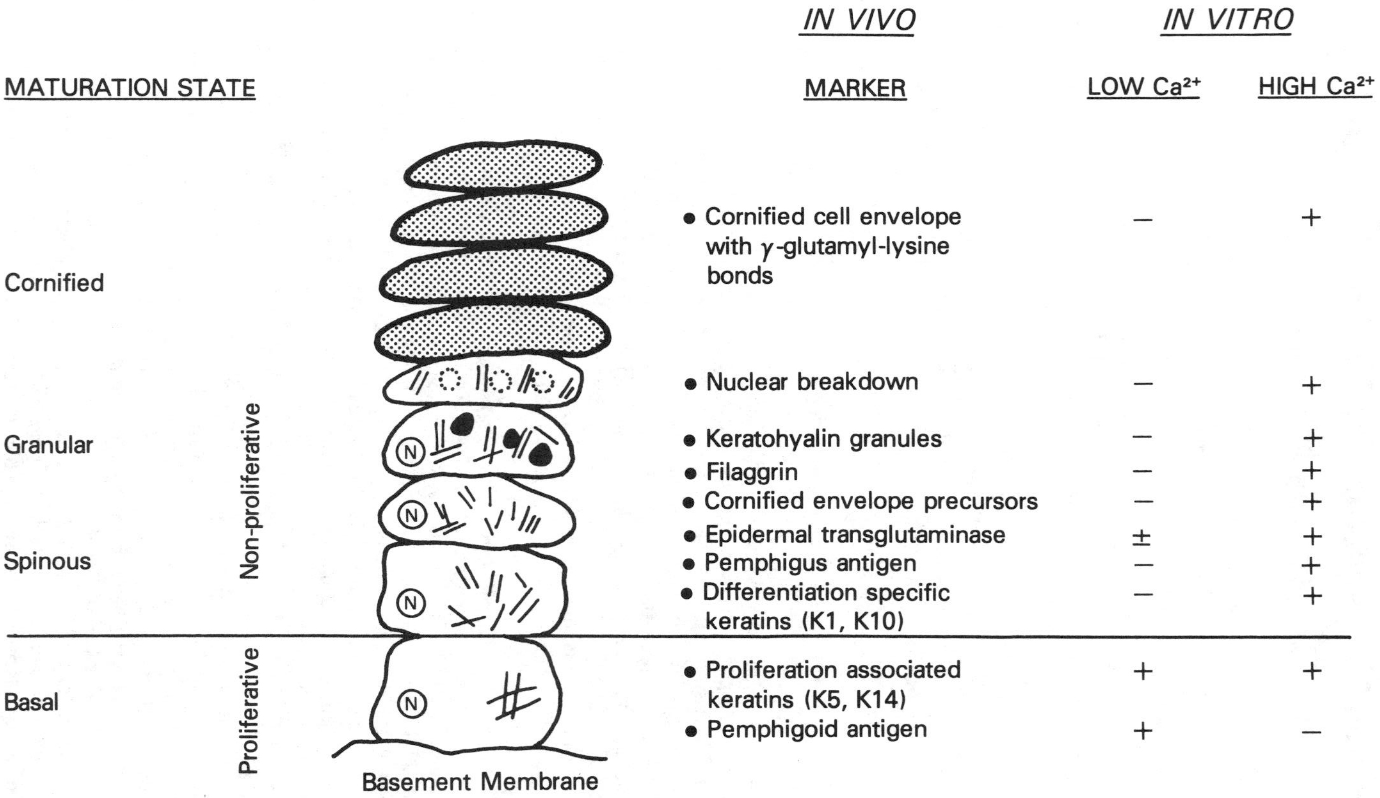

FIGURE 1. Schematic depiction of the terminal differentiation of mammalian epidermis *in vivo* and *in vitro*. Markers expressed in specific Ca_o of the culture medium are shown as + or −, but + does not necessarily indicate expression by all cells (see text). Low Ca^{2+} is 0.02–0.09 mM and high Ca^{2+} is >0.1 mM.

corneocytes analogous to those of the stratum corneum *in vivo*.[10] However, only 10–20% of keratinocytes in 1.0 mM Ca^{2+} express K1 or K10[11] and filaggrin,[12] markers expressed by most or all suprabasal cells *in vivo*. Thus Ca_o may not regulate differentiation *in vivo* or at least may not be the only signal to regulate the process.

Analysis of Ca^{2+} content in epidermis *in vivo* indicates that a Ca^{2+} gradient exists.[13,14] Total and free Ca^{2+} are low in the basal cell and first suprabasal cell layers relative to serum and dermal content. Ca^{2+} in the granular cell layer is extraordinarily high. The processes by which a tissue can regulate and maintain a Ca^{2+} gradient are not clear. Nevertheless, the existence of this gradient in the epidermis supports the potential importance of Ca_o as a physiological regulator of epidermal differentiation *in vivo*.

Pathways Involved in Ca^{2+}-induced Epidermal Differentiation

Evidence supporting the validity of the cell culture model to study epidermal differentiation was provided by reports that the tumor promoter 12-0-tetradecanoyl-phorbol-13-acetate (TPA) could induce epidermal differentiation *in vivo* and in keratinocytes cultured in 0.05 mM Ca^{2+} medium.[15–17] Since the biological activities associated with TPA responses are mediated through the activation of protein kinase C, a phospholipid and Ca^{2+} dependent enzyme present in virtually all cells of all multicellular organisms,[18,19] this enzyme would appear to play a role in the differentiation response. A link to the Ca^{2+} pathway was implicated since the pattern of changes in epidermal protein synthesis and protein phosphorylation which occur within one or two hr after exposure to either Ca^{2+} or TPA were found to be similar.[20] Additional studies showed that exogenous bacterial phospholipase C and exogenous diacylglycerol could mimic the action of TPA on epidermal differentiation.[21] Together, these results suggested a pathway regulating epidermal differentiation which involves exposure to increasing concentrations of extracellular Ca^{2+} activating an endogenous phospholipase C, a Ca^{2+} requiring enzyme.

Phospholipase C activation is associated with the rapid metabolism of cellular phosphatidylinositol (PI) yielding two intracellular second messengers, diacylglycerol and inositol trisphosphate (IP3).[22] Diacylglycerol is the endogenous activator of protein kinase C while IP3 mobilizes intracellular Ca^{2+} (Ca_i) from bound stores. To determine if a change in Ca_o could stimulate PI metabolism by activation of cellular phospholipase C, we prelabeled cultures with [^{3}H]inositol in 0.05 mM Ca^{2+} medium and added Ca^{2+} to 1 mM.[23] Inositol phosphates (IPs), the water soluble metabolites of phosphatidylinositol, increased within 2 min. There was a corresponding decrease in radiolabeled phosphoinositides suggesting that these were the source of the increased IPs. After 3 hr in 1 mM Ca^{2+} medium, each of the IPs remained increased to 130–140% of control levels. Two Ca^{2+} ionophores, A23187 and ionomycin, also increased IP levels in cells maintained in 0.05 mM Ca^{2+}, suggesting that a rise in Ca_i is important in the increased turnover of phosphatidylinositol. Phorbol esters stimulated phosphatidylcholine metabolism but not phosphatidylinositol turnover.[23] In concert with ionomycin, phorbol esters became more potent inducers of differentiation suggesting that both protein kinase C activation, elevation of Ca_i and PI turnover were important components of the signal for epidermal differentiation. These results demonstrate that the second messenger system for Ca^{2+}-mediated keratinocyte differentiation may be through a direct effect on phospholipase C activity.

To assess the Ca_i response to a change in Ca_o, digital image technology was employed by loading cells with a calcium sensitive probe, Fura 2, and measuring a change in intracellular fluorescence.[24] When Ca_o is increased from 0.05 mM to 1.2

mM, there is a 10–20-fold increase in Ca_i which occurs within a few minutes and is sustained for at least 30 min. The magnitude of this response is greatest in the presence of serum, but individual cells show substantial heterogeneity in both time course and magnitude of response. In contrast, serum-free conditions reduce the Ca_i increase to <5-fold, but the cells respond homogeneously. Serum-free conditions also produce a sustained rise in Ca_i in response to an increase in Ca_o. Such results suggest that exogenous factors found in serum could influence the differentiation response in individual cells. Since both Ca_i and PI metabolism are increased simultaneously by a change in Ca_o and both are sustained and tightly linked to the differentiation response, a rise in Ca_i is likely to be an intracellular signal responsible for the induction of terminal differentiation in keratinocytes.

A Specific Ca_o Regulates the Expression of Certain Differentiation Markers

The sequential expression of differentiation-specific markers is characteristic of the terminal differentiation of the epidermis.[25] Keratin markers such as K1 (67 kd) and K10 (59 kd) are expressed early in the differentiation program as cells begin their

TABLE 1. Influence of Ca_o on the Expression of Specific Differentiation Markers in Cultured Mouse Keratinocytes

	Concentrations of Ca^{2+} in Culture Medium (mM)							
	Expression at 24 Hr				Expression at 48 Hr			
Marker	0.05	0.1	0.5	1.0	0.05	0.1	0.5	1.0
K1	−	+ + +	+	+	−	+ + +	+ +	+
K10	−	+ +	+	+	−	+ + +	+	+
Filaggrin	−	+	−	ND[a]	−	+ + +	+	ND
CE	−	+	−	ND	−	+ + +	+	ND

[a]Not determined.

maturation in the basal or first suprabasal layer.[25,26] Cornified envelope precursors and filaggrin are expressed later as cells enter the granular cell layer[27,28] (see Roop et al., in preparation). Monospecific antibodies have been made to unique sequences of K1, K10, filaggrin and the major protein component of the cornified envelope (here called anti-CE). These antibodies were used as probes to study the Ca^{2+} requirements for differentiation since they recognize maturation stage dependent markers. Both immunofluorescence analysis of cultured cells and Western blotting of cell extracts were used to evaluate the expression of these markers in cultured cells.

When basal cells grown in 0.05 mM Ca^{2+} were switched to 1.0 mM Ca^{2+}, only 10–15% of cells expressed keratins K1 and K10[11] and fewer expressed CE and filaggrin by immunofluorescence analysis. Western blots of such cell extracts do not reproducibly detect K1 and K10 expression. When basal keratinocytes (0.05 mM Ca^{2+}) were switched to 0.10 mM Ca^{2+}, K1 was readily detected on Western blots by 24 hr and K10 staining was intense by 48 hr (TABLE 1). These marker proteins were reduced or not detectable when cells were switched instead to 0.5 mM or 1 mM Ca^{2+}. Western blots using antibodies to detect expression of filaggrin and CE showed a similar requirement for Ca_o of approximately 0.1–0.12 mM, although these bands were not seen prior to 48 hr. The expression of these differentiation specific genes diminished at approximately

0.2–0.3 mM Ca^{2+}. The number of positive cells was greatly increased by immunofluorescence analysis. When transcripts for K1 were analyzed by RNA slot blots, K1 message was abundant at 18 hr in 0.1 mM Ca_o but not detectable in cells maintained in 1.0 mM for that time. Assuming these specific changes in Ca_o reflect analogous changes in Ca_i, the intracellular Ca^{2+} milieu may dictate the expression of the differentiation phenotype of normal keratinocytes and this may be under transcriptional control. These studies also demonstrate the importance of maintaining a Ca^{2+} gradient *in vivo* for the proper regulation of epidermal homeostasis.

From our studies, we can conclude that one function for Ca_o is to control epidermal inositol lipid metabolism. A primary role of inositol lipid metabolism for keratinocyte differentiation could have direct clinical significance for some skin diseases. For example, patients chronically treated with Li^+ for manic-depressive illness suffer from a psoriatic-like skin disease. Chronic Li^+ treatment causes a decrease in brain inositol levels, but skin levels of inositol have not been examined.[22,29] Psoriasis is a hyperproliferative disease in which keratinocyte differentiation is impaired. Further studies seem warranted to assess the possible relationship between agents which are known to influence cellular inositol metabolism and impaired responsiveness to endogenous regulators of keratinocyte differentiation.

SUMMARY

In mouse and human epidermis, the Ca^{2+} environment of the basal cell layer is substantially below serum Ca^{2+}, while that of the granular cell layer is unusually high. Reduction of extracellular Ca^{2+} concentration (Ca_o) in the medium of keratinocyte cultures maintains a basal cell phenotype while serum Ca^{2+} concentrations induce terminal differentiation. Measurements of intracellular Ca^{2+} (Ca_i) by the use of Fura 2 and digital imaging technology reveal that Ca_i increases 10–20-fold in response to an increase in Ca_o and remains elevated. Concomitant with the rise in Ca_i is an increase in the metabolism of phosphatidylinositol (PI) to yield inositol phosphates and diacylglycerol. PI metabolism is also stimulated by calcium ionophores suggesting that a rise in Ca_i is directly responsible. The consequent increase in diacylglycerol and Ca_i would activate protein kinase C, an event known to trigger epidermal differentiation. Specific Ca_o and Ca_i determine the expression of individual markers of keratinocyte differentiation *in vitro*. These findings may account for the importance of the Ca^{2+} gradient for maintaining regulated growth and differentiation of the epidermis *in vivo*.

REFERENCES

1. RHEINWALD, J. G. & H. GREEN. 1977. Nature **265:** 421–424.
2. FUCHS, E. & H. GREEN. 1983. Cell **19:** 1033–1042.
3. ROOP, D. R., P. HAWLEY-NELSON, C. K. CHENG & S. H. YUSPA. 1983. Proc. Natl. Acad. Sci. USA **80:** 716–720.
4. RICE, R. H. & H. GREEN. 1979. Cell **18:** 681–694.
5. YUSPA, S. H. 1985. *In* Interrelationship Among Aging, Cancer and Differentiation. B. Pullman, P. O. P. T'so & E. L. Schneider, Eds.: 67–81. D. Reidel. Dordrecht.
6. HENNINGS, H., D. MICHAEL, C. CHENG, P. STEINERT, K. HOLBROOK & S. H. YUSPA. 1980. Cell **19:** 245–254.
7. YUSPA, S. H. 1985. *In* Methods in Skin Research. D. Skerrow & C. Skerrow, Eds.: 213–249. John Wiley. Sussex.
8. STANLEY, J. R. & S. H. YUSPA. 1983. J. Cell Biol. **96:** 1809–1814.
9. HENNINGS, H. & K. A. HOLBROOK. 1983. Exp. Cell Res. **143:** 127–142.

10. NAGAE, S., U. LICHTI, L. M. DE LUCA & S. H. YUSPA. 1987. J. Invest. Dermatol. 89: 51–58.
11. ROOP, D. R., H. HUITFELDT, A. KILKENNY & S. H. YUSPA. 1987. Differentiation 35: 143–150.
12. DALE, B. A., J. A. H. SCOFIELD, H. HENNINGS, J. R. STANLEY & S. H. YUSPA. 1983. J. Invest. Dermatol. 81: 90s–95s.
13. MALMQUIST, K. G., L. E. CARLSON, B. FORSLIND, G. M. ROOMANS & K. R. AKSELSSON. 1984. Nucl. Instrum. Methods Phys. Res. Sec. B 3: 611–617.
14. MENON, G. K., S. GRAYSON & P. M. ELIAS. 1985. J. Invest. Dermatol. 74: 508–512.
15. REINERS, J. J., JR. & THOMAS J. SLAGA. 1983. Cell 32: 247–255.
16. YUSPA, S. H., T. B. BEN, H. HENNINGS & U. LICHTI. 1980. Biochem. Biophys. Res. Commun. 97: 700–708.
17. YUSPA, S. H., T. BEN & H. HENNINGS. 1983. Carcinogenesis 4: 1413–1418.
18. BLUMBERG, P. M. 1988. Cancer Res. 48: 1–8.
19. NISHIZUKA, Y. 1986. Science 233: 305–312.
20. WIRTH, P. J., S. H. YUSPA, S. S. THORGEIRSSON & H. HENNINGS. 1987. Cancer Res. 47: 2831–2838.
21. JENG, A. Y., U. LICHTI, J. E. STRICKLAND & P. M. BLUMBERG. 1985. Cancer Res. 45: 5714–5721.
22. BERRIDGE, M. J. 1984. Biochem. J. 220: 345–360.
23. JAKEN, S. & S. H. YUSPA. 1988. Early signals for keratinocyte differentiation: Role of Ca^{2+}-mediated inositol lipid metabolism in normal and neoplastic epidermal cells. Carcinogenesis 9: 1033–1038.
24. WILLIAMS, D. A., K. E. FOGARTY, R. Y. TSEIN & F. S. FAY. 1985. Nature 318: 558.
25. ROOP, D. R., C. K. CHENG, R. TOFTGARD, J. R. STANLEY, P. M. STEINERT & S. H. YUSPA. 1985. Ann. N.Y. Acad. Sci. 455: 426–435.
26. SUN, T., R. EICHNER, W. G. NELSON, S. C. G. TSENG, R. A. WEISS, M. JARVINEN & J. WOODCOCK-MITCHELL. 1983. J. Invest. Dermatol. 81: 109s–115s.
27. HAYDOCK, P. V. & B. M. DALE. 1986. J. Biol. Chem. 261: 12520–12525.
28. ROTHNAGEL, J. A., T. MEHREL, W. W. IDLER, D. R. ROOP & P. M. STEINERT. 1987. J. Biol. Chem. 262: 15643–15648.
29. HOKIN-NEAVERSON, M. & K. SADEGHIAN. 1984. J. Biol. Chem. 259: 4346–4352.

Structure and Function of Growth Inhibitory Epidermal Pentapeptide

KJELL ELGJO[a] AND KARL L. REICHELT[b]

[a]Institute of Pathology
and
[b]Pediatric Research Institute
The National Hospital (Rikshospitalet)
0027 Oslo 1, Norway

The epidermal thickness is different in the various regions of the body, and from one species to another. But for a given region in a given species the normal thickness is remarkably constant throughout adult life. The corollary of this commonplace observation is that the cells that are lost from the surface are continually replaced by an equal number of new cells. Thus, the local control of cell renewal must be able to register quantitatively the need for new cells at any given time. A growth regulation of this type is best explained in terms of a negative feedback control. A stimulatory signal from already desquamated cells is physically impractical.

Most available *in vivo* data support the concept that regulation of normal cell renewal basically works according to a negative feedback principle. Although such a regulation could involve a large number of interacting factors it presupposes the production of one or more mitosis inhibitors by cells that belong to the same tissues in which they act. In such a system, loss of inhibitor-producing cells is followed by a corresponding fall in inhibitor concentration and hence, a compensatory increase in cell production.

Earlier experiments have shown that crude or partially purified skin extracts reversibly inhibit epidermal cell proliferation *in vivo* and *in vitro*.[1] Experiments with such extracts indicated that proliferating epidermal cells were inhibited at several phases in the cell cycle,[2] and that the effect was confined to keratinizing epithelia.[3] Such experiments cannot, however, prove the existence of specific growth inhibitors, and we therefore started a series of experiments to purify and characterize the putative inhibitor(s).

Purification Procedures

Water extracts of mouse skin prepared as previously described[4] represented the starting material. After trying different fractionation techniques we found that the following procedures were best suited:[5] Sephadex G25; C-18 reversed phase (Bondapak C-18/Porasil B 37–75 microns); cation exchange on Dowex 50; anion exchange on Dowex 1; Fractogel MG 2000, HPLC separation on Partisil M-9, and HPLC rechromatography of the active fraction. At each step of purification the fractions were tested for possible effects on mouse epidermal mitoses (G_2-M cell flux) by means of Colcemid.[6] These purification procedures yielded two inhibitory peptides: pGlu-Glu-Asp-Ser-GlyOH, and pGlu-GlyOH.[7] Most of the following experiments were made with the pentapeptide.

Biological Effects

In vitro, the isolated pentapeptide inhibited cell proliferation in primary mouse epidermal cells at a low and restricted dose range (10^{-8}–10^{-10} M).[8] After prolonged exposure the number of cornified envelopes was increased, indicating that the pentapeptide could modify terminal differentiation. Higher concentrations of pentapeptide were needed to obtain similar effects in a transformed cell line. *In vivo* too, the isolated pentapeptide was effective only when given at a low dose (10^{-11}–10^{-13} mol/mouse).

A synthetic pentapeptide with identical structure was made in 1984 (Peninsula Laboratories, Belmont, CA). The synthetic pentapeptide was tested *in vivo* together with a number of analogs.[7] Only the pentapeptide and a synthetic dipeptide identical to the one that had been found in the extracts (pGlu-GlyOH) were inhibitory over a limited, low-dose range, like the biological inhibitor. In later experiments, synthetic peptides have been used.

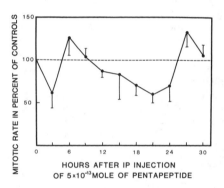

FIGURE 1. Hairless mice were injected ip at 0900 with 10^{-13} mol of the epidermal pentapeptide. In each cage of 8 mice, 4 were given the peptide; the remaining 4 animals served as controls and received only the solvent. The mitotic rate was estimated at 3-hr intervals by means of Colcemid (0.15 mg). Colcemid-arrested mitoses were counted in 25 successive high power vision fields of interfollicular epidermis. Each time point represents groups of 4 treated animals and 4 controls. The results are given as percentages of the control values with 1 SD, as estimated from the SD of both the treated groups and the controls.

Epidermal Cell Proliferation after a Single Treatment

Synthetic pentapeptide (10^{-13} mol) was given intraperitoneally (ip) to groups of hairless female mice, 7–8 weeks old, and the mitotic rate (MR) and the labeling index (LI) were estimated at 3-hr intervals during the following 30 hr. FIGURE 1 shows that the treatment was followed by an initial reduction of the MR in accordance with previous experiments. The inhibition was followed by a short overshoot, and subsequently by a long-lasting but moderate reduction of the MR. At 24 hr a second overshoot was observed, but at 30 hr after treatment the MR had returned to the normal range.

FIGURE 2 shows that the LI was initially reduced at 6 hr after treatment, but no overshoot was seen after this first, short period of reduced LI. After returning to the normal range at 9 hr a second and long-lasting inhibition of epidermal DNA synthesis was observed, followed by a relatively short overshoot at 27 hr. At 30 hr the LI was back to the normal range again.

A single ip treatment with 10^{-13} mol of pentapeptide thus resulted in one short and one longer-lasting period of reduced epidermal cell proliferation. The overshoot seen after termination of the second period of inhibition was not of such a magnitude and duration that it was likely to compensate fully for the reduced rate of cell proliferation.

FIGURE 2. Groups of hairless mice were treated with 10^{-13} mol of epidermal pentapeptide as described in the legend to FIGURE 1. Thirty min before sacrifice all mice received 30 μCi ^{3}H-thymidine. Labeled cells were counted in 25 successive high power vision fields of interfollicular epidermis. Each time point represents groups of 4 treated animals and 4 controls. The results are given as percentages of the control values with 1 SD, as estimated from the SD of both the treated groups and the controls.

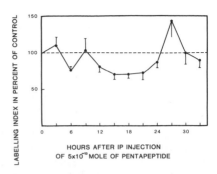

HOURS AFTER IP INJECTION
OF 5x10^{-10} MOLE OF PENTAPEPTIDE

At the present time we have no means to determine how long the synthetic pentapeptide is present at an effective concentration in epidermis. Therefore, we cannot tell whether the long-lasting effect *in vivo* is due to a slow degradation of synthetic pentapeptide, or to a short inhibitory signal that simultaneously affects susceptible cells in different phases of the cell cycle. On the basis of earlier *in vitro* data[8] we regard the latter alternative as the more likely one. If correct, this interpretation would suggest that the initial effect was due to a transient inhibition of cells in the G_2, M, and S phases, whereas the second and long-lasting period of decreased cell proliferation was related to an effect on cells in late G_1 phase.

The regulation of epidermal cell proliferation is obviously a local phenomenon. To create a more physiological situation we therefore applied the pentapeptide topically as a 0.02% cream, using a water-miscible cream base. Groups of female hairless mice, 7–8 weeks old, were treated once with the cream. In each cage, four animals were treated with the pentapeptide cream, and the remaining four mice used as controls and treated with the cream base. The pentapeptide cream-treated mice were kept separated from the controls for about 30 min after the application. The MR was followed during the following 5 days. FIGURE 3 shows that a single application of the pentapeptide cream was followed by oscillations in the MR with troughs at 3, 12, 36,

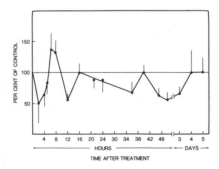

TIME AFTER TREATMENT

FIGURE 3. Epidermal mitotic rate after a single application of a 0.02% cream. 0.02% pentapeptide cream was applied once to the back skin of the mice, and the mitotic rate estimated at intervals by means of Colcemid during the following 5 days. Four mice in a cage of 8 were treated with the pentapeptide cream, and the remaining 4 used as controls and treated with the cream base. Each time point represents a group of 4 treated and 4 control animals. The results are expressed as percentages of the control values with 1 SD, as estimated from the SD of the treated group and of the controls.

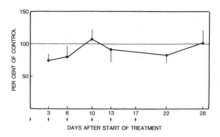

FIGURE 4. Epidermal mitotic rate after repeated applications of a 0.02% cream. 0.02% pentapeptide cream was applied twice weekly to the back skin of mice, and the mitotic rate estimated at intervals by means of Colcemid. Four mice in cages of 8 were treated with the pentapeptide cream, and the reamining 4 used as controls and treated with the cream base. Days of treatment and of Colcemid experiments are indicated by *arrows*. Each time point represents groups of 4 treated and 4 control animals. The results are expressed as percentages of the control values with 1 SD, as estimated from the SD of the treated group and of the controls.

and 45–48 hr after treatment. The MR was still reduced on day 3 but normal on day 4. Excpet for a short overshoot at 6–8 hr, the mitotic rate never exceeded the control between the trough values.

In another experiment, 0.02% pentapeptide cream was applied twice weekly. Here too, the controls received only cream base. One group of mice, 4 treated animals and 4 controls, were injected with Colcemid (0.15 mg) and killed on the days when the rest of the mice were treated with the cream. FIGURE 4 shows the reaction to the repeated applications. As in the separate single-application experiment described above, the mitotic rate was reduced on day 3 following the first treatment. Later, no statistically significant ($p > 0.05$ with the Wilcoxon rank sum test) alteration in the MR could be observed.

We also examined whether repeated applications of the pentapeptide cream was followed by a decreased epidermal thickness. TABLE 1 shows that repeated applications of the cream resulted in a transient hyperplasia that was less pronounced in the mice

TABLE 1. Epidermal Thickness after Cream Treatment[a]

Days After First Treatment	Cells per 25 Vision Fields of Epidermis		
	Peptide	Cream Base	Effect
2	2615 ± 106	2741 ± 147	95%
3	2689 ± 126	2881 ± 15	93%
6	2472 ± 95	2757 ± 18	89%[b]
10	1885 ± 236	2024 ± 25	93%
13	2476 ± 385	2719 ± 36	90%
22	2095 ± 122	2128 ± 6	98%

[a]Groups of mice were treated twice weekly by topical application of a 0.02% pentapeptide cream, or with the cream base. In each cage, 4 mice were given pentapeptide cream and the remaining 4 the cream base. The animals examined on days 2 or 3 had been treated only once. Epidermal cells were counted per 25 vision fields (eye-piece 12.5, objective 40) of interfollicular epidermis (= 12.5 mm). The effect is expressed as percentage of cream base-treated epidermis. The counts are given with 1 SEM. The mean number of epidermal cells per 25 vision fields in a group of 4 untreated mice of the same age and sex was 2112 ± 50.

[b]$p < 0.05$ with Wilcoxon rank sum test.

that had received the pentapeptide cream, but the difference was significant only on day 6 ($p < 0.05$). Five days after terminating the treatment the epidermal thickness was normal again.

The lack of a sustained effect of the repeated treatments could possibly be related to earlier observations that ip injection of a large, ineffective dose of pentapeptide abolishes the inhibitory effect of a smaller dose given 1 hr later.[7] A state of refractoriness thus seems to be induced after treatment with the pentapeptide, but we do not know whether this phenomenon is due to altered receptor kinetics, or to a compensatory increase in the concentration of growth factors that counteract the pentapeptide-induced inhibition.

We also tried to treat groups of mice with a 0.02% dipeptide (pGlu-GlyOH) cream, but except for a single time point this treatment had no inhibitory effect. We do not know if this result was due to incomplete diffusion of the dipeptide into the epidermis, to inactivation of the dipeptide by the cream base, or to a wrong dose. After ip injection the pentapeptide and the dipeptide give the same inhibition of epidermal mitoses at the same dose level,[7] but different doses of the dipeptide could be needed when given topically as a cream.

Why Is Cell Proliferation Only Partially Inhibited?

The experiments described above emphasize one problem that we have no unambiguous answer to at the present time: Epidermal cell proliferation is never reduced by more than 40–50% after treatment with either the biological or the synthetic pentapeptide. This restricted response could be accounted for in several ways. Proliferating epidermal cells represent a heterogeneous population with resting, or slowly cycling, cells in several phases of the cell cycle.[9] Different susceptibility to the inhibitor of the different subpopulations could result in only a partial inhibition of cell proliferation. On the other hand, it is quite possible that the isolated inhibitor is just one of many factors that regulate cell renewal in epidermis. An increased concentration of inhibitory pentapeptide could lead to altered active concentrations of stimulatory growth factors. Alternatively, the pentapeptide could be dependent some other factor to exert its effect, and a limited supply of such a second messenger would be the determining factor for the efficacy of the pentapeptide. Lastly, possible cell surface receptors for the pentapeptide might have a response pattern that precludes a 100% inhibition of proliferating cells.

Interaction with Intracellular Messenger Systems

Earlier experiments with crude or partially purified skin extracts have demonstrated that the effect is enhanced by adrenalin,[10] and abrogated by pretreatment with the beta-receptor blocking agent propranolol.[11] Preliminary *in vivo* experiments have indicated that pretreatment of mice with 5 μg propranolol abolishes the inhibitory effect of synthetic pentapeptide given 1 hr later (unpublished data). It is therefore possible that the inhibitory effect of the pentapeptide is influenced by, or dependent on a cAMP-related intracellular messenger system. The pentapeptide itself does not, however, alter adenylyl cyclase activity in a cell-free preparation of epidermal cells (H. Attramadal, 1987, personal communication). The curvilinear dose-response relationship is suggestive of surface receptor kinetics like those described by Monod *et al.* in 1965.[12] These problems obviously need further investigation.

Relationship to Other Inhibitory Peptides

The pentapeptide isolated from skin extracts bears a strong structural resemblance to the hemoregulatory pentapeptide identified in 1982 by Paukovits and Laerum.[13] These pentapeptides have the amino acids pyroglutamate, glutamate, and aspartate in common, and both have pyroglutamate at the N-terminal end. Recently, a mitosis inhibiting tripeptide, pGlu-His-GlyOH, was isolated from mouse intestinal extracts.[14] This peptide reversibly inhibits cell proliferation in the large bowel epithelium. Also, a mitosis inhibiting peptide has been isolated from normal liver[15] that has pyroglutamate at the N-terminal end, but the full structure is not yet known.

In spite of their similarities, the inhibitory peptides mentioned above are tissue specific at their respective active dose levels. Since it is reasonable to assume that all peptides of this category are broken down by a common pyroglutamyl peptidase, it would not be surprising if the tissue specificity would be less apparent at dose levels that would involve substrate competition for the available pyroglutamyl peptidase. This has not been investigated yet. Lastly, the growth inhibitory peptides referred to above have structural similarities to several peptide hormones, like thyrotropin-releasing hormone (TRH) and gonadotropin-releasing hormone (Gn-RH). It is tempting to speculate that such structural similarities could imply that cell communication in some respects is based on a group of closely related signal substances that monitor cell function and cell division.

REFERENCES

1. BULLOUGH, W. S., E. B. LAURENCE, O. H. IVERSEN & K. ELGJO. 1967. The vertebrate epidermal chalone. Nature **214:** 578–580.
2. ELGJO, K. & O. P. F. CLAUSEN. 1983. Proliferation-dependent effect of skin extracts on mouse epidermal cell flux at the G_1-S, S-G_2 and G_2-M transitions. Virchows Arch. B. **42:** 143–151.
3. NOME, O. 1975. Tissue specificity of the epidermal chalones. Virchows Arch. B. **19:** 1–25.
4. HENNINGS, H., K. ELGJO & O. H. IVERSEN. 1969. Delayed inhibition of epidermal DNA synthesis after injection of an aqueous skin extract. Virchows Arch. B. **3:** 45–53.
5. REICHELT, K. L., K. ELGJO & P. D. EDMINSON. 1987. Isolation and structure of epidermal mitosis inhibiting pentapeptide. Biochem. Biophys. Res. Commun. **146:** 1493–1501.
6. ELGJO, K. & K. L. REICHELT. 1984. Purification and characterization of a mitosis inhibiting epidermal peptide. Cell Biol. Int. Rep. **8:** 379–382.
7. ELGJO, K., K. L. REICHELT, P. D. EDMINSON & E. MOEN. 1986. Endogenous peptides in epidermal growth control. *In* Biological Regulation of Cell Proliferation. R. Baserga, P. Foa & E. E. Polli, Eds.: 259–265. Raven Press. New York, NY.
8. ELGJO, K., K. L. REICHELT, H. HENNINGS, D. MICHAEL & S. H. YUSPA. 1986. Purified epidermal pentapeptide inhibits proliferation and enhances terminal differentiation in cultured mouse epidermal cells. J. Invest. Dermatol. **87:** 555–558.
9. CLAUSEN, O. P. F., B. KIRKHUS, E. THORUD, A. SCHÖLBERG, E. MOEN & A. CROMARTY. 1986. Evidence of mouse epidermal subpopulations with different cycle times. J. Invest. Dermatol. **86:** 266–270.
10. BULLOUGH, W. S. & E. B. LAURENCE. 1964. Mitotic control by internal secretion: The role of the chalone-adrenalin complex. Exp. Cell Res. **33:** 176–194.
11. ELGJO, K. 1975. Epidermal chalone and cyclic AMP: An *in vivo* study. J. Invest. Dermatol. **64:** 14–18.
12. MONOD, J., J. WYMAN & J.-P. CHANGEUX. 1965. On the nature of allosteric transitions: A plausible model. J. Mol. Biol. **12:** 88–118.
13. PAUKOVITS, W. & O. D. LAERUM. 1982. Isolation and synthesis of a hemoregulatory peptide. Z. Naturforsch. **37c:** 1297–1300.

14.   SKRAASTAD, Ö., T. FOSSLI, P. D. EDMINSON & K. L. REICHELT. 1988. Purification and characterisation of a mitosis inhibitory tripeptide from mouse intestinal extracts. Epithelia **1:** 107–119.
15.   PAULSEN, J. E., K. L. REICHELT & A. K. PETTERSEN. 1987. Purification and characterization of a growth inhibitory hepatic peptide. Virchows Arch. B. **54:** 152–154.

The Epidermal G1-Chalone: An Endogenous Tissue-Specific Inhibitor of Epidermal Cell Proliferation

K. HARTMUT RICHTER, MATTHIAS CLAUSS,
WERNER HÖFLE, RUBEN SCHNAPKE,
AND FRIEDRICH MARKS

German Cancer Research Center
Institute of Biochemistry
Im Neuenheimer Feld 280
D-6900 Heidelberg 1, Federal Republic of Germany

Cellular proliferation is generally assumed to be under the positive control of certain hormones and growth factors. Such factors are especially suited to explain the proliferative behavior of cells growing *in vitro*. Several of these growth factors including EGF,[1] TGFα,[2] keratinocyte growth factor[3] and interleukin I[4] have been shown to stimulate skin keratinocytes in culture. Evidence for a physiological role in skin has been as yet only provided for EGF and the closely related TGFα. Both have been found to accelerate the re-epithelization of skin wounds[5] and EGF may be involved also in cornea wound-healing[6] as well as in eyelid opening and incisor eruption.[7] Thus, EGF and TGFα are probably involved in wound healing and developmental processes, whereas none of the known growth factors has ever definitely been shown to play a role in normal epidermal growth regulation, *ie.*, everyday tissue regeneration.

A characteristic feature of intact nondisturbed epidermis and most other tissues is a proliferative steady state. From the cybernetic point of view, the simplest method by which cell division in steadily renewing tissues is controlled in a tissue-specific way is a negative feedback mechanism: The mature (dying) cells produce a growth-inhibitory signal which acts on the proliferative compartment. In the early sixties it was found that an aqueous extract from skin inhibited cellular proliferation in skin *in vivo* and *in vitro*.[8,9] In this extract, distinguishable inhibitory activities called "chalones" were identified which act on sebaceous glands and interfollicular epidermis.[10] The latter activity was subfractionated into two entities with different points of attack in the cell cycle, *i.e.*, the G1-S and the G2-M transition.[11,12] These inhibitory activities were assumed to be due to two different antiproliferative tissue-derived factors which were tentatively called epidermal G1- and G2-chalone. Tissue-specific antiproliferative factors of the chalone type were also found in other tissues.[13] The discovery of TGFβ to be a powerful antiproliferative agent for epidermis and other epithelia[14] has recently focussed new interest on the problem of negative growth control.

Isolation, Bioassay and Physicochemical Properties of Epidermal G1-Chalone

The epidermal G1-chalone inhibits epidermal DNA synthesis. It is extracted from lyophilized skin powder with water. Due to its extraordinary stability the epidermal G1-chalone can be efficiently purified by a simple procedure[15] consisting of ethanol precipitation, extensive digestion with pronase, coprecipitation of chalone and polyanions using cetylpyridinium chloride, and phenol extraction.

Recently, we have developed another simple isolation procedure which omits pronase digestion. This procedure is based on a fractionated extraction of the skin extract (ethanol precipitate) with organic solvents. After nonactive material has been removed with n-butanol, acetone, ethylacetate, chloroform and chloroform-methanol (1:1) the active fraction is extracted with chloroform-methanol-water (5:5:1) and further purified by chromatography on a sephacryl 300 column using phosphate-buffered saline (PBS) plus 0.2% Triton X-100 as an eluant. The chalone is eluted as a mixed chalone-Triton X-100 micelle and separated from the detergent by extensive dialysis against water.

SDS polyacrylamide electrophoresis reveals 2 major bands in the active fractions at 8 and 10 kD (with polypeptides as references) which are made visible only by silver staining following dichromate oxidation. Silver staining alone results only in a very faint pattern, whereas Coomassie blue does not reveal any pattern on the gels. Higher amounts of material applied onto the polyacrylamide gel result in a smeared staining without discrete bands reaching from the start down to the 4-kD region. From a sephacryl 300 column, chalone activity is eluted with the void volume at neutral pH (FIG. 1). The activity is shifted to the 10-kD region (using cytochrome C as a marker) upon chromatography in 0.1 M formic acid (FIG. 1). Since this shift is reversible when the pH is raised again it is probably due to desaggregation-aggregation of the active material.

G1-chalone activity can be assayed either *in vivo*[15] (i.p. injection into adult mice and pulse-labelling of epidermal DNA 16 hr later) or *in vitro* using a rat-tongue epithelial cell line.[16] Assuming a molecular weight of 10 kD half-maximal inhibitory activity (ID_{50}) *in vitro* is found at a chalone concentration of 2×10^{-13} M. *In vivo* the ID_{50} is approximately 0.2 pmol/g body weight. The purification procedure and the staining properties as well as the stability of the inhibitor against heat, proteolytic digestion and denaturing agents are not consistent with a simple peptide or polypeptide structure of the epidermal G1-chalone.

Other possibilities that can easily be excluded based on the physicochemical behavior of the factor and on degradation experiments are nucleic acids, simple oligo- and polysaccharides and simple lipids and glycolipids. The chalone can be extracted from a SDS-solution by means of phenol, whereas in the absence of SDS it is found in a precipitate formed at the water-phenol interphase. This behavior and its tendency to form aggregates and mixed micelles with Triton X-100 indicate an amphipathic nature of the factor.

Preliminary analytical data indicate that highly enriched chalone fractions contain 5% carbohydrate and 60% amino acids. Thus, the factor may be a protease-resistant rather lipophilic glycopeptide.

Chalone activity is completely destroyed by 1 N NaOH (1 hr/37°C) and upon incubation with an exoglycosidase mixture (FIG. 2). Attempts to inactivate the factor with a single glycosidase have been hitherto unsuccessful. The chalone is partially inactivated by periodate. No sulfhydryl or disulfide residues were found in the chalone fraction and the activity is not diminished under reductive conditions such as treatment with mercaptoethanol.

Biological Properties of the Epidermal G1-Chalone (Reviewed by Marks et al.[17-21])

The most striking properties of the epidermal G1-chalone are its high biological activity and its pronounced specificity. Both G1-chalone production and chalone

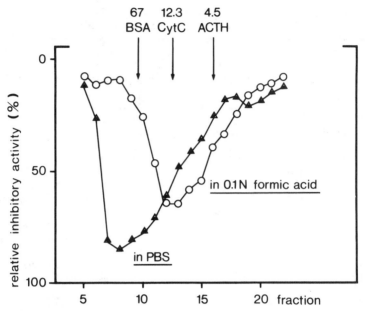

FIGURE 1. Gel permeation chromatography of epidermal G1-chalone on sephacryl 300. 30 g ethanol precipitate of an aqueous pig-skin extract[15] (ID_{50} = 400 ng/ml) were successively extracted with n-butanol, acetone, ethylacetate, chloroform and chloroform/methanol 1:1 (3 times with 300 ml each at room temperature, mixing by means of Branson B15-sonifier for 5 min each). The extracts were discarded. From the residue the chalone was extracted 6 times with 300 ml chloroform-methanol-water (5:5:1) each. After evaporation of the solvent the chalone-active material was dissolved by stepwise extraction with 6 portions of 50 ml water each using sonification for mixing. The aqueous extract was filtered through a 0.2-μm filter. 70 μg (200 μl) enriched chalone (ID_{50} = 1 ng/ml) were applied on a column containing 6 ml sephacryl 300. The column was eluted with Dulbecco's PBS (—▲—). In a second experiment the chalone preparation was heated in 0.5 M of formic acid at 56°C for 10 min before application onto the column. Elution was carried out with 0.1 M formic acid (—○—). 400 μl-fractions were collected and tested for chalone activity (inhibition of DNA labelling in RTE2 keratinocytes *in vitro* REF. 16).

responsiveness seem to be restricted to keratinizing epithelia. Other tissues, even closely related ones such as sebaceous glands or hair follicles and nonepidermal cells, neither respond to the factor nor seem to contain epidermal G1-chalone activity. Ontogenetic studies as well as experiments with epidermal carcinomas indicate that chalone production is closely linked to epidermal differentiation, *i.e.*, the process of keratinization, while in mice, chalone responsiveness of epidermis develops after birth. *In vivo* as well as in cell culture, the inhibitory effect of the epidermal G1-chalone is fully reversible within 1 day and levels off at 40–70% inhibition of DNA synthesis over a concentration range of at least 3 orders of magnitude.[16] This indicates that only a certain population of cells, rather than all cells of the proliferative compartment of epidermis, is chalone sensitive. Cytoflow-fluorometric analysis using partially synchronized RTE keratinocytes show that the inhibitory effect on epidermal cell proliferation exhibits a distinct point of attack in the late G1-phase of the cell cycle. Other cycle phases are not influenced (unpublished results). The factor is not species specific since chalone preparations derived from pig skin act on mouse epidermis and rat tongue epithelium. Chalone responsiveness and epidermal growth pattern show an inverse correlation, *i.e.,* mouse epidermis becomes transiently desensitized to the inhibitory effect of chalone injections when it is developing hyperplasia due to chemical or mechanical traumatization.

Since hyperplasia is the visible result of a temporary disturbance of tissue homeostasis, this indicates that the epidermal G1-chalone plays an important role in homeostatic control. This conclusion is strongly supported by the observation that under certain circumstances, such as skin massage or treatment with certain nonirritating agents such as the phorbol ester 4-0-methyl-TPA, a strong hyperproliferative response can be evoked in mouse epidermis which does *not* result in epidermal hyperplasia. Under these conditions where the control of tissue homeostasis is obviously maintained epidermis is not desensitized to G1-chalone. Attempts to treat experimental skin cancer with epidermal G1-chalone have proved unsuccessful.

Relationship between Epidermal G1-Chalone and Other Endogeneous Inhibitors of Epidermal Cell Proliferation

Epidermal G1-chalone is not the only cell growth inhibitor found in skin. As already mentioned above a still ill-defined "sebaceous gland chalone" may exist.[10]

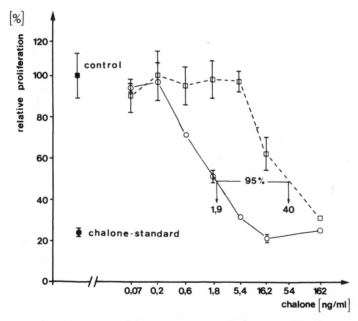

FIGURE 2. Inactivation of epidermal G1-chalone by treatment with exoglycosidases. 100 μg enriched chalone was dissolved in 1 ml 40 mM acetate pH 5, 2.5 mM octylglucoside containing 100 units penicillin, 100 μg streptomycin and 5 mg exoglycosidase mixture from Turbo cornutus (Seikagaku Kogyo, Tokyo, Japan). After incubation for 4 hr at 37°C with (−□−) or without exoglycosidases (−○−) chalone activity was tested using rat tongue epithelial cells in culture as previously described.[16] Under the conditions employed the enzyme mixture by itself did not show any effect on DNA synthesis.

Furthermore, for interfollicular epidermis, G1-chalone activity could be clearly distinguished from another activity acting on mitosis (G2-chalone).[11,12] Finally, transforming growth factor β (TGFβ) has recently been shown to be a potent inhibitor of epithelial cell proliferation including keratinocyte proliferation.[14]

The literature on epidermal G2-chalone is somewhat bewildering. Initially it was described as a high-molecular weight protein (>20 kDa) which acts synergistically with adrenalin in inhibiting epidermal mitosis in a highly tissue-specific and cell cycle-specific manner.[17,21] Recently, a pentapeptide pyroGlu-Glu-Asp-Ser-GlyOH with G2-chalone activity has been isolated from skin.[22]

Whether or not a relationship exists between the high molecular weight G2-chalone described in the older literature and the inhibitory pentapeptide remains an open question. Moreover, the role of adrenalin and glucocorticoid hormone in G2-chalone action and epidermal growth control has not achieved much attention in the last 15–20 years. What is known is that the well-established antimitotic effect of catecholamines on epidermis is mediated by cyclic AMP via a β-adrenergic mechanism[17] and that glucocorticoids prevent agonist-induced down-regulation of the β-adrenergic signal transduction machinery.[22]

According to a series of preliminary experiments carried out in our laboratory, a close relationship between the antiproliferative pentapeptide and G1-chalone does not exist. Thus, we found the pentapeptide to be inactive under conditions of the *in vivo* assay for G1-chalone (i.p. injection followed by pulse-labelling of epidermal DNA,[15] see TABLE 1) and 5 orders of magnitude less active in inhibiting epidermal DNA synthesis *in vitro* (using RTE cells[16]). Moreover, G1-chalone is strictly cell cycle specific, whereas the pentapeptide has been found to act on both DNA synthesis and mitosis.[24] It cannot yet be ruled out, of course, that the pentapeptide is a breakdown product of G1-chalone with reduced specificity. To answer this question one has to wait for the elucidation of G1-chalone structure.

Like epidermal G1-chalone, TGFβ has been found to inhibit epithelial cell proliferation with a point of attack near the G1-S transition.[25] Moreover, suprabasal epidermal cells are able to express TGFβ-mRNA at least when stimulated by the phorbol ester TPA.[26] Even the molecular weight of the TGFβ monomer (12.5 kDa) appears to be rather close to that of the major bands found upon SDS-polyacrylamide gel electrophoresis of G1-chalone. Thus, TGFβ seems to fulfil several requirements for an epidermal chalone proper. On the other hand, in contrast to G1-chalone, i.p. injection of TGFβ did not result in any inhibition of epidermal DNA labelling *in vivo* (TABLE 1). In the *in vitro* test system (RTE cells[16]) the TGFβ dose (4×10^{-11} M) required for half-maximal inhibition of DNA synthesis was approximately 2 orders of magnitude higher than that of G1-chalone (2×10^{-13} M). Moreover, both epidermal G1-chalone and TGFβ differ profoundly in physiochemical behavior and tissue specificity. No TGFβ band at 12.5 kDa could be seen upon SDS polyacrylamide gel electrophoresis of highly purified chalone as compared with a control using authentic TGFβ as a reference. In addition, incubation with an anti-TGFβ antibody (R & D Systems, Minneapolis, MN) did not reduce G1-chalone activity, whereas it completely neutralized TGFβ activity (unpublished results). Taken together, these results rule out an identity or any close structural relationship between TGFβ and chalone. On the other hand, a *functional* relationship between chalone and TGFβ may well exist. Thus, the chalone could be a local specific mediator of the antiproliferative action of TGFβ or TGFβ could somehow interfere with the mechanism of action of G1-chalone. Another possibility would be that TGFβ replaces G1-chalone and its regulative functions on cell proliferation under emergency situations such as wound healing, etc. There is indeed evidence for TGFβ to play a role as a "wound hormone."[27] Moreover, autocrine production of TGFβ seems to be induced under conditions where epidermis becomes desensitized for epidermal G1-chalone, *i.e.,* hyperplastic transformation by phorbol ester TPA.[26] Thus, on the cellular level, TGFβ and epidermal G1-chalone may well act in an analogous fashion or even share a similar mechanism of action. Such a proposal is supported by the observation that both TGFβ and G1-chalone (and to a lesser extent also the inhibitory pentapeptide) not only inhibit DNA synthesis of keratinocytes *in vitro* but at the same time stimulate terminal differentiation of primary neonatal mouse epidermal cells grown *in vitro*[28] (4 × MEM, 10% fetal calf serum, 0.08 mM CaCl$_2$) as shown by a dose-dependent increase of transglutaminase activity (assayed according to REF. 29). Similar as for inhibition of DNA synthesis the transglutaminase-inducing efficacy of the three factors was found to be in the order G1-chalone (4 × 10^{-13}) >TGFβ (3 × 10^{-12} M) >inhibitory pentapeptide (3 × 10^{-10} M). A stimulatory effect of the inhibitory pentapeptide on terminal differentiation of cultured mouse epidermal cells has been recently shown also by other authors.[30]

TABLE 1. Effect of G1-Chalone, Inhibitory Pentapeptide
PyroGlu-Glu-Asp-Ser-GlyOH and TGFβ on DNA Synthesis in Mouse Epidermis *In Vivo*[a]

Inhibitor	Dose/Mouse	DNA Labelling	
		cpm/μ DNA	%
None	—	27 ± 8	100
Pentapeptide[b]	10^{-13} Mol	29 ± 3	107
	10^{-12} Mol	29 ± 4	107
	10^{-11} Mol	32 ± 6	119
TGFβ[c]	50 ng	27 ± 13	100
	100 ng	29 ± 5	107
G1-Chalone[d]	300 μg[d]	9 ± 2	30

[a]The inhibitors were dissolved in 0.2 ml isotonic NaCl solution and intraperitoneally injected into 7-week-old female NMRI mice (30 g). 16 hr later pulse labelling of epidermal DNA was carried out and specific radioactivity of DNA was determined according to REFERENCE 12. N = 5 (animals), ± S.D.
[b]The dose of the pentapeptide was chosen so that according to the literature[24] an inhibitory effect could be expected.
[c]Porcine TGFβ purchased from R & D Systems, Minneapolis, MN.
[d]The semipurified chalone preparation (ethanol precipitate of pig-skin extract) used in this experiment is equivalent to 50–100 ng highly purified epidermal chalone. The preparation does not contain other inhibitory activity beside chalone.[15]

However, despite these similarities in their biological effects, neither the inhibitory pentapeptide nor epidermal G1-chalone are able to compete with TGFβ for the binding to the TGFβ receptor of epidermal cells (unpublished results; for receptor assay, see REF. 31).

SUMMARY

An apparently macromolecular factor is isolated from aqueous skin extracts which inhibits DNA synthesis *in vivo* and *in vitro* with high efficacy (ID$_{50}$ *in vivo* 0.2 pmol/g, *in vitro* 0.2 pM) and in a highly specific manner showing a point of attack in the late G1-phase of the cell cycle (epidermal G1-chalone). Preliminary characterization indicates an unusual highly amphipathic structure consisting of amino acids and carbohydrate. Despite its apparent molecular weight of ≈ 10 kD the chalone is stable against denaturing agents and most enzymes, including proteases. An inverse correlation between chalone responsiveness of mouse epidermis *in vivo* and the development of hyperplasia due to injury indicates an important role of the factor in the regulation of tissue homeostasis. According to its physicochemical and biological properties the epidermal G1-chalone appears not to be related to other endogeneous inhibitors of epidermal cell proliferation such as the pentapeptide pyroGlu-Glu-Asp-Ser-GlyOH and transforming growth factor β (TGFβ).

REFERENCES

1. RHEINWALD, J. G. & H. GREEN. 1977. Nature 265:421–424.
2. COFFEY, R. J., R. DERYNCK, J. N. WICOX, T. S. BRINGMAN, A. S. GOUSTIN, H. L. MOSES & M. R. PITTELKOW. 1987. Nature **328**: 817–820.
3. GILCHREST, B. A. 1983. J. Invest. Dermatol. **81** (Suppl. 1): 184s–189s.
4. RISTOW, H. J. 1987. Proc. Natl. Acad. Sci. USA **84**: 1940–1944.
5. SCHULTZ, G. S., M. WHITE & R. MITCHELL. 1987. Science **235**: 350–352.
6. SAVAGE, R. C. & S. COHEN. 1973. Exp. Eye Res. **15**: 361–366.
7. COHEN, S. 1962. J. Biol. Chem. **237**: 1555–1562.
8. BULLOUGH, W. S. & E. B. LAURENCE. 1964. Exp. Cell Res. **33**: 176–194.
9. IVERSEN, O. H., E. AANDAHL & K. ELGJO. 1965. Acta Pathol. Microbiol. Scand. **64**: 506–510.
10. BULLOUGH, W. S. & E. B. LAURENCE. 1970. Cell Tissue Kinet. **3**: 291–300.
11. ELGJO, K., H. HENNINGS & W. EDGEHILL. 1971. Virchows Arch. B. **10**: 342–347.
12. MARKS, F. 1971. Hoppe-Seyler's Z. Physiol. Chem. **352**: 1273–1274.
13. HOUCK, J. C., Ed. 1976. Chalones. North Holland.
14. MASSAGUE, J. 1987. Cell **49**: 437–438.
15. MARKS, F. 1974. Hoppe-Seyler's Z. Physiol. Chem. **356**: 1989–1992.
16. RICHTER, K. H., A. JEPSEN & F. MARKS. 1984. Exp. Cell Res. **150**: 68–76.
17. MARKS, F. 1976. *In* Chalones. J. C. Houck, Ed.: 173–227. North Holland. Amsterdam.
18. MARKS, F. 1981. *In* Biology of Skin Cancer (Excluding Melanomas). O. D. Laerum & O. H. Iversen, Eds. UICC Technical Report Series, Vol. 63: 16–21.
19. MARKS, F. & K. H. RICHTER. 1984. Br. J. Dermatol. **11,** Suppl. 27: 58–63.
20. MARKS, F., M. CLAUSS, A. TIEGEL & K. H. RICHTER. 1986. *In* Biological Regulation of Cell Proliferation. R. Baserga, P. Foa, D. Metcalf & E. E. Polli, Eds.: 267–274. Raven Press. New York, NY.
21. HONDIUS-BOLDING, W. & E. B. LAURENCE. 1968. Europ. J. Biochem. **5**: 191–198.
22. REICHELT, K. L., K. ELGJO & P. D. EDMINSON. 1987. Biochim. Biophys. Res. Commun. **146**: 1493–1501.
23. IIZUKA, H. & A. OHKAWARA. 1983. J. Invest. Dermatol. **80**: 524–528.
24. ELGJO, K., K. L. REICHELT, P. D. EDMINSON & E. MOEN. 1986. *In* Biological Regulation of Cell Proliferation. R. Baserga, P. Foa, D. Metcalf & E. E. Polli, Eds.: 259–266. Raven Press. New York, NY.
25. SHIPLEY, G. D., M. R. PITTELKOW, J. J. WITTE, R. E. SCOTT & H. L. MOSES. 1986. Cancer Res. **46**: 2068–2071.
26. AKHURST, R. J., F. FEE & A. BALMAIN. 1988. Nature **331**: 363–365.
27. ROBERTS, A. B., M. B. SPORN, R. K. ASSOIAN, J. M. SMITH, N. S. ROCHE, L. M. WAKEFIELD, U. I. HEINE, L. A. LIOTTA, U. FALANGER, J. H. KETERI & A. FANCI. 1986. Proc. Natl. Acad. Sci. USA **83**: 4167–4171.
28. FÜRSTENBERGER, G., M. GROSS, J. SCHWEIZER, I. VOGT & F. MARKS. 1986. Carcinogenesis **8**: 1745–1753.
29. THACHER, S. M. & R. H. RICE. 1983. Cell **40**: 685–695.
30. ELGJO, K., K. L. REICHELT, H. HENNINGS, D. MICHAEL & S. H. YUSPA. 1986. J. Invest. Dermatol. **87**: 555–558.
31. M. ROGERS. 1988. Ph.D. thesis. University of Heidelberg.

Keratinocytes Produce and Are Regulated by Transforming Growth Factors

MARK R. PITTELKOW,[a] ROBERT J. COFFEY, JR.,[b,c] AND
HAROLD L. MOSES[b]

[a]*Department of Dermatology*
Mayo Clinic
Rochester, Minnesota 55905
and
Departments of [b]*Cell Biology and* [c]*Medicine*
Vanderbilt University
Nashville, Tennessee 37232

INTRODUCTION

The epidermis is a stratified epithelium that continually undergoes cell renewal and loss. Keratinocytes constitute the greatest mass of the epidermis and are highly replicative cells. In the normal state, the epidermis stringently regulates keratinocyte proliferation, growth arrest and differentiation. Cell and tissue control of keratinocyte growth, stratification and differentiation are complex. A variety of positive and negative regulators have been implicated in epidermal growth control. However, other presently unknown factors may also be involved in the regulation of keratinocyte proliferation. This paper reviews the biochemical, biological, and molecular properties of the transforming growth factors type alpha and beta in the control of epidermal growth and their possible roles in cutaneous morphogenesis, tissue homeostasis and skin disease.

Transforming Growth Factors and Autocrine/Paracrine Regulation

The transforming growth factors type α and β (TGF-α and TGF-β) were initially isolated and purified from virally transformed mesenchymal cells.[1,2] Conditioned medium from these cell lines stimulate anchorage-independent colony formation and morphologic alteration of nontransformed mesenchymal cells. The biological factor(s) isolated from transformed cells was called sarcoma growth factor (SGF). Distinct biological activities that constituted SGF are separable. Individually, these factors fail to simulate the pronounced alterations induced by SGF. When SGF is reconstituted, however, its ability to induce a transformed phenotype is restored. The polypeptides exerting these effects have been purified, identified biochemically and designated TGF-α and TGF-β (previously called TGFI and TGFII respectively).[3–5] Since transformed cells not only produce these factors, but, in turn, respond to them, the concept of autocrine secretion (cellular self stimulation) was developed to account for the possible role of these transforming factors in the altered growth behavior of neoplasia.[6] However, subsequent discoveries have necessitated a revision and expansion of the model for autocrine factors in neoplasia to include autocrine and paracrine (neighboring or locally mediated cell response) secretion in the growth regulation of normal, nontransformed, diploid cells and tissues. These concepts evolved as a result of studies showing that cultured keratinocytes synthesize and secrete TGF-α and

TGF-β.[7-9] Further investigation has revealed that TGF-α and TGF-β regulate proliferation, growth arrest and differentiation of keratinocytes in an autocrine/ paracrine fashion.

Transforming Growth Factor Type Alpha (TGF-α)

TGF-α was initially identified in conditioned medium from cultured, Maloney MSV-transformed 3T3 cells.[1,2,4,5] Partially purified material was capable of inducing the transformed phenotype in nontransformed cells by initiating a phenotypic change in morphology and promoting anchorage-independent cell growth. Purification of one of the growth factors from conditioned culture medium showed that it specifically bound to epidermal growth factor (EGF) membrane receptors. However, this growth factor was distinct from EGF. Further investigation demonstrated that this receptor binding factor was TGF-α, a unique growth factor that constituted part of the activity detected in conditioned medium from the viral transformed mouse mesenchymal cells. Additional studies comparing EGF and TGF-α showed that EGF could substitute for the transforming properties of TGF-α in soft agar growth assays.

Structural Characteristics of TGF-α

The structural properties of TGF-α have been elucidated at the protein and gene level[10-12] (reviewed in REFS. 13, 14). Mature TGF-α is a 5.6-Kd polypeptide consisting of a single chain of 50 amino acids. Larger molecular weight species of TGF-α have been detected in cultured tumor cell lines and fresh tumor tissues. In addition, TGF-α has been detected in a variety of nonneoplastic cells and tissues, including human placenta,[15] mouse and rat embryos,[16,17] cultured bovine pituitary cells[18] and human and murine keratinocytes.[8,9]

The sequence of rat TGF-α shows a significant homology to both human and mouse EGF.[10-12] The amino acid sequence of TGF-α is highly conserved among species. Like EGF, TGF-α is synthesized from a precursor. The open reading frame of the cloned human TGF-α gene is sufficient to encode a protein of 160 amino acids of which residues 40–89 encode native TGF-α.[10] Transcripts of 4.5–4.8 kilobases have been detected in a cell line of human renal cell carcinoma, from which a cDNA clone was originally isolated.[10] A number of other fresh human tumor tissues and cultured tumor cells contain TGF-α mRNA. Carcinomas predominately express this growth factor transcript. The human gene locus for TGF-α is located on the short arm of chromosome 2.[19]

The mature, 50 amino acid molecule appears to be one of the major secreted species of TGF-α. Other higher molecular weight forms a TGF-α may be secreted or cell associated.[20,21] Eleven to nineteen–Kd precursors are secreted and these species are N-glycosylated. Even larger forms identified as TGF-α by immunochemical and receptor binding assays have been detected in medium or cell extracts.[22,23] The 160 amino acid precursor for TGF-α, pro-TGF-α, is synthesized as an integral membrane glycoprotein with the mature TGF-α sequence located in the extracellular domain.[20] The NH_2-terminal signal sequence of 23 uncharged amino acids is followed by a domain of about 74 amino acids that includes the 50 amino acid, mature TGF-α sequence. Sequentially, the transmembrane domain of 23 amino acids is followed by the cysteine-rich COOH-terminal domain of 39 amino acids. Elastaselike enzymes appear to cleave the extracellular domain at the COOH terminus of the mature TGF-α

sequence.[24] This proteolytic step creates a 17–19-Kd glycosylated species of TGF-α that is a soluble and bioactive molecule. Further cleavage of the NH_2 terminus yields mature TGF-α.

Present evidence would indicate that secreted forms of TGF-α are multiple and can be glycosylated to variable extent. Less is known of the biological activity and processing mechanisms that regulate the membrane-associated precursor, pro-TGF-α. Transmembrane and cytoplasmic portions of the precursor molecule also may have regulatory roles in TGF-α protein activation, receptor-ligand binding, processing and possibly, transcriptional control. Ultimate delineation of the biosynthesis, processing and biological function of TGF-α will likely require the use of cell lines tranfected with TGF-α expression vectors, normal cell types that produce or are induced to express TGF-α mRNA and protein, and transgenic mice.

Overview of Biological Activities of TGF-α

As its name denotes, TGF-α influences growth and cellular morphology and induces a transformed phenotype in susceptible cells. This biological activity allowed purification of the TGFs and subsequently, separation of TGF-α and TGF-β. It is now recognized that TGF-α has a broad range of biological activities. Based on its homology to EGF, TGF-α would be expected to exhibit many effects similar to EGF. In this respect, several polypeptides bear homology to EGF and are included in the EGF family of growth factors. TGF-α and vaccinia virus growth factor (VVGF) are among this class.[25,26]

Bioassays developed to quantitate and characterize EGF activity, in turn, have been adapted to measure the bioactivity and relative potency of TGF-α.

EGF was first detected and purified using a bioassay of eyelid opening in the newborn mouse.[27] In parallel studies recombinant human TGF-α causes accelerated eyelid opening in newborn mice.[28] Murine and human EGF were compared to human TGF-α, and all three growth factors exhibit essentially identical potency in this bioassay. In other experiments, synthetic rat TGF-α accelerated incisor eruption and eyelid opening in newborn mice but retarded growth rates of body weight and hair.[29] Significantly fewer hairs as well as a shorter and finer coat were produced by daily subcutaneous injection of TGF-α, similar to the results of previous studies with EGF.

TGF-α and EGF exert biological effects on nonepithelial tissues. TGF-α has been shown to markedly enhance resorption of bone in organ culture, stimulate formation of osteoclastlike cells in long-term human marrow cultures and induce a significant rise in plasma calcium concentration when administered *in vivo*.[30–32] The potency of TGF-α in these studies was severalfold greater or markedly (10–100-fold) increased compared to EGF.

In addition to causing increased calcium resorption, TGF-α has been shown to inhibit parathyroid hormone (PTH) responsive-adenylate cyclase in osteoblastlike cells.[33] EGF also exerts this effect. This evidence suggests that these growth factors may modulate PTH receptor activation and may be associated with decreased bone formation as well as increased absorption.

TGF-α has been shown to exert potent regulatory effects on other hormone-responsive cell types, similar to those displayed by EGF. TGF-α and EGF attenuate estrogen biosynthesis by granulosa cells. This occurs via neutralization of follicle stimulating hormone action and alteration of cAMP generation.[34]

Interestingly, for selected comparisons of *in vitro* and *in vivo* responses to EGF and TGF-α, *in vivo* bioassays may reveal differences in biological activities of these two growth factors that may not be detectable *in vitro*. An example of this difference has been shown for angiogenesis using the hamster cheek pouch bioassay.[35]

TGF-α is produced by a variety of carcinoma cell lines and tumor tissues. Angiogenesis is a recognized feature in the pathology of tumor growth and also is seen in physiologic processes, including wound healing. TGF-α production may favor neoplastic growth by promoting neovascularization.

Purified, recombinant human TGF-α is more potent than natural mouse EGF in stimulating angiogenesis following subcutaneous injection into the hamster cheek pouch. TGF-α is 3–10-fold more active in eliciting morphologic evidence of neovascularization and stimulating mitogenicity of capillary endothelial cells *in vivo*. By contrast, no significant differences are apparent in displacement of receptor-ligand binding by TGF-α and EGF *in vitro* or in the relative potency or maximal stimulation of mitogenesis in several cultured cell types, including pulmonary artery and lung microvascular endothelial cells. Several mechanisms could account for the quantitative differences in activity. Additional studies demonstrate that the two peptides have similar kinetics of release. However, TGF-α may have a lower sensitivity to protease digestion *in vivo,* or decreased clearance that would accentuate or prolong bioavailability. Another possibility is that EGF and TGF-α have separate receptors on the cell membrane. Previous studies have suggested that TGF-α interacts with a separate receptor on rat kidney cells,[36] but an antibody to EGF receptor prevents mitogenesis by TGF-α and EGF.[37] Perhaps preferential receptor binding or events subsequent to ligand-receptor activation such as internalization or ligand recycling may explain these disparate observations. Alternatively, indirect effects of these growth factors may be differentially regulated. Other angiogenic factors may be preferentially induced by TGF-α or cooperate with TGF-α to stimulate endothelial proliferation and neovascularization to a greater degree.

A significant finding has been the isolation of TGF-α from embryonic tissue. TGF-α may play a critical role in embryonic development or indirectly in the angiogenic, proliferative and differentiative reactions occurring in the maternal tissues that support and nourish the embryo. Initial studies isolated TGF-α-like material from mouse embryos.[38,39] More detailed analyses suggested that TGF-α was synthesized by early gestation embryos of mouse.[40] However, *in situ* hybridization subsequently localized TGF-α mRNA expression to maternal decidual tissue, and found the highest expression immediately adjacent to the embryo.[41] Angiogenesis and proliferation within decidua may be regulated by TGF-α. In addition, paracrine or endocrine effects on adjacent maternal tissues or the embryo also could be exerted by TGF-α. Further studies will be required to understand the function of TGF-α in the physiology of pregnancy.

Since initial isolation from transformed mesenchymal cells, TGF-α and TGF-β have been shown to exhibit biological activities that implicate their importance in cellular transformation and malignant neoplasia. TGF-α protein has been identified in a variety of human tumor cell lines including squamous and mammary carcinomas, melanomas and some sarcomas.

A survey of a large number of tumor cell lines and fresh tissues demonstrates that TGF-α mRNA is not transcribed in hematopoietic cell lines but is expressed in multiple cell lines derived from solid tumors, especially carcinomas.[42,43] Renal cell carcinomas and squamous cell carcinomas (head and neck, lung) preferentially express high levels of TGF-α mRNA. Melanoma and mammary carcinomas also express transcripts for TGF-α. More restricted expression is seen in sarcoma cell lines, though a fibrosarcoma and osteosarcoma cell line accumulate markedly enhanced levels of TGF-α mRNA. Representative cell lines also synthesize TGF-α protein including cell-associated and secreted species. Many of these cell lines and tumors coordinately express EGF receptor mRNA. These findings suggest that tumor cells are capable of responding to increased autocrine growth factor production by enhanced expression of the pertinent receptor for the growth factor.

A mouse mammary epithelial cell line has been developed as a tumor model to determine whether transformation by an activated C-Harvey ras proto-oncogene enhances TGF-α production and results in acquisition of growth in soft agar and tumorigenesis when injected into nude mice.[44] The results of this study show that transformation by an activated ras proto-oncogene causes increased production of biologically active TGFs. The H-ras transformed cells grow autonomously and fail to respond to exogenous EGF or TGF-α. The number and affinity of EGF receptors also is reduced, likely because of the increased capacity of these cells to secrete TGF-α. Interestingly, infection of Balb/MK-2, a mouse epidermal keratinocyte cell line, with Harvey or Kirsten murine sarcoma viruses eliminates the EGF requirement of these

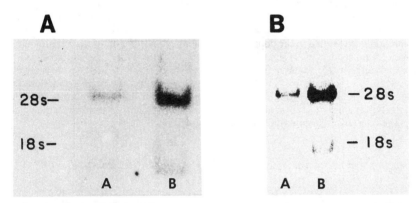

FIGURE 1. Northern blot analysis of poly-A + RNA from HK. Cells were grown to confluency in complete MCDB 153 medium and switched to medium lacking growth factors for two days. Panels A and B, lanes A: RNA isolated from HK cultured in absence of growth factors for 2 days; lanes B: RNA isolated from HK restimulated with EGF, 10 ng/ml for 4 hr (Panel A) or TGF-α, 10 ng/ml for 4 hr (Panel B). 2 μg of RNA was loaded in each lane. Hybridization with TGF-α cDNA probe and autoradiography was performed. Migration of 28s and 18s rRNA markers are indicated.

cells for adherent, monolayer clonal growth.[45] Viral ras oncogene or activated cellular ras proto-oncogenes may diminish or eliminate epithelial cell dependence on exogenous growth factors and induce tumorigenesis by enhancing production of TGF-α.

Biological activity of TGF-α has been characterized in several cell lines retrovirally transfected by TGF-α cDNA expression vectors. Nontransformed Fischer rat fibroblasts (Rat-1) cells synthesize and secrete significant amounts of TGF-α upon transfection. Loss of anchorage-dependent growth and tumor formation in nude mice results, and soft agar colony formation can be prevented by anti-TGF-α monoclonal antibodies.[46] Tumor formation occurred in 3/9 control animals and 9/9 animals inoculated with TGF-α transfected Rat-1 cells. Tumor growth was similar in both groups of tumor-bearing animals and was not rapidly progressive. Normal rat kidney (NRK) cells transfected with the rat TGF-α gene form colonies in soft agar in the presence of TGF-β.[47] NIH3T3 cells have been transfected with the TGF-α gene placed under the control of either a metallothionen promoter or a retroviral long terminal repeat. These transfectants express significant quantities of TGF-α, yet they have failed to form transformed foci. Sublines expressing high levels of TGF-α were not tumorigenic in nude mice. TGF-α expression and secretion by transfectants resulted in increased saturation density in culture that could be partially inhibited by TGF-α antibody.[48] In conjunction with the previous findings, these observations support the proposition that TGF-α is not a direct-acting oncogene but, rather, a gene product that enhances growth in culture and indirectly "promotes" growth and tumorigenicity of already malignant cells.

Further support for the role of TGF-α in tumor growth has been shown for pancreatic cancer cells that overexpress EGF receptor.[49] In each of five cell lines that exhibit markedly enhanced expression of EGF receptor, TGF-α mRNA also is synthesized. TGF-α was 10 to 100-fold more potent than EGF in promoting anchorage-independent growth at very low seeding densities. The evidence suggests that TGF-α production and concomitant overexpression of EGF receptors establishes a superagonist autocrine cycle for certain cancer cells to obtain a growth advantage.

The concept that TGF-α enhances tumorigenicity coincides with the "promotion" phase of two-stage or multistage carcinogenesis wherein the initiated, malignant cell is preferentially expanded in the population by tumor promoters, such as phorbol esters.[50] Enhanced expression of TGF-α is an event that occurs within the initiated cell or surrounding cells (if the tissue is capable of expressing TGF-α) to generate a growth advantage for the initiated cell during subsequent stages of tumor promotion and progression. Not only is availability of TGF-α important, but EGF receptor quantity and ligand-receptor activation also are closely linked elements in the overall biological response that governs neoplasia and tumor progression.

Production and Biological Activities of TGF-α: Keratinocytes and Epidermis

TGF-α is synthesized by carcinomas, but several normal cell types also produce this polypeptide. Human and murine keratinocytes and normal bovine pituitary cells in culture have been shown to secrete TGF-α.[8,9,18] These unexpected findings conclusively demonstrate that production of TGF-α is not exclusively coupled to cell transformation and neoplasia. Neonatal and adult human epidermal keratinocytes in culture synthesize and secrete TGF-α into the medium. TGF-α mRNA is expressed in cultured human keratinocytes (HK) propagated in serum-free medium containing EGF, insulin, pituitary extract and supplements.[8] Interestingly, confluent cultures of HK deprived of exogenous growth factors express low or undetectable levels of TGF-α mRNA and do not secrete detectable amounts of TGF-α into the medium (FIG. 1). EGF or TGF-α induces expression of TGF-α mRNA within several hours of addition to cultured HK. Sustained accumulation of TGF-α mRNA and synthesis of TGF-α protein is produced by continued treatment with EGF.

Autoinduction of TGF-α by HK likely serves to regulate basal-level expression of this mitogenic polypeptide in normal epidermis. Immunoreactive TGF-α is detected by several antibodies and localizes to multiple layers of the epidermis. It is not known which layers of epidermis synthesize TGF-α most actively. Furthermore, it remains to be determined whether posttranslational modifications such as glycosylation and proteolytic cleavage that create multiple cell-associated and secreted species of TGF-α are important processes that regulate cellular function and restrict biological activity to epidermis and neighboring cell types in appendages and dermis.

TGF-α and EGF are potent mitogens for clonal growth of cultured HK. Two classes of growth factors are required for efficient proliferation of HK at low seeding densities. The EGF family of growth factors (EGF or TGF-α) and insulin or insulinlike growth factor (IGF-1) both are necessary for initiating clonal development.[51] If either factor is absent, no colonies form (FIG. 2). Once a critical colony mass is generated, however, exogenous growth factors no longer are required for further propagation of the colony (unpublished data). In early stages of clonal expansion TGF-α and EGF are equipotent mitogens. EGF and TGF-α equivalency is calculated by competitive ligand-receptor binding in A431 cells or mink lung epithelial cells. A recent study has shown that large colony (megacolony) growth of HK departs from exponential, radial outgrowth in the absence of exogenous growth factors. EGF and TGF-α restore logarhythmic expansion, and TGF-α is approximately 1.3-fold more potent than EGF in this bioassay.[52] This effect appears to be related to the ability of HK located at the peripheral of the megacolony to migrate outward. In this respect, TGF-α may accelerate cell migration by improving cellular locomotion, enhancing synthesis of cell matrix or adhesion molecules or lessening cohesive associations among neighboring cells. Certainly, combinations of these factors and likely other effects interact to create this biological response. These observations may be of biological significance in

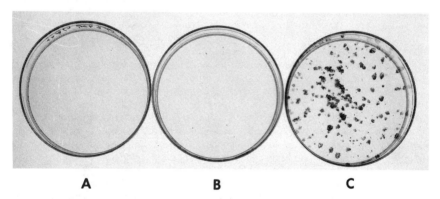

| A | B | C |

FIGURE 2. Effect of EGF and insulin on clonal growth of HK. 500 cells were seeded into standard MCDB 153 medium (lacking growth factors) and (**A**) 10 ng/ml EGF, (**B**) 5 μg/ml insulin, and (**C**) 10 ng/ml EGF + 5 μg/ml insulin. 10 days later plates were fixed, stained and photographed.

epidermal wound healing and imply that exogenous growth factors of the EGF family are able to augment endogenous TGF-α expression. These factors regulate re-epithelialization, dermal healing responses (*e.g.,* transient neovascularization) and maintain epidermal integrity.

Detection of TGF-α in maternal decidua also suggests that TGF-α serves as an important polypeptide mediator in local tissue responses during embryonic development. Levels of TGF-α in embryonic tissues, exclusive of decidua, have not been quantitated precisely. Presently, it is not known at which stage in development epidermis and possibly other ectodermal and mesenchymal tissues begin to express TGF-α. Our previous studies have demonstrated that TGF-α mRNA and protein are synthesized in newborn foreskin epidermis. If TGF-α regulates local epidermal growth, the initial, single-layered structure of presumptive epidermis (4 to 6 weeks gestation) would be expected to produce TGF-α. Other specialized embryonic structures, such as the periderm, or maternal decidual tissues may supply TGF-α to epidermis during early stages of development. Periderm and epidermis have been shown to express EGF receptors and therefore are capable of responding to TGF-α.

Neonatal and adult epidermis express EGF receptor as well.[53] Normally, EGF receptors predominate in the basal layers of epidermis and levels of receptor decrease significantly coincident with stratification and keratinization.

EGF receptor antibodies and direct visualization of ligand-receptor binding using radiolabeled EGF have shown that these receptors also are located in appendageal structures, such as eccrine gland and may function in ion transport as well as in epithelial growth and differentiation.[54] Localization of TGF-α within appendageal structures has not been thoroughly evaluated, but undoubtedly will yield important information regarding the role of this growth factor in eccrine development and function, ontogeny and cyclic regeneration of the hair germ, and formation of dermal papilla and hair.

The hair germ is composed of epithelial cells that proliferate, grow downward and envelop the dermal papilla with each cycle of the active (anagen) phase of hair development. In anagen, epithelial cells of the hair germ begin to mitose and expanding buds reinvest the dermal papilla. Onset of hair germ proliferation precedes the increase in dermal vascularization observed during the anagen phase. Enhanced production and bioavailability of TGF-α may contribute to the proliferative response and neovascularization during this phase of the hair cycle. Smaller, secreted forms of TGF-α may diffuse locally and penetrate into subjacent tissues.

Other epithelial tissues respond to TGF-α *in vivo*. Lobuloalveolar development of the mouse mammary gland *in vivo* is stimulated by local, slow-release pellets containing TGF-α.[55] The response is mediated locally since contralateral lobuloalveolar gland development does not occur. Mammary glands appear to be approximately 5-fold more sensitive to TGF-α than EGF, and TGF-α activity does not require priming the animal with steroid hormones. Maximal responses occur after 4–5 days of growth factor exposure. Possibly endogenous TGF-α acts on postnatal mammary tissue during critical periods of morphogenesis to induce development of lobuloalveolar epithelium and stroma.

Cutaneous wound healing and hyperproliferative skin disease are representative models of perturbed keratinocyte growth that undoubtedly will delineate the importance of TGF-α in epidermal physiology and pathobiology. As previously mentioned, TGF-α enhances HK outgrowth greater than EGF. Chemically synthesized TGF-α and purified VVGF have been compared to EGF in their ability to enhance re-epithelialization of partial thickness burns of pigs.[56] TGF-α and VVGF facilitate epithelial wound healing significantly greater than EGF-treated or untreated wounds. Psoriasis is an archetypal disease of benign epidermal hyperproliferation. The germinative population of epidermal keratinocytes is significantly expanded, the average duration of the epidermal cell cycle is dramatically decreased and differentiation is altered remarkably. Among the distinctive abnormalities of differentiation is the persistent expression of EGF receptors in upper layers of involved, lesional epidermis.[57] Recent evidence supports our initial hypothesis that TGF-α mRNA and protein is significantly increased in psoriatic epidermis. TGF-α misregulation and overproduction coupled to persistent expression of EGF receptor in upper epidermis may be a central mechanism that perpetuates the psoriatic pathology of epidermal hyperproliferation, altered differentiation and tortuous, dermal capillary growth.

Transforming Growth Factor Type Beta (TGF-β)

TGF-β was originally identified in conditioned medium of transformed cells. It is now well recognized that the vast majority of normal mammalian cell types and tissues produce TGF-β. Not only have the sources of TGF-β been more completely character-ized, but the structural identities and biological functions of the family of TGF-β molecules have been studied extensively.

Structural Characteristics of TGF-β and Its Receptor

TGF-β was originally purified to homogeneity from human platelets and placenta, bovine kidney and virally transformed rat cells (for reviews see REFS. 58, 59). In subsequent investigation, a family of polypeptide factors related to TGF-β has been discovered. The initial form of TGF-β to be characterized was a 25-Kd disulfide-linked homodimer. Additional forms of TGF-β have been identified and named accordingly, TGF-β_1 (formerly TGF-β) and TGF-β_2. Representative heterodimeric forms combine β_1 and β_2 subunits. Marked conservation of the polypeptide sequence among species implies that TGF-β_1 and its molecular family serve crucial functions in organisms and cells expressing these factors. Other members of the family appear less related and have more specialized functions. These include the inhibins/activins, mullerian inhibitor substance and the gene product of decapentaplegic gene complex that modulate embryogenesis and morphogenesis in drosophila.

cDNA for TGF-β_1 of several mammalian species and human TGF-β_2 have been cloned.[60,61] The gene for TGF-β_1 is transcribed into a 2.5-Kb mRNA. TGF-β_2 is transcribed into 4.1, 5.1 and 6.5-Kb mRNAs, with different size messages likely related to differential RNA splicing and alternative polyadenylation.

Several normal tissues provide abundant sources of TGF-β. Platelets store TGF-β_1 in granules and serum contains TGF-β_1 that is released from platelets during coagulation. Platelets and many cells, including keratinocytes, produce TGF-β_1 in a latent form that is irreversibly activated by acid treatment.[62] Based on cDNA analysis, the precursor is encoded within a 390 residue open reading frame where the mature TGF-β_1 subunits is encoded by residues 279–390. The precursor can be activated by acid, SDS, urea and proteolytic enzymes such as plasmin and cathepsin D.[63] Activation may occur by dissociation or proteolytic digestion from a precursor molecule or TGF-β-binding protein complex.

TGF-β_2 has been identified in porcine platelets (along with TGF-β_1), bovine bone (formerly called cartilage-inducing factor B),[64] human glioma cells,[61] a prostatic adenocarcinoma cell line (PC-3)[65] (that also produces TGF-β_1) and BSC-1 cells (formerly called growth inhibitor).[66] Recently, the term polyergin (polyfunctional regulator of growth) has been proposed for TGF-β_2.[66] TGF-β_2, like TGF-β_1, also is synthesized as a large precursor protein.[59,61]

The TGF-β receptor is different from other growth factor receptors and is widely distributed among most, but not all cell types.[67] Radioreceptor assays have demon-strated ligand-receptor dissociation constants ranging from 1 to 140 pM and receptor number per cell from 600 to 80,000.[68] At least two types of TGF-β receptors have been proposed on the basis of chemical cross-linking studies. Differential receptor affinities for TGF-β_1 and TGF-β_2 appear to exist. No direct kinase or other enzymatic activities have been reported for the TGF-β receptor to date, though protein kinase C may be activated by TGF-β in selected tissues.[69]

Overview of Biological Activities of TGF-β

TGF-β has complex effects on cell proliferation. As new biological activities are discovered, it is clear that the term TGF-β inadequately describes the multiple seemingly diverse effects of this polypeptide. Cataloging the many functions of TGF-β

and developing a unifying concept to define its actions are challenging. Proliferation or antiproliferation has been described for many normal and neoplastic cells in culture. Other biological activities unrelated to mitogenesis also are modulated by TGF-β. This paper only briefly reviews studies showing that TGF-β exerts potent effects on tissues *in vivo* and modulates biological functions at the cellular and molecular levels.

TGF-β stimulates proliferation of a variety of fibroblastic cell types in monolayer culture. Evidence suggests that the "mitogenic" activity of TGF-β is mediated indirectly via other gene products that encode defined growth factors. AKR-2B cells express *c-sis* following TGF-β treatment, and PDGF appears to be the direct mitogen for cell proliferation.[70] Based on the initial demonstration of anchorage-independent cell proliferation, TGF-β must be regarded as a bonafide growth factor. Other normal cell types such as osteoblasts and schwann cells also proliferate in response to TGF-β.[59,71] By contrast, TGF-β induces growth arrest in a variety of cultured cells including epithelial cells, T and B lymphocytes, and other mesenchymal cells such as late embryonic or adult fibroblasts.[3,58] However, growth of early embryonic fibroblasts in monolayer culture or human foreskin fibroblasts in soft agar is stimulated by the addition of TGF-β_1.[72] Interestingly, keloid-derived fibroblasts have reduced growth factor requirements and have enhanced growth response to the combination of EGF and TGF-β.[73] By contrast, TGF-β inhibits monolayer growth of normal dermal fibroblasts. Keloid-derived fibroblasts mimic early fetal fibroblast growth induced by TGF-β, and keloids may be composed of a distinct subpopulation of fibroblasts that occur in early wounds and fetal tissue. In other studies, TGF-β inhibits differentiation of adipocytes and myoblasts. However, the mitogenic activity of insulin and glucocorticoids for adipocytes is retained, suggesting that effects of TGF-β on differentiation are not strictly coupled with loss of proliferation.

In general, proliferative or antiproliferative responses exerted by TGF-β are dependent on several factors. Among these factors are the derivation and type of cell (*i.e.,* specific embryologic germ layers and mammalian species) and the maturity of the organism at the stage of cell donation (early or late fetal, neonatal or adult). Response also depends on monolayer or soft agar culture and addition of other growth factors and hormones. The degree of neoplastic transformation also determines the response to TGF-β and, in some cases, this response is coupled to induction of autocrine growth factor production.

Many of the biological effects of TGF-β are not directly linked to mitogenesis. Perhaps the most prominent responses are regulation of extracellular matrix production and degradation. It has been recognized that potent biochemical changes induced by TGF-β are translated into biological responses. Furthermore, these advances guide future research aimed at elucidating the biochemical and molecular mechanisms that initiate, transmit and modulate TGF-β signaling.

TGF-β strongly enhances formation of collagen and fibronectin and to a lesser extent other extracellular matrix proteins such as thrombospondin.[58,59,74] The type I collagen gene contains promoter sites that are activated indirectly by TGF-β. TGF-β also enhances expression of receptor sites on cells that recognize and bind extracellular matrix proteins.[75] Normal extracellular matrix is organized into a three-dimensional meshwork that is continually remodeled. Pathologic conditions such as malignancy are know to alter extracellular matrix synthesis and degradation. Proteases and protease inhibitors control metabolism of the extracellular matrix. Normal lung fibroblasts and their neoplastic counterparts produce plasminogen activator (PA) and PA-inhibitor (PA-I).[76] TGF-β causes transformed cells to transiently enhance PA-I mRNA expression while normal cells express PA-I mRNA and protein for a prolonged duration after TGF-β treatment. The net proteolytic effect on extracellular matrix is greater for the malignant cell type, creating a potential for tumor invasion and metastasis.

In vivo studies have confirmed the biological potency of TGF-β. TGF-β has been shown to regulate cell growth and enhance extracellular matrix production and accelerate healing of cutaneous wounds. For example, mammary end buds of mouse are highly mitotic, multilayered epithelia that enlarge as the ductal tree grows. TGF-β_1 reversibly inhibits bud proliferation and enlargement when released from capsules implanted adjacent to end buds.[77] Proliferation is restored upon removal of TGF-β-containing capsules. In other studies, TGF-β injected subcutaneously into newborn mice causes formation of granulation tissue.[78] Neovascularization and fibrosis is produced. Also, incisional wounds in rats heal more quickly and wound tensile strength is significantly increased by TGF-β application.[79] More pronounced cellularity (macrophages and fibroblasts) is observed in TGF-β-treated wounds. TGF-β has been shown to increase fibroblast chemotaxis in culture and may account for this response.[80]

Biological Activities of TGF-β: Keratinocytes and Epidermis

TGF-β is produced and secreted in an inactive form by normal HK and squamous carcinoma (SC) cell lines.[81] Acid-activated material from conditioned medium of HK and SC cells competes with radiolabeled TGF-β-ligand for cell receptor binding. Cultured HK express mRNA for TGF-β_1. BALB/MK, a murine keratinocyte cell line, also expresses TGF-β mRNA and secretes TGF-β in an inactive form.[9] TGF-β_1 is a potent, reversible inhibitor of HK and BALB/MK proliferation.[7,9] TGF-β at 10 ng/ml induces cell cycle-specific, growth arrest in the G_1 phase. At subconfluence, TGF-β_1-related growth arrest does not induce terminal differentiation of HK. Growth factor removal (EGF and insulin) and increase of medium calcium concentration promotes irreversible loss of proliferation and terminal differentiation.[81] These studies show that TGF-β is produced by cultured HK in an inactive form and inhibition of proliferation in HK can be reversibly exerted by TGF-β. Modulation of terminal differentiation is dependent on additional factors including cell density, growth factor and calcium concentrations.

In normal human and mouse epidermis basal levels of TGF-β mRNA, detected by Northern or *in situ* hybridization, are low or undetectable.[8,82] Following treatment with the potent, tumor-promoting phorbol ester, phorbol myristic 12,13 acetate (PMA) mouse epidermis markedly induces TGF-β_1 mRNA accumulation. The response peaks within 6 hr and decreases thereafter, though a second, but less intense, stimulation of TGF-β_1 mRNA production occurs at 48–72 hr. This was confirmed by *in situ* hybridization. TGF-β hybridization localized to suprabasal layers of epidermis. It is not known, however, if TGF-β protein is synthesized following phorbol ester stimulation or whether activation of latent TGF-β_1 occurs in epidermis. Hair follicle cells also show enhanced expression of TGF-β mRNA upon PMA stimulation. Upper follicle cells preferentially express this transcript compared to hair bulb cells. Marked induction of TGF-β by phorbol esters may be a mechanism by which epidermis is capable of producing a negative growth modulator to counteract proliferative signals triggered by TGF-α. TGF-β could dampen the intense mitogenic activity in basal and suprabasal cells produced by phorbol ester application. Within several days of a single application of PMA, epidermal growth returns to normal and TGF-β expression diminishes. Another, not mutually exclusive effect of TGF-β mRNA production may be to accelerate keratinization since this response is also observed in epidermis after PMA treatment. Regardless of this speculation, TGF-β undoubtedly plays an important role in modulation of growth and differentiation of HK and in regulation of extracellular matrix formation. Based upon *in vitro* and preliminary *in vivo* studies, a more definitive understanding of TGF-β effects on skin is expected to follow.

One example is the recent localization of TGF-β_1 protein in developing mouse embryo.[83] TGF-β_1 antibody stains mesenchyme or tissues derived from mesenchyme such as cartilage, bone and connective tissue. Staining appears greatest during mitogenesis. Intense staining is also seen in mesenchyme adjacent to developing epithelium, such as hair follicles. TGF-β_1 protein is minimal or negligible in overlying epithelium. Nonetheless, TGF-β_1 production may serve to promote mesenchymal growth and differentiation, as well as acting as an inductive, paracrine mediator in epithelial development. Adult mammalian epidermis also has the capacity to express TGF-β during stages of hyperplasia caused by phorbol esters. Conceivably, epidermal wounding or other perturbed states of growth and differentiation may also express TGF-β as part of the pathologic process or as a regulator to restore epidermal homeostasis and tissue integrity.

REFERENCES

1. DeLarco, J. E. & G. Todaro. 1978. Proc. Natl Acad. Sci. USA **75:** 4001–4005.
2. Todaro, G. J. & J. E. DeLarco. 1978. Cancer Res. **38:** 4147–4154.
3. Moses, H., R. Tucker, E. Leof, R. J. Coffey, J. Halper, G. D. Shipley. 1985. *In* Growth Factors and Transformation, Cancer Cells. J. Feramisco, B. Ozanne & C. Stiles, Eds. Vol. **3:** 59–64. Cold Spring Harbor Laboratory. New York, NY.
4. Sporn, M. B. & A. B. Roberts. 1985. Nature **313:** 745–747.
5. Goustin, A. S., E. B. Leof, G. D. Shipley & H. L. Moses. 1986. Cancer Res. **46:** 1015–1029.
6. Sporn, M. B. & G. J. Todaro. 1980. N. Engl. J. Med. **303:** 878–880.
7. Shipley, G. D., M. R. Pittelkow, J. J. Wille, Jr., R. E. Scott & H. L. Moses. 1986. Cancer Res. **46:** 2068–2971.
8. Coffey, R. J., R. Derynck, J. N. Wilcox, T. S. Bringman, A. S. Goustin, H. L. Moses & M. R. Pittelkow. 1987. Nature **328:** 817–820.
9. Coffey, R. J., N. J. Sipes, C. C. Bascom, R. Graves-Deal, C. Y. Pennington, B. E. Weissman & H. L. Moses. 1988. Cancer Res. **48:** 1596–1602.
10. Derynck, R., A. B. Roberts, M. E. Winkler, E. Y. Chen & D. V. Goeddel. 1984. Cell **38:** 287–297.
11. Marquardt, H., M. W. Humkapiller, L. E. Hood & G. J. Todaro. 1984. Science **223:** 1079–1082.
12. Winkler, M. E., T. S. Bringman & B. J. Marks. 1986. J. Biol. Chem. **261:** 13838–13843.
13. Derynck, R. 1986. J. Cell. Biochem. **32:** 293–304.
14. Derynck, R., A. Rosenthal, P. B. Linquist, T. S. Bringman & D. V. Goeddel. 1986. Cold Springs Harbor Symp. Quant. Biol. **51:** 649–655.
15. Stromberg, K., D. A. Pigott, J. E. Ranchalis & D. R. Twardzki. 1982. Biochem. Biophys. Res. Commun. **106:** 354–361.
16. Matrisian, L. M., M. Pathak & B. E. Magun. 1982. Biochem. Biophys. Res. Commun. **107:** 761–769.
17. Twardzki, D. R., J. E. Ranchalis & G. J. Todaro. 1982. Cancer Res. **42:** 590–593.
18. Samsoondar, J., M. S. Kobrin & J. E. Kudlow. 1986. J. Biol. Chem. **261:** 14408–14413.
19. Brissenden, J. E., R. Derynck & U. Franke. 1985. **45:** 5593–5597.
20. Bringman, T. S., P. B. Lindquist & R. Derynck. 1987. Cell **48:** 429–440.
21. Dart, L. L., D. M. Smith, C. A. Meyers, M. B. Sporn & C. A. Frolik. 1985. Biochemistry **24:** 5925–5931.
22. Hazarika, P., R. L. Pardue, R. Earls & J. R. Dedman. 1987. Biochemistry **26:** 2067–2070.
23. Teixido, J. & J. Masague. 1988. J. Biol. Chem. **263:** 3924–3929.
24. Ignotz, R. A., B. Kelley, R. J. Davis & J. Massague. 1986. Proc. Natl. Acad. Sci. USA **83:** 6307–6311.

25. BROWN, J. P., D. R. TWARDZIK, H. MARQUARDT & G. J. TODARO. 1985. Nature **313:** 491–492.
26. CARPENTER, G. & J. G. ZENDEGUI. 1986. Exp. Cell Res. **164:** 1–10.
27. COHEN, S. & G. CARPENTER. 1975. Proc. Natl. Acad. Sci. USA **72:** 1317–1321.
28. SMITH, J. M., M. B. SPORN, A. B. ROBERTS, R. DERYNCK, M. E. WINKLER & H. GREGORY. 1985. Nature **315:** 515–516.
29. TAM, J. P. 1985. Science **229:** 673–675.
30. TASHJIAN, A. H., JR., E. F. VOELKEL, M. LAZZANO, F. R. SINGER, A. B. ROBERTS, R. DERYNCK, M. E. WINKLER & L. LEVINE. 1985. Proc. Natl. Acad. Sci. USA **82:** 4535–4538.
31. TAKAHASHI, N., B. R. MacDONALD, J. HON, M. E. WINKLER, R. DERYNCK, G. R. MUNDY & G. D. ROODMAN. 1986. J. Clin. Invest. **78:** 894–898.
32. TASHJIAN, A. H., JR., E. F. VOEKLEL, W. LLOYD, R. DERYNCK, M. E. WINKLER & L. LEVINE. 1986. J. Clin. Invest. **78:** 1405–1409.
33. GUTIERREZ, G. E., G. R. MUNDY, R. DERYNCK, E. L. HOWLETT & M. S. KATZ. 1987. J. Biol. Chem. **262:** 15845–15850.
34. ADASHI, E. Y., C. E. RESNICK & D. R. TWARDZIK. 1987. J. Cell Biochem. **33:** 1–13.
35. SCHREIBER, A. B., M. E. WINKLER & R. DERYNCK. 1986. Science **232:** 1250–1253.
36. MASSAGUE, J., M. P. CZECH, K. IWATA, J. E. DELARCO & G. J. TODARO. 1982. Proc. Natl. Acad. Sci. USA **79:** 6822–6826.
37. CARPENTER, G., C. M. STOSCHECK, J. E. PRESTON & J. E. DELARCO. 1983. Proc. Natl. Acad. Sci. USA **80:** 5627–5631.
38. PROPER, J. J., C. L. BJORNSON & H. L. MOSES. 1980. J. Cell Physiol. **110:** 169–174.
39. TWARDZIK, D. R., J. R. RANCHALIS & G. J. TODARO. 1982. Cancer Res. **42:** 590–593.
40. TWARDZIK, D. R. 1985. Cancer Res. **45:** 5413–5416.
41. HAN, V. K., E. S. HUNTER, R. M. PRATT, J. G. ZENDEGUI & D. C. LEE. 1987. Mol. Cell. Biol. 2335–2343.
42. DERYNCK, R., D. V. GOEDDEL, A. ULLRICH, J. U. GUTTERMAN, R. D. WILLIAMS, T. S. BRINGMAN & W. H. BERGER. 1987. Cancer Res. **47:** 707–712.
43. COFFEY, R. J., JR, A. S. GOUSTIN, A. M. SODERQUIST, G. D. SHIPLEY, J. WOLFSHOL, G. CARPENTER & H. L. MOSES. 1987. Cancer Res. **47:** 4590–4594.
44. SALOMON, D. S., I. PERROTEAU, W. R. KIDWELL, J. TAM & R. DERYNCK. 1987. J. Cell. Physiol. **130:** 397–409.
45. WEISSMAN, B. & S. A. AARONSON, 1985. Mol. Cell. Biol. **5:** 3386–3396.
46. ROSENTHAL, A., P. G. LINDQUIST, T. S. BRINGMAN, D. V. GOEDDEL & R. DERYNCK. 1986. Cell **46:** 301–309.
47. WATANABE, S., E. LAZAR & M. B. SPORN. 1987. Proc. Natl. Acad. Sci. USA **84:** 1258–1262.
48. FINZI, E., T. FLEMING, O. SEGATTO, C. Y. PENNINGTON, T. S. BRINGMAN, R. DERYNCK & S. A. AARONSON. 1987. Proc. Natl. Acad. Sci. USA **84:** 3733–3737.
49. SMITH, J. J., R. DERYNCK & M. KORC. 1987. Proc. Natl. Acad. Sci. USA 7567–7570.
50. YUSPA, S. H. & M. C. POIRER. 1988. Adv. Cancer Res. **50:** 25–70.
51. WILLE, J. J., JR., M. R. PITTELKOW, G. D. SHIPLEY & R. E. SCOTT, 1984. J. Cell. Physiol. **121:** 31–44.
52. BARRANDON, Y. & H. GREEN. 1987. Cell **50:** 1131–1137
53. NANNEY, L. B., J. A. McKANNA, C. M. STOSCHECK, G. CARPENTER & L. E. KING. 1984. J. Invest. Dermatol. **82:** 165–169.
54. NANNEY, L. B., M. MAGID, C. M. STOSCHECK & L. E. KING. 1984. J. Invest. Dermatol. **83:** 385–393.
55. VONDERHAAR, B. K. 1987. J. Cell. Physiol. **132:** 581–584.
56. SCHULTZ, G. S., M. WHITE, R. MITCHELL, G. BROWN, J. LYNCH, D. R. TWARDZIK & G. J. TODARO. 1987. Science **235:** 350–352.
57. NANNEY, L. B., C. M. STOSCHECK, M. MAGID & L. E. KING. 1986. J. Invest. Dermatol. **86:** 260–265.
58. KESKI-OJA, J., E. B. LEOF, R. M. LYONS, R. J. COFFEY, JR. & H. L. MOSES. 1987. J. Cell. Biochem. **33:** 95–107.
59. SPORN, M. B., A. B. ROBERTS, L. M. WAKEFIELD & B. deCROMBRUGGHE. 1987. J. Cell. Biol. **105:** 1039–1045.

60. DERYNCK, R., J. A. JARRETT, E. Y. CHEN, D. H. EATON, J. R. BELL, R. K. ASSOIAN, A. B. ROBERTS, M. B. SPORN & D. V. GOEDDEL. 1985. Nature **316:** 701–705.
61. DEMARTIN, R., B. HAENDLER, R. HOFER-WARBINEK, H. GANGITSCH, M. WRANN, H. SCHLUSENER, J. M. SEIFERT, S. BODMER, A. FONTANA & E. HOFER. 1987. EMBO J. **6:** 3673–3677.
62. LAWRENCE, D. A., R. PIRCHER, C. KRYCEVE-MARTINERIE & P. JULLIEN. 1984. J. Cell. Physiol. **121:** 184–188.
63. LYONS, R. M., J. KESKA-OJA & H. L. MOSES. 1988. J. Cell. Biol. **106:** 1659–1665.
64. SEYEDIN, P. R., P. R. SEGARINI, D. M. ROSEN, A. Y. THOMPSON, H. BENTZ & J. GRAYCAR. 1987. J. Biol. Chem. 1946–1949.
65. LIOUBIN, M. N., T. IKEDA & H. MARQUARDT. 1987. Biochemistry **26:** 2406–2410.
66. HANKS, S. K., R. ARMOUR, J. H. BALDWIN, F. MALDONADO, J. SPIESS & R. W. HOLLEY. 1988. Proc. Natl. Acad. Sci. USA **85:** 79–82.
67. TUCKER, R. F., E. L. BRANUM, G. D. SHIPLEY, R. J. RYAN & H. L. MOSES. 1984. Proc. Natl. Acad. Sci. USA **81:** 6757–6761.
68. WAKEFIELD, L. M., D. M. SMITH, T. MASUI, C. C. HARRIS & M. B. SPORN. J. Cell. Biol. **105:** 965–975.
69. MARKOVAC, J. & G. W. GOLDSTEIN. 1988. Biochem. Biophys. Res. Commun. **150:** 575–582.
70. LEOF, E. B., J. A. PROPER, A. S. GOUSTIN, G. D. SHIPLEY, P. E. DICORLETTO & H. L. MOSES. 1986. Proc. Natl. Acad. Sci. USA **83:** 2453–2457.
71. CENTRELLA, M., T. L. MCCARTHY & E. CANALIS. 1987. J. Biol. Chem. **262:** 2869–2874.
72. HIU, D. J., A. J. STRAIN, S. F. ELSTOW, I. SWENNE & R. D. MILNER. 1986. J. Cell. Physiol. **128:** 322–328.
73. RUSSELL, S. B., K. M. TRUPIN, S. RODRIGUEZ-EATON, J. D. RUSSELL & J. S. TRUPIN. 1988. **85:** 587–591.
74. PENTTINEN, R. P., S. KOBAYASHI & P. BORNSTEIN. 1988. Proc. Natl. Acad. Sci. USA **85:** 1105–1108.
75. IGNOTZ, R. A. & J. MASSAGUE. 1987. Cell. **51:** 189–197.
76. KESKI-OJA, J., R. RAGHOW, M. SAWDEY, D. J. LOSKUTOFF, A. E. POSTLEWAITE, A. H. KANG & H. L. MOSES. 1988. J. Biol. Chem. **263:** 3111–3115.
77. GILBERSTEIN, G. B. & C. W. DANIEL. 1987. Science **237:** 291–293.
78. ROBERTS, A. B., M. B. SPORN, R. K. ASSOIAN, J. M. SMITH, N. S. ROCHE, L. M. WAKEFIELD, U. I. HEINE, L. A. LIOTTA, V. FALANGA, J. H. KEHRL & A. S. FAUCI. 1986. Proc. Natl. Acad. Sci. USA **83:** 4167–4171.
79. MUSTOE, T. A., G. F. PIERCE, A. THOMASON, P. GRAMATES, M. B. SPORN & T. F. DEUEL. 1987. Science **237:** 1333–1336.
80. POSTLETHWAITE, A. E., J. KESKI-OJA, H. L. MOSES & A. H. KANG. 1987. J. Exp. Med. **165:** 251–256.
81. WILKE, M. S., B. M. HSU, J. J. WILLE, JR., M. R. PITTELKOW & R. E. SCOTT. 1988. Am. J. Pathol. **131:** 171–181.
82. AKHURST, R. J., F. FEE & A. BALMAIN. 1988. Nature **331:** 363–365.
83. HEINE, U. I., E. F. MUNDZ, K. C. FLANDERS, L. R. ELLINGSWORTH, H.-Y. P. LAM, N. L. THOMPSON, A. B. ROBERTS & M. D. SPORN. 1987. J. Cell. Biol. **105:** 2861–2876.

Immunology and the Skin

Current Concepts

B. J. LONGLEY, I. M. BRAVERMAN,
AND RICHARD L. EDELSON

Department of Dermatology
Yale University School of Medicine
LCI 500
P.O. Box 3333
New Haven, Connecticut 06510-8059

INTRODUCTION

The skin is the main interface between the organism and the environment, the route by which we first encounter many infectious and noxious agents, and has long been a vehicle for study and manipulation of the immune system. The earliest concepts of hypersensitivity evolved through the study of skin responses, and one of the most successful campaigns of medicine, the eradication of smallpox, was accomplished using skin vaccination. Yet the skin is still regarded by many biologists as a passive barrier. We are just beginning to understand the complexity of its immune functions, but it is clear that skin cells take an active role in the processing and presentation of antigen to the central lymphoid compartments. Beyond this, the skin has many characteristics which suggest that it can function as a relatively autonomous immunologic organ.[1,2] The epidermis, the most superficial part of the skin, may have its own immune surveillance system, and may be a site of postthymic differentiation of lymphoid cells. The resident cells with known immunologic capabilities in the epidermis include antigen presenting cells, keratinocytes, and cells with many of the characteristics of T-cells. In this overview we will highlight some of the well established histologic, cellular and biochemical characteristics of the epidermis and suggest ways in which these units may function together to make the skin a unique and semiautonomous immunologic organ. Although the supporting tissue of the skin, the dermis, has the same active immunologic components found in most supporting tissues, such as endothelial cells and macrophages, we will concentrate on the cells of the epidermis which are more tissue specific.

The immunologically active cells of the epidermis can be divided into two groups: bone marrow derived cells (Langerhans cells (LC) and T-lymphocytelike cells) and keratinocytes. LCs are considered members of the macrophage lineage and are the best studied of the bone marrow derived cells resident in the epidermis.[3] They are antigen presenting cells which phagocytose foreign antigen, and carry it to the central lymphoid organs.[4] The main T-lymphocytelike cell described in normal murine skin is a dendritic cell which represents less than 5% of the total epidermal cells and which was first recognized as a Thy-1$^+$ dendritic epidermal cell (Thy-1 DEC).[5,6] Thy-1 DECs express the protein Thy-1, a surface antigen characteristic of murine T-cells, and because they express the gamma-delta T-cell antigen receptor they are also known as epidermal gamma-delta cells. The properties of the Thy-1 DEC are less well understood than those of the LC, and the equivalent cell in human epidermis has not yet been conclusively identified. The other major immunologically active epidermal cell recognized is the keratinocyte. Keratinocytes are the most abundant cells in the epidermis and have been shown to be potent producers of immunologically active cytokines *in vitro*.[9] The role of these epidermally-derived cytokines locally and systemically has not yet been elucidated, although their effects are likely to be profound: a more complete understanding of the capabilities of these cells may show important interactions between keratinocytes, bone marrow derived cells, and other cell types in health and disease.

The Thymic Epithelium and Lymphoid Development

In order to comprehend the immunologic capabilities of the skin, it is useful to review the structure and function of another immune organ, the thymus, because it shares many anatomic, biochemical, and functional properties with the epidermis. With that background, we will then review both generally accepted and controversial features of the different populations of epidermal cells. The similarity of the epidermis and thymus is underscored by the relationship between the development of normal epidermis and thymus in mice. Congenitally athymic mice are known as "nude mice" because they lack hair, a normal epidermal appendage. Genes for the normal development of the epidermis and for the athymic state appear to be identical or closely linked in these animals, because the traits have never been separated despite extensive breeding of these mice.[2]

From an immunologic viewpoint, the thymus contains at least two very important cellular components: keratinizing epithelium and the thymocytes which infiltrate it. The thymic epithelium expresses major histocompatability complex (MHC) class I and II molecules.[10,11] Although some cytokines produced by thymic epithelium *in vitro* have been identified[12] this work has lagged behind similar research with epidermal cells. However, preliminary unpublished reports by Birchall and Kupper[13] indicate that thymic epithelial cells can produce all of the cytokines produced by epidermal keratinocytes. These cytokines include interleukins (ILs) 1 and 6, and hematopoietic growth factors: granulocyte colony stimulating factor (GCSF), monocyte colony-stimulating factor (MCSF) and granulocyte-monocyte colony-stimulating factor (GMCSF).[12,13] *In vitro*, the colony-stimulating factors produced by keratinocytes act on multiple and varied target cell types. They stimulate hematopoietic cells of different lineages to proliferate and contribute to the differentiation of less mature, pluripotent bone marrow-derived cells.[9,14] It appears that they can have similar effects on hematopoietic cells *in vivo*[15] contributing to the development of the host defense system.

The other major thymic cellular component, the thymocytes, develop membrane-localized receptors which recognize specific antigens during their maturation. A brief description of this process will help explain the markers used to describe T-lymphocyte populations in the skin. In peripheral blood these T-cell receptors for antigen (TCRs) most commonly consist of an alpha-beta chain heterodimer associated with the CD3 protein complex. This ensemble recognizes antigen presented in the context of MHC residues.[16,17] A second heterodimer, composed of a gamma and a delta chain can also associate with CD3, instead of the alpha and beta chain heterodimer. Although it is believed that this gamma-delta CD3 complex can also function to recognize antigens, this hypothesis has not yet been proved.[18,19]

The alpha, beta, gamma and delta chains are polymorphic, and during thymocyte differentiation the genes encoding these polypeptides rearrange to make use of different constant, variable and joining regions. Ontologically, gene rearrangement probably occurs first in the delta gene followed by the gamma, beta, and finally alpha genes.[19] The exact relationship between gamma-delta T-cells and alpha-beta T-cells has not been determined. Most adult thymocytes are alpha-beta TCR expressing cells, but many of them have nonproductive gamma gene rearrangements. Gamma-delta T-cells are a small percentage of adult thymocytes. Since gamma-delta cells appear in fetal thymuses before alpha-beta TCR cells, and since in the mouse they do not express the L3T4 and LYT2 surface antigens associated with mature T-helper and T-suppressor phenotypes respectively, these cells have been regarded as immature or less differentiated than alpha-beta T-cells. However, these cells may be mature products of a separate lineage or limb of hematopoietic development rather than less mature cells of the alpha-beta lineage. During the thymic maturation process, cortical thymocytes express terminal deoxynucleotidyl transferase (TdT), a nuclear enzyme which adds nucleotides to ends of DNA molecules without requiring a template.[10,20] TdT may help develop the diversity of the T-cell antigen repertoire by inducing variation in the sequence encoding the antigen recognition chains.

Most human cortical thymocytes also express surface antigens of the CD1 family, but no detectable major histocompatibility complex (MHC) gene products.[10] There is a growing body of evidence that the CD1a protein has structural similarities to MHC class I molecules,[21–23] and it has been speculated that CDI molecules may temporarily replace MHC I molecules during thymocyte development.[24] No murine equivalent of CD1a has been identified, limiting our understanding of the importance of this molecule in experimental systems. As thymocytes mature and pass to the thymic medulla, CD1 and TdT expression decreases and MHC I products are expressed.

Bone Marrow-Derived Cells of the Epidermis

Like the thymus, the epidermis is a mixed tissue, composed of epithelial and bone marrow-derived cells. The ratio of bone marrow-derived cells to epithelial cells is much lower in the skin than in the thymus, but because the skin is a much larger organ than the thymus there may be a larger absolute number of these cells associated with the skin than there are associated with the thymus. Two distinct populations of bone marrow-derived cells with dendritic morphology have been identified in normal murine epidermis. These are the Thy-1[+] dendritic epidermal cell (Thy-1 DEC) and LC.[3–6] The limited knowledge available to date indicates that Thy-1 DEC express the gamma and delta TCR chains in association with CD3 and do not express MHC class II surface antigen or the L3T4 and LYT2 surface proteins found in mature T-helper and T-suppressor phenotypes.[7,8,19] Thus they resemble early fetal thymocytes. Less is known about the in vivo function of the Thy-1 DECs than their phenotypes. There are data suggesting that the Thy-1 DECs may have an important role locally in the skin. They can induce down regulation of contact hypersensitivity.[25] Such a system might balance the stimulatory effect of LCs on cutaneous hypersensitivity. Thy-1 DECs show some natural killer activity in vitro,[26,27] so it has been postulated that they act as killer cells in immune surveillance of the epidermis.[18] Since the skin is constantly exposed to the carcinogenic influences of environmental chemicals, ultraviolent irradiation, and viruses, it would be important to have an intraepithelial immune system capable of identifying and destroying tumors or infected cells before they spread to the rest of the body. Resident cells with killer functions and the capability to recognize foreign antigen, such as the TCR bearing Thy-1 DEC, could play such a role. Another possibility is that the Thy-1 DECs are immature cells which are sensitized to antigen in the skin and mature into antigen specific clones. This sort of extrathymic differentiation (maturation) has been reported in cultured Thy-1 DECs which have apparently developed a more mature T-helper phenotype in culture.[27] This is a controversial point, as the authors recognize, because of the possibility of contamination alpha-beta T-cells during isolation of the Thy-1 DECs.[28] If this sort of maturation does occur *in vivo,* the cells might persist in the skin and induce local effects or they might migrate, like LCs, as messengers to the central compartments of the immune system. Although they have been sought, TCR alpha-beta cells have not been identified in normal murine epidermis, making the former scenario less likely.

The human equivalent of the murine Thy-1 DEC has not yet been conclusively identified. The human Thy-1 molecule is expressed predominantly on neural cells and only on a small subpopulation of lymphocytes, unlike the case in mice, so it is not a helpful marker in looking for these cells.[29,30] The human gamma and delta genes and their protein products have been identified,[17–19] and one group has published preliminary reports of dendritic CD1a+, CD3+, gamma-delta cells in human epidermis.[31] Furthermore, in the skin of patients with cutaneous T-cell lymphoma, Braverman and Kupper[32] have identified dendritic cells which contain Birbeck granules and which simultaneously express CD1a and CD2. CD2 is the sheep erythrocyte receptor, the classical marker of T-lymphocytes. Using monoclonal antibodies, they have also demonstrated expression of the human alpha-beta TCR on a subset of dendritic epidermal cells in patients with this disease, showing that the human CD1a+ dendritic epidermal cell population may contain cells of the T-cell lineage. In humans, a variable percentage of the epidermal CD1a+ population contains Birbeck granules, the identifying marker of "Langerhans" cells.[33,34] Thus the coexpression of CD2 and possibly TCR in CD1a+ cells may seem paradoxical. The relationship between the CD2+, CD1a+ dendritic epidermal cells, the gamma-delta TCR cells, and human "Langerhans" cells is not clear, but it is possible that the human CD1a+ dendritic epidermal cell population performs the functions of both murine LC and Thy-1 DEC. Whether these functions would be in separate subpopulations of the CD1a+ cells or whether CD1a+ cells are pluripotent or undifferentiated awaits experimental data.

CD1a+ cells constitute from 1 to 4 percent of the total cells of the epidermis and appear to represent most of the bone marrow-derived cells of the epidermis. When removed from the skin and cultured *in vitro,* human CD1a+ cells do not proliferate and lose expression of CD1a, even in the presence of keratinocytes.[35] CD1a cells do proliferate in situ.[36] They are very rare in normal peripheral blood, and there is no evidence that CD1a+ cells preferentially home to the skin. Hence, one very plausible explanation of this system is that precursors from the peripheral blood migrate to the epidermis, produce prodigious quantities of CD1a locally, and may proliferate further *in situ.* An observation supporting this possibility is the finding that cells from one patient with cutaneous T-cell lymphoma have been induced to express CD1a when cocultivated with epidermal cells.[38] Therefore, like the thymic microenvironment, the epidermis appears to take an active role in the induction of the thymocytelike CD1a phenotype. Furthermore, in both murine and human systems, cocultivation of normal autologous peripheral blood lymphocytes with keratinocytes (the mixed lymphocyte epidermal reaction, MLER) induces TdT expression by a variable percentage of the lymphocytes.[38] Since TdT appears to be involved in rearrangement of TCR genes during thymocyte differentiation, this observation suggests that keratinocytes can induce genotypic as well as phenotypic changes and supports the thesis of extrathymic differentiation.

Beyond the similarities between epidermis and thymus, the skin has cells with a well defined role as antigen presenting cells. These cells, LCs, are clearly established as the farthest outposts of the afferent arm of the immune system, existing only 50 microns or so from the external environment.[3,39] They represent 2 to 4 percent of epidermal cells and are capable of stimulating a systemic immune response to antigens encountered in the skin. In mice, LCs are a distinct population which have cytoplasmic Birbeck granules demonstrable by electron microscopy, and express MHC class II (Ia) but not the Thyl antigen or CD3 and its associated TCR chains.[39,27,28] Their ability to sensitize lymphocytes has been demonstrated *in vitro:* if exposed to GMCSF[40] they mature into the most potent antigen presenting cells known.[41,42] In guinea pigs, these cells have been shown to carry cutaneously applied antigen to regional lymph nodes.[4]

As previously discussed, the lineage of human LCs may be intertwined with a human equivalent of the Thy-1[+] dendritic epidermal cells. Like murine LCs, human LCs contain cytoplasmic Birbeck granules but they also express the surface antigens CD1a and CD4, and the MHC class II antigens: HLA-DR, DP, and DQ.[33,34,43,44] In addition Braverman and Kupper have shown that they may coexpress CD1a and CD2 in cutaneous T-cell lymphoma.[32] MHC class II antigens probably bind antigen for presentation by these human LCs,[16] and it is attractive to postulate the CD1a molecules may also directly contribute to antigen presentation.[23] Because human LCs have not been successfully propogated *in vitro*, their functions are less well characterized than those of murine LCs. Removal of human LCs from epidermal suspensions using anti-CD1a or HLA-DR antibody and complement decreases the proliferative response of autologous peripheral blood lymphocytes in the MLER.[45] The distribution of human LC has been found by immunohistochemical studies to vary in contact hypersensitivity reactions[46] and in different disease states, particularly leprosy[47,48] and HIV infection associated dermatoses.[49] Because of their experimentally well defined role as antigen presenting cells, it seems likely that the LC have a central role in the pathogenesis of these inflammatory skin conditions.

Keratinocytes and Cytokines

Besides having lymphoid populations, the skin and thymus both have keratinizing epithelium. Keratinocytes produce a number of immunologically active peptides *in vitro,* and there is increasing evidence that they produce them *in vivo* as well.[15] One of the first epidermal hormones to be identified was called thymopoietin, because it was first isolated from thymus.[50] Thymopoietin regulates early stages of T-cell differentiation.[51] At approximately the same time, keratinocytes were shown to produce a T-cell stimulating substance[52,53] which was subsequently shown to be identical to interleukin-1.[54] It soon became clear that keratinocytes can produce a number of cytokines *in vitro* and probably *in vivo*.[9] Two of these polypeptides are identical to IL-1 alpha and IL-1 beta produced by macrophages. Another message, with 20–30% sequence identity to IL-1 has been isolated from keratinocytes, and been called IL-1 kappa, but a corresponding protein has not yet been identified.[55] Keratinocyte-derived IL-1 activity may function mainly as an autocrine growth factor for keratinocytes and as an inducer of inflammatory response to tissue injury; however, systemic effects and effects on the bone marrow-derived cells resident in the skin are possible.[56] Keratinocytes do not appear to make IL-2 or gamma interferon, products of the TH 1 (T-helper inducer) CD4+ T-lymphocyte subgroup, but they may make most other interleukins including IL-3 and IL-6, beta interferon, as well as colony stimulating factors MCSF, GCSF and GMCSF.[13] Since these cytokines stimulate growth and differentiation of hematopoietic cells *in vitro*,[9,13] Kupper[57] has proposed that they promote lymphocyte growth and maturation in the skin.

Overall these findings suggest that development of CD3 expressing "T-lymphocytes" can be induced by the epidermal microenvironment, either postthymically or perhaps even independently of the thymus. If this hypothesis is proved, the name "T-lymphocytes" meaning "thymus-derived lymphocyte" would not be accurate for these cells. Similar cells expressing the CD3 associated gamma-delta heterodimer in the gut epithelium have been called intraepithelial lymphocytes (IELs).[58] Since the thymus is also an epithelial organ, the term IEL could also be applied to thymocytes.

Regardless of the nomenclature, thymocytes and IELs are distinct from B-lymphocytes which do not require an epithelial microenvironment and which do not express CD3 associated heterodimers.

CONCLUSIONS

In summary, the skin is a unique immunologic organ populated by potent immunostimulatory cells capable of initiating a systemic response to cutaneously applied antigens. The epidermis has most of the cellular and hormonal components present in the thymus and may be involved in extrathymic maturation of lymphoid cells. T-cells normally present in the skin may provide an *in situ* surveillance system which guards against the development of malignancies and prevents the spread of infection beyond the confines of the epidermis.

REFERENCES

1. STREILEIN, J. W. 1983. Skin-associated lymphoid tissues (SALT): Origins and functions. J. Invest. Dermatol. **80:** 12s–15s.
2. EDELSON, R. & J. FINK. 1985. The immunologic function of skin. Sci. Am. **252:** 46–53.
3. WOLFF, K. & G. STINGL. 1983. The Langerhans cell. J. Invest. Dermatol. **80:** 175–215.
4. SILBERBERG-SINAKIN, I., G. J. THORBECKE, R. I. BAER, S. A. ROSENTHAL & V. BEREZOWS-KY. 1976. Antigen-bearing Langerhans cells in skin, dermal lymphatics and in lymph nodes. Cell. Immunol. **25:** 137–151.
5. TSCHACHLER, E., G. SCHULER, J. HUTTERER, H. LEIBL, K. WOLFF & G. STINGL. 1983. Expression of Thy-1 antigen by murine epidermal cells. J. Invest. Dermatol. **81:** 282–285.
6. BERGSTRESSER, P. R., R. E. TIGELAAR, J. H. DEES & J. W. STREILEN. 1983. Thy-1 antigen-bearing dendritic cells populate murine epidermis. J. Invest. Dermatol. **81:** 286–288.
7. KUZIEL, W. A., A. TAKASHIMA, M. BONYHADI, P. R. BERGSTRESSER, J. P. ALLISON, R. E. TIGELAAR & P. W. TUCKERS. 1987. Regulation of T-cell receptor gamma-chain RNA expression murine Thy-1+ dendritic epidermal cells. Nature **328:** 263–266.
8. STINGL, G., F. KONONG, H. YAMADA, W. M. YOKOYAMA, E. TSCHACHLER, J. A. BLUESTONE, G. STEINER, L. A. SAMELSON, A. M. LEW, J. E. COLIGAN & E. M. SHEVACH. 1987. Thy-1+ dendritic epidermal cells express T3 antigen and the T-cell receptor gamma chain. Proc. Natl. Acad. Sci. USA **84:** 4586–4590.
9. KUPPER, T. S. 1988. Interleukin 1 and other human keratinocyte cytokines: Molecular and functional characterization. Adv. Dermatol. **3:** 293–308.
10. JANOSSY, G., J. A. THOMAS, F. J. BOLLUM, S. GRANGER, G. PIZZOLO, K. F. BRADSTOCK, L. WONG, A. MCMICHAEL, K. GANESHAGURU & A. V. HOFFBRAND. 1980. The human thymic microenvironment: An immunohistologic study. J. Immunol. **125:** 202–212.
11. HSU, S. & E. S. JAFFE. 1985. Phenotypic expression of T lymphocytes in thymus and peripheral lymphoid tissues. Am. J. Pathol. **121:** 69–78.
12. LEE, T. P., J. KURTZBERG, S. J. BRANDT, J. E. NEIDEL, B. F. HAYNES & K. H. SINGER. 1988. Human thymic epithelial cells produce granulocyte and macrophage colony-stimulating factors. J. Invest. Dermatol. **90:** 580.
13. BIRCHAL, N. & T. KUPPER. Personal communication.
14. SIEFF, C. A. 1987. Hematopoietic growth factors. J. Clin. Invest. **79:** 1549–1557.
15. CLARK, S. C. & R. KAMEN. 1987. The human hematopoietic colony-stimulating factors. Science **236:** 1229–1237.
16. ALLEN, P. M., G. R. MATSUEDA, R. J. EVANS, J. B. DUNBAR, JR., G. R. MARSHALL & E. R. UNANUE. 1987. Identification of the T-cell and Ia contact residues of a T-cell antigenic epiope. Nature **327:** 713–715.
17. CLEVERS, H., B. ALARCON, T. WILEMAN & C. TERHORST. 1988. The T cell receptor/CD3 complex: A dynamic protein ensemble. Annu. Rev. Immunol.

18. JANEWAY, C. A. JR., B. JONES, A. HAYDAY. 1988. Specificity and function of T cells bearing gamma:delta receptors: An hypothesis. Immunology Today **9:** 73–76.

19. CHIEN, Y. H., M. IWASHIMA, D. A. WETTSTEIN, K. B. KAPLAN, J. F. ELLIOTT, W. BORN & M. M. DAVIS. 1987. T-cell receptor delta gene rearrangements in early thymocytes. Nature **330:** 24–31.

20. BOLLUM, F. J. 1979. Terminal deoxynucleotidyl transferase: Biological studies. Adv. Enzymol. **203:** 347–374.

21. COTNER, T., H. MASHIMO, P. KUNG, P. GOLDSTEIN & J. STROMINGER. 1981. Human T cell surface antigens bearing a structural relationship to HLA antigens. Proc. Natl. Acad. Sci. USA **78:** 3858–3862.

22. VAN DE RIJN, M., P. G. LERCH, R. W. KNOWLES & C. TERHORST. 1983. The thymic differentiation markers T6 and M241 are two unusual class I antigens. J. Immunol. **131:** 851–855.

23. LONGLEY, J., J. KRAUS, M. ALONSO & R. EDELSON. Molecular cloning of CD1a (T6) a human epidermal dendritic cell marker related to class I MHC molecules. J. Invest. Dermatol. In press.

24. GAMBON, F., M. KREISLER & F. DIAZ-ESPADA. 1988. Correlated expression of surface antigens in human thymocytes. Evidence of class I HLA modulation in thymic maturation. Eur. J. Immunol. **18:** 153–159.

25. CRUZ, P. D., JR., J. NIXON-FULTON, R. E. TIGELAAR & P. R. BERGSTRESSER. 1988. UV-irradiated Ia+ and Thy-1+ epidermal cells induce down-regulation of contact hypersensitivity. J. Invest. Dermatol. **90:** 552.

26. ROMANI, N., G. STINGL, E. TSCHACHLER, W. D. WITMER, R. M. STEINMAN, E. M. SHEVACH & G. SCHULER. 1985. The Thy-1 bearing cell of murine epidermis: a distinctive leukocyte perhaps related to natural killer cells. J. Exp. Med. **161:** 1368–1383.

27. CAUGHMAN, S. W., S. M. BREATHNACH, S. O. SHARROW, D. A. STEPHANY & S. I. KATZ. 1986. Culture and characterization of murine dendritic Thy-1+ epidermal cells. J. Invest. Dermatol. **86:** 615–624.

28. STINGL, G., K. C. GUNTER, E. TSCHACHLER, H. YAMADA, R. I. LECHLER, W. M. YOKOYAMA, G. STEINER, R. N. GERMAIN & E. M. SHEVACH. 1987. Thy-1+ dendritic epidermal cells belong to the T-cell lineage. Proc. Natl. Acad. Sci. USA **84:** 2430–2434.

29. WILLIAMS, A. F. & J. GAGNON. 1982. Neuronal cell Thy-1 glycoprotein; homology with immunoglobulin. Science **216:** 696–703.

30. McKENZIE, J. L., & J. W. FABRE. 1981. Human Thy-1: Unusual localization and possible functional significance in lymphoid tissues. J. Immunol. **126:** 843–849.

31. GROH, V., H. YOKOZEKI, W. M. YOKOYAMA, C. A. FOSTER, F. KONING, M. B. BRENNER & G. STINGL. 1988. T-Cell receptors on dendritic cells of the human epidermis. J. Invest. Dermatol. **90:** 565 and personal communications.

32. BRAVERMAN, I. & T. KUPPER. 1988. Both the epidermal dendritic cell population and the majority of epidermotropic cells in cutaneous T-cell lymphoma (CTCL) bear both CD1 and T-cell markers. J. Invest. Dermatol. **90:** 549 and personal communications.

33. FITHIAN, E., P. KUNG, G. GOLDSTEIN, M. RUBENFELD, C. FENOGLIO & R. EDELSON. 1981. Reactivity of Langerhans cells with hybridoma antibody. Proc. Natl. Acad. Sci. USA **78**(4): 2451–2454.

34. TAKEZAKI, S., S. L. MORRISON, C. L. BERGER, G. GOLDSTEIN, A. C. CHU & R. L. EDELSON. 1982. Biochemical characterization of a differentiation antigen shared by human epidermal Langerhans cells and cortical thymocytes. J. Clin. Immunol. **2:** 128S–134S.

35. CZERNICLEWSKI, J., M. DOMARCHEZ & M. PRUNIEVOS. 1984. Human Langerhans cells in epidermal cell culture, *in vitro* skin explants and skin grafts onto "nude" mice. Arch. Dermatol. Res. **276:** 288–292.

36. CZERNIELEWSKI, J., P. VAIGOT & M. PRUNIERAS. 1985. Epidermal Langerhans cells—a cycling cell population. J. Invest. Dermatol. **84:** 424–426.

37. CHU, T., C. BERGER, J. MORRIS & R. EDELSON. 1987. Induction of an immature T-cell phenotype in malignant helper T cells by cocultivation with epidermal cell cultures. J. Invest. Dermatol. **89:** 358–361.

38. RUBENFELD, M. R., A. E. SILVERSTONE, D. M. KNOWLES, J. P. HALPER, A. DO SOSTOA,

C. M. FENOGLIO & R. L. EDELSON. 1981. Induction of lymphocyte differentiation by epidermal cultures. J. Invest. Dermatol. **77:** 221–224.

39. STINGL, G., S. I. KATZ, E. M. SHEVACH, A. S. ROSENTHAL & I. GREEN. 1978. Analogous functions of macrophages and Langerhans cells in the initiation of the immune response. J. Invest. Dermatol. **71:** 59–64.

40. WITMER-PACK, M. D., W. OLIVIER, J. VALINSKY, G. SCHULER & R. M. STEINMAN. 1987. Granulocyte/macrophage colony-stimulating factor is essential for the viability and function of cultured murine epidermal Langerhans cells. J. Exp. Med. **166:** 1484–1498.

41. SCHULER, G. & R. M. STEINMAN. 1985. Murine epidermal Langerhans cells mature into potent immunostimulatory dendritic cells *in vitro*. J. Exp. Med. **161:** 526–546.

42. INABA, K., G. SCHULER, M. D. WITMER, J. VALINSKY, B. ATASSI & R. M. STEINMAN. 1986. Immunologic properties of purified epidermal Langerhans cells. J. Exp. Med. **164:** 605–613.

43. RALFKIAER, E., H. STEIN, T. PLESNER, K. HOUS-JENSEN & D. MASON. 1984. *In situ* immunological characterization of Langerhans cells with monoclonal antibodies: Comparison with other dendritic cells in skin and lymph nodes. Virchows Arch. A. **403:** 401–412.

44. SONTHEIMER, R. D., P. STRASTNY & G. NUNEZ. 1986. HLA-D region antigen expression by human epidermal Langerhans cells. J. Invest. Dermatol. **87:** 707–710.

45. CZERNIELEWSKI, J. M., D. SCHMITT, M. R. FAURE & F. THIVOLET. 1983. Functional and phenotypic analysis of isolated human Langerhans cells and indeterminate cells. Br. J. Dermatol. **108:** 129–138.

46. SILBERBERG, I. 1971. Ultrastructural studies of Langerhans cells in contact sensitive and primary irritant reactions to mercuric chloride. Clin. Res. **19:** 715.

47. LONGLEY, J., A. HAREGEWOIN, T. YEMANEBARHAN, T. WARNDORFF VAN DIEPEN, J. NSIBAMI, D. KNOWLES, K. A. SMITH & T. GODAL. 1985. *In vivo* responses to mycobacterium leprae: Antigen presentation interleukin-2 production, and immune cell phenotypes in naturally occurring leprosy lesions. Int. J. Lepr. **53:**(3): 385–394.

48. LONGLEY, J., A. HAREGEWOIN, W. DE BEAUMONT, K. A. SMITH & T. GODAL. 1986. Lepronin stimulates interleukin-2 production and interleukin-2 receptor expression *in situ* in lepromatous leprosy patients. Symposium on the Immunology of Leprosy, Oslo. Lep. Rev. **57**(Suppl. 2): 189–198.

49. GIELEN, V., D. SCHMITT, C. DEZUTTER-DAMBUYANT, J. F. NICOLAS & J. THIVOLET. 1987. AIDS and Langerhans cells: CD1, CD4 and HLA class II antigen expression. J. Invest. Dermatol. **89:** 324.

50. CHU, A. C., G. GOLDSTEIN, J. A. K. PATTERSON, C. L. BERGER, S. TAKESAKI & R. EDELSON. 1982. Thymopoietin-like substance in human skin. Lancet **8301:** 766–767.

51. SCHEID, M. P., G. GOLDSTEIN & E. A. BOYSE. 1978. The generation and regulation of lymphocyte populations. J. Exp. Med. **147:** 1727–1740.

52. LUGER, T. A., B. M. STADLER, S. I. KATZ & J. J. OPPENHEIM. 1981. Epidermal cell derived thymocyte activating factor. J. Immunol. **127:** 1493–1498.

53. SAUDER, D. N., C. CARTER, S. I. KATZ & J. J. OPPENHEIM. 1982. Epidermal cell production of thymocyte activation factor (ETAF). J. Invest. Dermatol. **79:** 34–39.

54. KUPPER, T. K., D. BALLARD, A. O. CHUA *et al.* 1986. Human keratinocytes contain MRNA indistinguishable from monocyte interleukin 1 and mRNA: Keratinocyte epidermal cell-derived thymocyte activating factor is identical to interleukin 1. J. Exp. Med. **164:** 2095.

55. BELL, T. V., C. B. HARLEY, D. STETSKO & D. N. SAUDER. 1987. Expression of mRNA homologous to interleukin-1 in human epidermal cells. J. Invest. Dermatol. **88:** 375–379.

56. GILCREST, B. A. & D. N. SAUDER. 1984. Autocrine growth stimulation of human keratinocytes by epidermal cell derived thymocyte activating factor (ETAF): Implications for cellular aging. Clin. Res. **32:** 613.

57. KUPPER, T. S. 1988. The role of epidermal cytokines. *In* The Immunophysiology of Cells and Cytokines. E. Shevach & J. Oppenheim, Eds. Oxford Univ. Press, New York, NY. In press.

58. GOODMAN, T. & L. LEFRANCOIS. 1988. Expression of the gamma-delta T-cell receptor on intestinal CD8+ intraepithelial lymphocytes. Nature **30:** 855–858.

Amino Acid Sequence of Thymopoietin Isolated from Skin

TAPAN K. AUDHYA AND GIDEON GOLDSTEIN

Immunobiology Research Institute
Route 22 East
P.O. Box 999
Annandale, New Jersey 08801-0999

INTRODUCTION

Thymopoietin is a linear polypeptide hormone secreted by epithelial cells of the thymus.[1,2] The complete amino acid sequence has been determined for bovine (49 amino acids) and human (48 amino acids) thymopoietin[3,4] and these show extensive homology including identity at residues 32–36. The synthetic pentapeptide corresponding to these amino acids (thymopentin, Arg-Lys-Asp-Val-Tyr) appears to be the active site in that it has the biological activities of the native polypeptide.[5]

Thymopoietin is pleiotropic in that it has demonstrable activities on the nicotinic cholinergic neuromuscular transmission,[6,7] early T cell differentiation,[8,9] functions of mature T cells,[10–12] and certain endocrine secretions.[13] The main emphasis in studies of thymopoietin/thymopentin has been on immunoregulatory effects, as evidenced in both experimental and clinical studies.[14–17]

An immunoassay to thymopoietin detected thymopoietin in thymus extracts but also detected immunoreactive material in spleen and lymph node extracts (skin extracts were not tested in these early studies).[4] The thympoietin-reactive materials in spleen and lymph node have been purified and amino acid sequences determined. The thympoietin substance in spleen was originally termed splenin[4] but we now rename it thysplenin to avoid confusion with the previously described splenic extract splenin.[18] Thysplenin is identical to thymopoietin except for two amino acid changes including one at position 34 in the active site, and this change alters the functional repertoire of the molecule.[19,20] The amino acid sequence of the thymopoietin-reactive substance in lymph node, termed thylymphin, is not yet completed, but already reveals a pattern of stretches of similarity and stretches of disimilarity of sequence with thymopoietin and thysplenin. Thus, there appear to be at least three distinct but related genes encoding thymopoietin, thysplenin and thylymphin.

The presence of thymopoietin in skin was first suggested from functional experiments in which supernatants from keratinocyte cultures were shown to induce differentiation of T cells in a manner similar to thymopoietin.[21] Subsequent studies utilizing antibodies to thymopoietin established that thymopoietin-reactive material was present in basal keratinocytes,[22] although these studies did not establish whether the material was thympoietin, thysplenin, thylymphin or yet another related gene product. We now provide experimental evidence, culminating in isolation and amino acid sequencing, that the material extracted from basal keratinocytes is identical to thymopoietin extracted from the thymus.

233

MATERIALS AND METHODS

Materials

Fresh fetal bovine skin was obtained on ice from Max Cohen Inc., Livingston, NJ.

Radioimmunoassay of Thymic Thymopoietin (TP) and Skin Thymopoietin (skTP)

The purification of skTP was monitored by using a RIA for TP as described previously.[4]

Isolation of skTP

The isolation methods used are summarized as a flow sheet in TABLE 1. Briefly, 20 kg of fetal bovine skin was extracted in 100 mm ammonium bicarbonate, as used previously to extract thysplenin.[4] Subsequent steps involved gel filtration on TSK 200 SW column (21.5 × 600 mm) (Hewlet-Packard, Avondale, PA), hydrophobic chromatography on Synchropak H-propyl column (Biorad, Richmond, CA) and anion exchange by FPLC on MonoQ HR 5/5 column (Pharmacia, Piscataway, NJ).

Amino Acid Sequence

Automated sequence analysis was performed using a model 470A gas-phase sequencer (Applied Biosystems, Foster City, CA) with Polybrene as carrier and a

TABLE 1. Flow Sheet of Thymopoietin Isolation from Fetal Bovine Skin

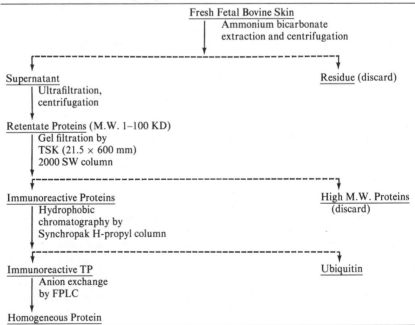

TABLE 2. Yields of Thymopoietin Isolated from Bovine Thymus and Skin

Purification Step	Total Protein	Immunoreactive Protein		Fold Purification	Recovery
		mg	%		
Thymic extract					
Ultrafiltered protein	262.5 g	162.4	0.1	1	100.0
Gel filtration	13.9 g	87.3	0.6	11	53.8
Hydrophobic chromatography	63.6 mg	28.7	45.1	752	17.7
Preparative FPLC	24.2 mg	23.1	95.2	1587	14.2
Skin extract					
Ultrafiltered protein	68.8 g	309.5	0.5	1	100.0
Gel filtration	2.9 g	76.9	2.6	60	25.0
Hydrophobic chromatography	181.2 mg	32.6	18.0	400	11.0
Preparative FPLC	1.4 mg	1.3	93.0	2070	0.4

standard single-coupling single-cleavage program, as described previously.[4] C-terminal analysis was performed using carboxy peptidase Y, as described previously.[4] Amino acid analysis was performed with a Liquimat III amino acid analyzer (Liquimat Corp., Boston, MA) as described previously.[4]

Neuromuscular Assay

Mice were injected intraperitoneally and tested electromyographically 18–24 hours later, as described previously.[23]

Cyclic GMP Assay

CEM cells were grown as described previously and incubated with peptide for two minutes. The process was arrested by addition of 10 percent TCA and cyclic GMP was extracted and measured by RIA as described previously.[12]

RESULTS

Purfication

Twenty kg (wet weight) of skin was extracted to yield 68.8 gm of soluble protein of which 309.5 mg (0.45 percent) was immunoreactive thymopoietin (TABLE 1). This yielded 1.3 mg of purified skTP after isolation. From these data 20 kg (wet weight) of skin contains 0.002 percent of skTP. The yields from skin and thymus are contrasted in TABLE 2. While skin actually contained more thymopoietin than thymus per 20 kg wet weight (309.5 mg versus 162.4 mg) the yields of skTP during isolation were lower.

Amino Acid Sequence

The amino acid sequence of thymopoietin isolated from skin was identical to that of thymopoietin isolated from thymus (FIG. 1). The first 37 amino acids were determined

	1	2	3	4	5	6	7	8	9	10	11	12	13	14	15	16	17	18	19	20	21	22	23	24	25
TSP	PRO	GLU	PHE	LEU	GLU	ASP	PRO	SER	VAL	LEU	THR	LYS	GLU	LYS	LEU	LYS	SER	GLU	LEU	VAL	ALA	ASN	ASN	VAL	THR
TP	PRO	GLU	PHE	LEU	GLU	ASP	PRO	SER	VAL	LEU	THR	LYS	GLU	LYS	LEU	LYS	SER	GLU	LEU	VAL	ALA	ASN	ASN	VAL	THR
sk TP	PRO	GLU	PHE	LEU	GLU	ASP	PRO	SER	VAL	LEU	THR	LYS	GLU	LYS	LEU	LYS	SER	GLU	LEU	VAL	ALA	ASN	ASN	VAL	THR

	26	27	28	29	30	31	32	33	34	35	36	37	38	39	40	41	42	43	44	45	46	47	48	49
TSP	LEU	PRO	ALA	GLY	GLU	GLN	ARG	LYS	GLU	VAL	TYR	VAL	GLU	LEU	TYR	LEU	GLN	HIS	LEU	THR	ALA	LEU	LYS	ARG
TP	LEU	PRO	ALA	GLY	GLU	GLN	ARG	LYS	ASP	VAL	TYR	VAL	GLU	LEU	TYR	LEU	GLN	SER	LEU	THR	ALA	LEU	LYS	ARG
sk TP	LEU	PRO	ALA	GLY	GLU	GLN	ARG	LYS	ASP	VAL	TYR	VAL	GLU	LEU	TYR	LEU	GLN	SER	LEU	THR	ALA	LEU	LYS	ARG

FIGURE 1. Amino acid sequences of bovine thysplenin (TSP) and thymic (TP) and skin (skTP) thymopoietin.

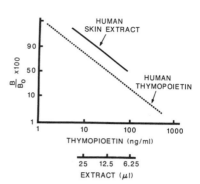

FIGURE 2. Radioimmunoassay for human thymopoietin showing parallel displacement curves for human thymopoietin isolated from thymus and a partially purified extract of human skin.

from gas-phase amino acid sequencing. Amino acids 43–49 were determined from C-terminal analysis. The amino acid compositions were identical so that we infer that residues 38–42 are also identical.

Radioimmunoassay

FIGURE 2 shows the displacement curves for human thymopoietin and an extract of human skin. The displacement curves are parallel.

Neuromuscular Assay

Skin thymopoietin caused a detectable change in neuromuscular transmission at a threshold dose of 0.1 μg/mouse as is found with thymic thymopoietin (TABLE 3).

Cyclic GMP Assay

Skin thymopoietin induced an intracellular elevation of cyclic GMP in CEM cells at a threshold dose of 0.1 μg/ml as is found with thymic thymopoietin (TABLE 3).

DISCUSSION

The present studies firmly establish that the substance in basal keratinocytes reactive with antibodies to thymopoietin is indeed identical to thymopoietin isolated

TABLE 3. Identity of Skin and Thymic Thymopoietin in Functional Assays and Contrast with Thysplenin

	Thysplenin	Thymic Thymopoietin	Skin Thymopoietin
Neuromuscular assay	−	+	+
Induction prothymocyte differentiation	+	+	+
cGMP stimulation in CEM T cell line	−	+	+

from the thymus. The amino acid sequence proved to be identical, confirming earlier studies showing antigenic and functional similarity in immunoassays and tests on neuromuscular transmission and cyclic GMP responsiveness of a thymopoietin-responsive human T cell line. The amounts of thympoietin contained per wet weight of tissue was greater in skin than thymus. This may be due to the fact that in young animals thymopoietin containing epithelial cells are only a minor component of the thymus which consists mainly of thymocytes and early T cells.

The morphological and tinctorial similarity of keratinocytes and certain epithelial cells in thymus has been recognized for a long time,[24] but it is nevertheless surprising and somewhat puzzling to find that basal keratinocytes produce thymopoietin. Neonatal thymectomy in mice produces a runting syndrome due to failure of development of the immune system[25]; prothymocytes cannot migrate to the thymus and undergo their normal differentiation to thymocytes and T cells within the thymus under the influence of thymopoietin. While this may be ascribed to the anatomical lack of the thymic epithelioreticulum it is clear that the thymopoietin in skin does not substitute for the lack of thymic thymopoietin. In adult animals the T cell system is established and thymectomy produces a series of changes related to lack of thymopoietin, since administration of thymopentin (Arg-Lys-Asp-Val-Tyr), an active synthetic frequent of thymopoietin, can largely restore these changes. These thymectomy-induced alterations include facilitation of neuromuscular transmission,[26,27] decrease of Thy-1$^+$ and Ly 1,2,3$^+$ T cells[28] and enhanced skin graft rejection based on the HY antigen in female C3H mice.[20]

All these experiments lead to the implication that circulating thymopoietin is largely produced by the thymus, with no significant contribution from basal keratinocytes in skin, despite the large amount of tissue this must represent. It appears, therefore, that the thymopoietin of basal keratinocytes may have a local apocrine role or be involved in modulating local immune reactions in the skin, a major first line of defense against the external environment. It is also of interest to note the association of hereditary athymia with lack of hair ("nude" mouse, nu/nu)[29] and speculate whether this may involve abnormalities of thymopoietin in the skin. Studies are in progress to attempt to delineate more precisely the apocrine or immunoregulatory role(s) of thymopoietin in skin.

SUMMARY

The isolation of thymopoietin-reactive material in fetal bovine skin was monitored by means of a radioimmunoassay to thymopoietin. The amino acid sequence of this material was determined to be identical with that of thymopoietin isolated from the thymus.

Experimental evidence suggests that thymopoietin in the circulation derives from the thymus and not from the skin, suggesting that the thymopoietin in keratinocytes has a local function, either apocrine and/or immunoregulatory.

ACKNOWLEDGMENTS

We thank Paula Feath for typing and Marilyn Sanders for editing the manuscript and Ron King, James Chen and Gary Campbell for expert technical assistance.

REFERENCES

1. GOLDSTEIN, G. 1974. Isolation of bovine thymin: A polypeptide hormone of the thymus. Nature 247: 11–14.

2. VIAMONTES, G. I., T. AUDHYA & G. GOLDSTEIN. 1986. Immunohistochemical localization of thymopoietin with an antiserum to synthetic cys-thymopoietin 28–39. Cell. Immunol. 100: 305–313.

3. AUDHYA, T., D. H. SCHLESINGER & G. GOLDSTEIN. 1981. Complete amino acid sequences of bovine thymopoietins I, II and III: Closely homologous polypeptides. Biochemistry 20: 6195–6200.

4. AUDHYA, T., D. SCHLESINGER & G. GOLDSTEIN. 1987. Isolation and complete amino acid sequence of human thymopoietin and splenin. Proc. Natl. Acad. Sci. USA 84: 3545–3591.

5. GOLDSTEIN, G., M. P. SCHEID, E. A. BOYSE, D. H. SCHLESINGER & J. VAN WAUWE. 1979. A synthetic pentapeptide with biological activity characteristic of the thymic hormone thymopoietin. Science 204: 1309–1310.

6. VENKATASUBRAMANIAN, K., T. AUDHYA & G. GOLDSTEIN. 1986. Binding of thymopoietin to acetylcholine receptor. Proc. Natl. Acad. Sci. USA 83: 3171–3174.

7. REVAH, F., C. MULLE, C. PINSET, T. AUDHYA, G. GOLDSTEIN & J.-P. CHANGEUX. 1987. Calcium dependent effect of thymopoietin on desensitization of the nicontinic acetylcholine receptor. Proc. Natl. Acad. Sci. USA 84: 3477–3487.

8. BASCH, R. S. & G. GOLDSTEIN. 1974. Induction of T cell differentiation in vitro by thymin, a purified polypeptide hormone of the thymus. Proc. Natl. Acad. Sci. USA 71: 1474–1478.

9. SCHEID, M. P., G. GOLDSTEIN & E. A. BOYSE. 1978. The generation and regulation of lymphocyte populations. Evidence from differentiative induction systems in vitro. J. Exp. Med. 147: 1727–1743.

10. BASCH, R. S. & G. GOLDSTEIN. 1975. Antigenic and functional evidence for in vitro inductive activity of thymopoietin (thymin) on thymocyte precursors. Ann. N.Y. Acad. Sci. 249: 290–299.

11. SUNSHINE, G. H., R. S. BASCH, R. G. COFFEY, K. W. COHEN, G. GOLDSTEIN & J. W. HADDEN. 1978. Thymopoietin enhances the allogeneic response and cyclic GMP levels of mouse peripheral, thymus-derived lymphocytes. J. Immunol. 120: 1594–1599.

12. AUDHYA, T., M. A. TALLE & G. GOLDSTEIN. 1984. Thymopoietin radioreceptor assay utilizing lectin-purified glycoprotein from a biologically responsive T cell line. Arch. Biochem. Biophys. 234: 167–177.

13. MALAISE, M. G., M. T. HAZEE-HAGELSTEIN, A. M. REUTER, T. VRINDS-GEVAERT, G. GOLDSTEIN, & P. FRANCHIMONT. 1987. Thymopoietin and thymopentin enhance the levels of ACTH, β-endorphin and β-lipotropin from rat pituitary cells in vitro. Acta Endocrinol. 115: 455–460.

14. LAU, C. Y. & G. GOLDSTEIN. 1980. Functional effects of thymopoietin 32–36 (TP5) on cytotoxic lymphocyte precursor units (CLP-U). 1. Enhancement of splenic CLP-U in vitro and in vivo after suboptimal antigenic stimulation. J. Immunol. 124: 1861–1865.

15. LAU, C. Y., E. Y. WANG & G. GOLDSTEIN. 1982. Studies of thymopoietin pentapeptide (TP5) on experimental tumors I. TP5 relieves immunosuppression in tumor-bearing mice. Cell. Immunol. 66: 217–232.

16. VEYS, E. M., H. MIELANTS, G. VERBRUGGEN, T. SPIRO, E. NEWDECK, D. POWER & G. GOLDSTEIN. 1984. Thymopoietin pentapeptide (thymopentin, TP5) in the treatment of rheumatoid arthritis. A compilation of several short- and longterm clinical studies. J. Rheumatol. 11: 462–466.

17. BOLLA, K., D. DJAWARI, E. M. KOKOSCHKA, J. PETRES, S. LIDEN, R. GONSETH, P. AMBLARD, M. G. BERMENGO, J. J. BONERANDI, A. CLAUDY, H. DEGREEF, J. DEMAUBEUGE, J. MEYNADIER, J. H. SAURAT, J. H. SCHOPF, W. HOBEL, J. P. CASTAIGNE & E. SUNDAL. 1985. Prevention of recurrences in frequently relapsing herpes labialis with thymopentin. Surv. Immunol. Res. 4 (Suppl. 1): 37–47.

18. UNGAR, G. 1945. Endocrine function of the spleen and its participation in the pituitary-adrenal response to stress. Endocrinology **37:** 329–340.
19. AUDHYA, T., M. P. SCHEID & G. GOLDSTEIN. 1984. Contrasting biological activities of thymopoietin and splenin, two closely related polypeptide products of thymus and spleen. Proc. Natl. Acad. Sci. USA **81:** 2847–2849.
20. GOLDBERG, E. H., G. GOLDSTEIN, D. B. HARMAN & E. A. BOYSE. 1984. Contrasting effects of thymopentin and splenopentin on the capacity of female mice to reject syngeneic male skin. Transplantation **38:** 52–55.
21. RUBENFELD, M. R., A. E. SILVERSTONE, D. M. KNOWLES, J. P. HALPER, A. DE SOSTOA, C. M. FENOGLIO & R. L. EDELSON. 1981. Induction of lymphocyte differentiation by epidermal cultures. J. Invest. Dermatol. **77:** 221–224.
22. CHU, A. C., J. A. K. PATTERSON, G. GOLDSTEIN, C. L. BERGER, S. TAKEZAKI & R. L. EDELSON. 1983. Thymopoietin-like substance in human skin. J. Invest. Dermatol. **81:** 194–197.
23. AUDHYA, T. & G. GOLDSTEIN. 1985. Thymopoietin and ubiquitin. *In* Methods in Enzymology. G. Di Sabato, J. L. Langone & H. Van Vunakis, Eds.: 279–291. Academic Press. New York, NY.
24. GOLDSTEIN, G. & I. R. MACKAY. 1969. The Human Thymus. William Heinemann. London.
25. MILLER, J. F. A. P. 1964. Effect of thymic ablation and replacement. *In* Thymus in Immunobiology. R. A. Good & A. E. Gabrielsen, Eds.: 436–443. Harper & Row (Hoeber). New York, NY.
26. GOLDSTEIN, G. & W. W. HOFMANN. 1969. Endocrine function of the thymus affecting neuromuscular transmission. Clin. Exp. Immunol. **4:** 181–189.
27. GOLDSTEIN, G. & W. W. HOFMANN. 1968. Electrophysiological changes similar to those of myasthenia gravis in rats with experimental auto-immune thymitis. J. Neurol. Neurosurg. Psychiatry **31:** 453–459.
28. SCHEID, M. P., G. GOLDSTEIN, U. HAMMERLING & E. A. BOYSE. 1975. Lymphocyte differentiation from precursor cells *in vitro*. Ann. N.Y. Acad. Sci. **249:** 531–540.
29. PANTELOURIS, E. M. 1968. Absence of thymus in a mouse mutant. Nature **217:** 370–371.

Biology and Molecular Biology of Epidermal Cell-Derived Thymocyte Activating Factor[a]

DANIEL N. SAUDER,[b] TRACY ARSENAULT,[c]
RODERICK C. MCKENZIE,[b] DAWN K. STETSKO,[b]
AND CALVIN B. HARLEY[c]

Departments of [b]Medicine and [c]Biochemistry
McMaster University Medical Centre
1200 Main Street West
Hamilton, Ontario, Canada L8N 3Z5

INTRODUCTION

The last 20 years have witnessed remarkable advances in our understanding of the importance of cell mediated immunity and humoral immunity working in complementary fashion to maintain the integrity of the body. Chemical messengers, generally lacking antigen specificity, play a vital role in the induction and expression of immunity. Lymphokines represent a major subgroup of these chemical messengers. The term "lymphokine" was first used by Dumonde *et al.* to describe "nonantibody" mediators of cellular immunity generated by lymphocyte activation.[1] Later, the definition of lymphokine was extended to include immunoregulatory products of cells other than lymphocytes. "Monokine" is the term used to describe immunoregulatory products of mononuclear phagocytes. "Cytokine" is a generic term applied to lymphokines, monokines or other cell products influencing the behavior of target cells.[2]

Throughout the 1960s, the existence of cytokines as discrete chemical entities was unsupported by chemical and molecular data. Then, in the late 1970s the development of discrete, reproducible and unequivocal assays for cytokine function allowed biochemical and molecular characterization of cytokine gene products. In addition, recombinant DNA techniques capable of directing the biosynthesis of lymphokine proteins in heterologous prokaryotic expression systems provided commercial and scientific impetus to produce sufficient quantities of homogeneous cytokine protein for testing in preclinical and clinical conditions.

Six years ago an epidermal cytokine termed "epidermal cell-derived thymocyte activating factor" (ETAF)[3,4] was identified. It is now felt that ETAF and related cytokines play a major role in skin immunity and perhaps in normal homeostasis.[5]

Over the past decade, it has also become clear that the skin can play a significant role in epidermal immunity and thus functions as the peripheral arm of the immune system.[5,6] Steilein used the designation "skin associated lymphoid tissue" or SALT to describe this immune function of skin.[7] The epidermal Langerhans cells with their antigen presenting ability act as a significant member of SALT[8] (reviewed in REFS. 9,

[a]Supported in part by the Canadian Medical Research Council, the Canadian Dermatology Foundation, the Ontario Ministry of Health and a grant from the Physicians' Services Incorporated.

10). Antigen presentation is thought to involve two signals which are presented to T cells by an antigen presenting cell. The first signal involves uptake, processing and presentation of the nominal antigen in association with Ia antigen and likely the T cell receptor. In addition, a second antigen nonspecific signal is also thought to be required. This signal appears to be mediated by interleukin-1 (IL-1).[11]

Since Langerhans cells are known to be potent stimulators of T cell activation and are capable of functioning as antigen presenting cells, it was hypothesized that Langerhans cells could also produce IL-1. These studies led to the discovery that Langerhans cell depleted cultures (keratinocytes) produce a factor with interleukin-1-like activity.[3] Since IL-1 was initially defined as a macrophage-derived cytokine, the above epidermal-derived factor was designated "epidermal cell-derived thymocyte activating factor" (ETAF).[3,4]

Biologic Properties of ETAF/IL-1

The cytokine IL-1 has pleotropic proinflammatory and immunoregulatory properties (reviewed in REFS. 12–14). IL-1 is an antigen nonspecific immune regulating glycopeptide originally described in activated macrophages but is now shown to be produced by a plethora of cell types including endothelial cells, neutrophils, fibroblast, epithelial cells, neutrophils, astrocytes, microglial cells and epidermal cells.[12–14] IL-1 mediates a wide range of biological activities including stimulation of thymocyte proliferation via induction of interleukin-2 (IL-2) release, stimulation of B lymphocyte maturation and proliferation, fibroblast growth factor activity and induction of acute-phase protein synthesis by hepatocytes.[15]

Immunoregulatory Effects of IL-1 and ETAF

In conjunction with the delivery of a processed antigen from an accessory cell, IL-1 provides the activation signal to a target T cell.[11] Subsequent induction of IL-2 and expression of IL-2 receptors in the target T cell underlie the T cell proliferative response. From the rapid onset of the kinetics of IL-1 production in mitogen stimulated monocytes, it is suggested that rapid *de novo* transcription of mRNA occurs upon challenge, since the IL-1 production can be inhibited with actinomycin D and cycloheximide.

In addition, IL-1, is noted for its effects on augmenting natural killer cell (NK) activities as well as B cell proliferation and maturation.[16] In the epidermis, Langerhans cells and keratinocytes respond to stimulation with bacterial antigens to release ETAF.[17] ETAF also potentiates the production of IL-2 in mitogen stimulated peripheral T cells to further enhance T cell clonal expansion.[3]

Inflammatory Effects of IL-1 and ETAF

IL-1 is proinflammatory.[18–23] It is chemotactic for monocytes, lymphocytes and neutrophils and causes neutrophil degranulation *in vitro*.[18–20] Subcutaneous injection of IL-1 results in neutrophil margination followed by exudation.[21,22] ETAF is chemotactic for neutrophils[19] and T lymphocytes.[20] An intradermal injection of IL-1 "prepares" the skin for a subsequent injection to elicit a local inflammatory skin reaction typical of the Schwartzman reaction.[21,22] Under the stimulation of IL-1, secondary inflammatory mediators such as prostaglandins and platelet activating factor (PAF) are produced by endothelial cells,[23] prostaglandins and colony stimulating factors are produced by fibroblasts, and histamine is produced from mast cells. These signify the potential multiplicity of IL-1 and ETAF as endogenous mediators in acute inflammatory response.

Systemically, IL-1 or keratinocyte-derived ETAF causes fever[19,24] and induces hepatocytes to synthesize and secrete plasma proteins which are inducible by trauma during the acute phase of inflammation.[15,25,26] It is now apparent that another cytokine termed "interleukin 6" (IL-6) (previously "β2 interferon" or "hepatocyte stimulating factor") is the major cytokine responsible for the hepatocyte stimulating factor activity.[27] Muscle proteolysis can be triggered by IL-1 or ETAF and is further enhanced by the febrile temperature and PGE_2 production.[28,29] The catabolic process potentially increases the amino acid pool available for reparative synthesis of more essential proteins. It is the release of these secondary mediators such as PGE_2 that is responsible for many of the pathologic manifestations of the IL-1 induced inflammatory response.

The proliferative effect of IL-1 in pathologic states can be extended to a number of cell types such as the fibroblasts, osteoblasts, glial cells and keratinocytes. Keratinocyte-derived ETAF acts as an autocrine cytokine to enhance the growth of keratinocytes[30,31] as well as fibroblasts which may be beneficial for wound healing.

IL-1 in Pathologic States

IL-1 is usually not detectable in normal skin biopsies using immunofluorescent techniques. However, significant activity can be found from culturing lesional skin from patients. For example, IL-1 can be detected in skin biopsies using immunofluorescence in skin lesions from patients with cutaneous T cell lymphoma. Using a monoclonal anti-IL-1β antibody, intense epidermal fluorescence was demonstrated in 10 out of 10 patients with cutaneous T cell lymphoma, whereas, minimal reactivity was seen in inflammatory dermatoses. It is an attractive hypothesis that hypersecretion of the chemotactic epidermal cytokine (IL-1 or ETAF) *in vivo* initiates "homing" and local proliferation of the malignant T cells.[32] IL-1 activity is also found to be increased in psoriatic epidermis.[33,34]

In addition to disease where ETAF/IL-1 is increased, certain therapies used in dermatology can alter ETAF/IL-1. Cultured epidermal cells or keratinocyte cell line irradiated with broad band UV display increased ETAF/IL-1 activity.[35] Mice given similar treatment also showed increased levels of acute phase proteins and neutrophilia together with a transient rise in ETAF activities in the serum.[36] Human volunteeers given one minimal erythema dose to the whole body resulted in increased serum IL-1.[37] However, ETAF activity was found to be decreased when keratinocytes were irradiated with narrow band UVA.[38] Thus, UV radiation can augment or inhibit ETAF/IL-1 depending on dosage and wavelength. In contrast to broad band UVB, narrow band UVA had a suppressive effect in ETAF/IL-1. This alteration in ETAF/IL-1 may have therapeutic implications for diseases affected by UV radiation.

ETAF/IL-1 Receptors

The biologic actions of polypeptide hormones are mediated by plasma membrane receptors. It has previously been demonstrated that purified radiolabeled IL-1 binds specifically to a receptor on the surface of fibroblasts as well as certain T cell lines[39,40] (reviewed in REF. 41).

Previous studies have shown that keratinocytes respond to ETAF/IL-1 by proliferating,[29,30] suggesting that keratinocytes would be suitable target cells for IL-1 receptor binding studies. Using iodinated IL-1α, it has been shown that keratinocytes possess high affinity receptors for IL-1[42] at approximately equivalent receptor density to that of human fibroblasts. IL-1α and IL-1β appear to occupy the same receptor.[43] Whether IL-1k, one member of the ETAF family (see below), competes for the same receptor has not yet been determined.

Molecular Biology of IL-1

Two distinct derivatives of IL-1 peptides have been identified. The predominant neutral form is IL-1β and the acidic form is IL-1α, which are encoded by two different genes.[14,44–48] These two molecules share only limited structural homology, but similar (if not identical), biological activities. The recombinant human IL-β is made up of 153 amino acid residues with a molecular weight of 17,500.[45] IL-1 is believed to be synthesized in a form of ≈30 kDa precursor which is then processed into a smaller 15–20 kDa active molecule. IL-1α and IL-1β appear to have equivalent affinities for the IL-1 receptor on T cells.[38,43]

Molecular Studies of ETAF

At first, it was impossible to detect ETAF by Western blotting with available polyclonal IL-1 antibody. Therefore, molecular studies were undertaken to determine the relationship between ETAF and IL-1α and IL-1β. For these studies, a synthetic oligonucleotide probe complementary to conserved regions of IL-1α and IL-1β mRNA from murine and human sequences was utilized to screen Northern transfers of RNA from the keratinocyte cell line COLO-16, an epidermal cell line,[49] and subsequently COLO-16 cDNA libraries. The IL-1β oligonucleotide probes hybridized strongly to 1.6-kb (IL-1) mRNA from lipopolysaccharide stimulated peripheral blood monocytes and to a similar size mRNA from COLO-16 cells, but not to RNA from negative control cells. A partial cDNA library made from COLO-16 mRNA primed with the IL-1β probe yielded positive clones of a size expected from IL-1 related mRNA. However, partial restriction site mapping and sequencing has indicated that this cDNA is distinct from monocyte IL-1β. On Northern transfers, the keratinocyte (COLO-16) IL-1-like clone hybridizes to the 1.6-kb RNA from COLO-16 cells but not the 1.6-kb RNA from monocytes. Most importantly, the clone is capable of hybrid selecting mRNA from COLO-16 cells which produces IL-1-like activity in the thymocyte assay following injection into frog oocytes.[50]

Cloning IL-1-like cDNAs from Epidermal Cells

Initially, a partial cDNA library was constructed using an IL-1β complimentary oligonucleotide[49] and screened using standard technology.[52,53] About 100,000 independent clones were obtained. Screening this library with the partial cDNA clones resulted in isolation of clones designated "D9" and "F8" which gave a positive signal when hybridized with an IL-1β cDNA probe (kindly supplied by Dr. P.E. Auron, Massachusetts Institute of Technology, Cambridge, MA).

These clones were initially assayed for biological activity using hybrid select/ oocyte injection technique. Briefly, plasmid DNA from the above clones was denatured, fixed to nitrocellulose, and then hybridized with poly (A) + keratinocyte RNA. The hybrid selected mRNA was then microinjected into Xenopus laevis oocytes. Fifty nanoliters of mRNA (dissolved in sterile H_2O at a concentration of 0.5–1 mg/ml)

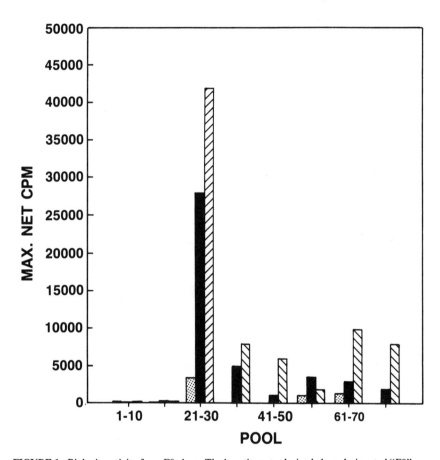

FIGURE 1. Biologic activity from F8 clone. The keratinocyte-derived clone designated "F8" was used to hybrid select keratinocytes mRNA which was then injected into Xenopus oocytes. Oocyte supernatant was fractionated on a TSK 2000 HPLC column and tested for IL-1 activity, using the thymocyte costimulator assay. *Dotted columns,* pPR selected RNA; *filled columns,* F8 selected RNA; *hatched columns,* poly (A) + macro RNA.

was injected into each oocyte. After injection the oocytes were placed at 4°C for 45 min, then transferred to a microlitre plate and incubated in fresh oocyte medium for 18 hr at 23°C. After incubation, oocytes were harvested, the supernatant concentrated 10× and chromatographed on a TSK 2000 size exclusion HPLC column to remove an IL-1 inhibitor. The material was then tested for the presence of ETAF/IL-1 using the thymocyte costimulator assay. The negative control consisted of poly (A) + keratino-

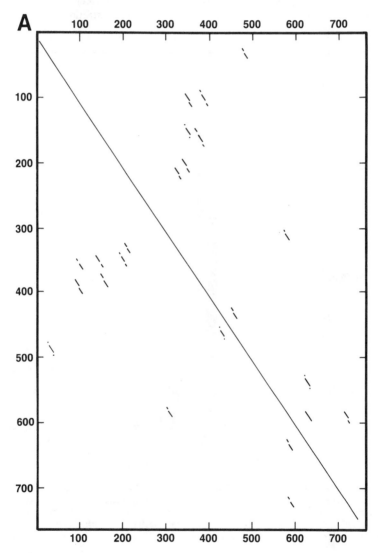

FIGURE 2. (**A**) Dot matrix comparison of IL-1β to IL-1β. Note homologous sequences yield straight line along the diagonal. (**B**) Dot matrix comparison of IL-1α to IL-1β. IL-1α has limited homology to IL-1β as shown by lack of straight lines on the diagonal.

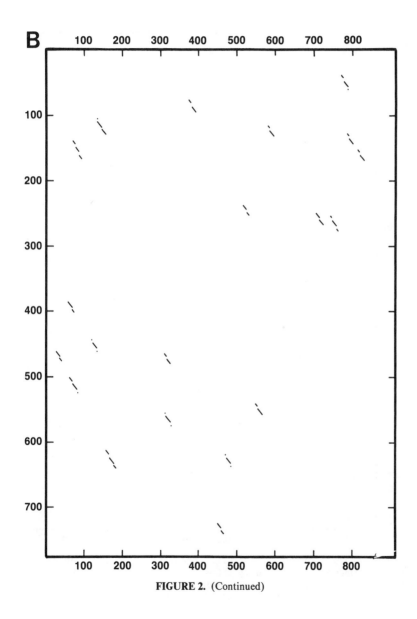

FIGURE 2. (Continued)

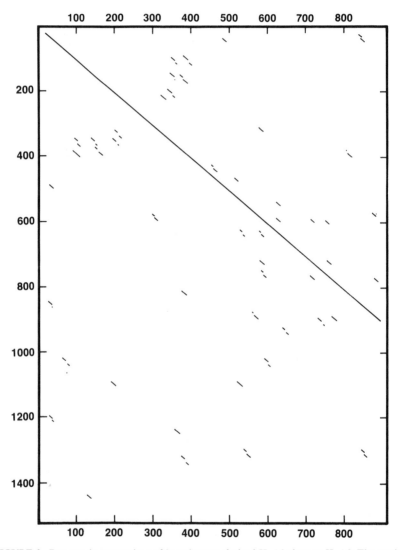

FIGURE 3. Dot matrix comparison of keratinocyte-derived IL-1β clone to IL-1β. The straight line on diagonal indicated strong homology. In this case, the full nucleotide sequence of IL-1β was compared to a partial sequence of keratinocyte IL-1β and hence the straight line is off the diagonal.

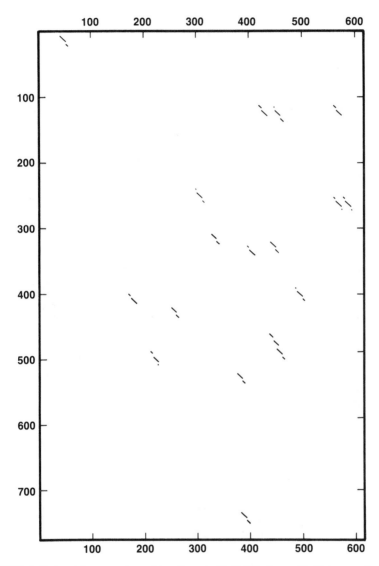

FIGURE 4. Dot matrix comparison of keratinocyte IL-1k/F8 clone with IL-1α revealing less than 30% homology. Similar results were found comparing IL-1k/F8 with IL-1β.

cyte RNA selected with the vector alone without the putative ETAF insert, and the positive control consisted of poly (A) + RNA from stimulated macrophages (FIG. 1).

Sequencing was performed by the Sanger dideoxy method[54] using the plasmid/ M13 phage hybrid pUC118/119.

Comparison of ETAF with IL-1α, IL-1β

To compare the above sequence generated from F8/D9, a dot matrix comparison was performed. Dot matrix comparison can be used to compare two nucleic acid or two protein sequences. Whenever a specified degree of similarity is reached between two sequences, (in the case of FIG. 2, ≥9 of 12 nucleotides) a "." is typed. If sequence "A" is totally homologous to sequence "B", a diagonal line is printed.

From the above oligo dT primed cDNA library, interleukin-1β cDNAs have also been identified. An interleukin-1β cDNA has been cloned and partially sequenced using dot matrix comparison (shown in FIG. 3) strong homology was identified between IL-1β and keratinocyte IL-1β. Thus, IL-1β cDNA has been cloned from keratinocytes. Screening for IL-1α clones is currently underway.

Interleukin-1k

When the clones designated "F8" were partially sequenced and compared to IL-1α or IL-1β, less than 30% homology was found (FIG. 4). Yet these clones were able to hybrid select mRNA that produced IL-1-like activity. In addition, preliminary data has shown that when F8 is transfected into a mammalian expression vector system, significant IL-1 activity was found. We propose that once the full molecular character-ization of the F8 clone is obtained, the molecule should be termed "IL-1k." Previous studies have shown keratinocytes contain mRNA homologous to IL-1α and IL-1β.[49–51] Thus, keratinocytes contain at least three members of the IL-1 family—IL-1α, IL-1β and IL-1k. ETAF, therefore, includes at least the above three molecules.

SUMMARY

ETAF/IL-1 has a multiplicity of divergent biological effects: enhancement of thymocyte proliferation, stimulation of cells in the hypothalamus to mediate fever, leukocyte chemotaxis, stimulation of hepatic synthesis of acute-phase proteins, aug-mentation of IL-2 production and keratinocyte proliferation. Until recently, it has not been possible to determine whether these divergent activities are mediated by closely related cytokines or separate cytokines. Now with the identification of IL-1α, IL-1β and IL-1k from keratinocytes, these studies will become possible. In either case, it is likely that ETAF/IL-1 plays an important role in local cutaneous and systemic inflammatory and immunological events.

REFERENCES

1. DUMONDE, D. C., R. A. WOLSTENCROFT, G. S. PANAYI, M. MATTHEW, J. MORLEY & W. T. HOWSON. 1969. Nature **224**: 38–42.
2. COHEN, S., P. E. BIGAZZI & T. YOSHIDA. 1974. Cell. Immunol. **12**: 150–159.
3. SAUDER, D. N., C. CARTER, S. I. KATZ & J. J. OPPENHEIM. 1982. Invest. Dermatol. **79**: 34–39.

4. LUNGER, T. A., B. M. STADLER, S. I. KATZ. 1981. J. Immunol. **127:** 1493–1498.
5. BREATHNACH, S. M. & S. I. KATZ. 1986. Hum. Pathol. **17:** 161–167.
6. EDELSON, R. L. & J. M. FINK. 1985. Sci. Am. **252:** 46–54.
7. STREILEIN, J. W. 1983. J. Invest. Dermatol. **80:** 12s–16s.
8. STINGL, G., S. I. KATZ, L. CLEMENT, I. GREEN & E. M. SHEVACH. 1978. J. Immunol. **121:** 2005–2013.
9. ROWDEN, G. 1981. The Langerhans Cell. CRC Crit. Rev. Immunol. **3:** 95–180.
10. CHOI, K. L. & D. N. SAUDER. 1987. Mech. Ageing & Dev. **39:** 69–79.
11. UNANUE, E. R. & P. M. ALLEN. 1984. Science **236:** 551–557.
12. DINARELLO, C. A. & J. W. MIER. 1986. Annu. Rev. Med. **37:** 173–178.
13. DINARELLO, C. A., J. G. CANNON, J. W. MIER, H. A. BERHNEIM, G. LO PRESTI, D. L. LYNN, R. N. LOVE, A. C. WEBB, P. E. AURON, R. C. REUBEN, A. RICH, S. M. WOLFF & S. D. PUTNEY. 1986. J. Clin. Invest. **77:** 1734–1739.
14. OPPENHEIM, J. J., E. J. KOVACS, K. MATSUSHIMA & S. K. DURUM. 1986. Immunology Today **7:** 45–56.
15. GAULDIE, J., D. N. SAUDER, K. P. W. J. MCADAM & C. A. DINARELLO. 1987. Immunology **60:** 203–207.
16. PILLAI, P. S., S. D. REYNOLDS, D. W. SCOTT, J. GAULDIE & D. N. SAUDER. 1987. J. Leukocyte Biol. **42:** 222–229.
17. SAUDER, D. N., C. A. DINARELLO & V. B. MORHENN. 1984. J. Invest. Dermatol. **82:** 605–607.
18. DINARELLO, C. A. 1985. J. Clin. Immunol. **5:** 287–297.
19. SAUDER, D. N., N. L. MOUNESSA, S. I. KATZ, C. A. DINARELLO & J. I. GALLIN. 1984. J. Immunol. **132:** 828–832.
20. SAUDER, D. N., M. M. MONICK & G. W. HUNNINGHAKE. 1985. J. Invest. Dermatol. **85:** 431–433.
21. GRANSTEIN, R. & D. N. SAUDER. 1987. Lymphokine Res. **6:** 187–193.
22. BECK, G. & G. S. HABICHT. 1986. J. Immunol. **136:** 3025–3031.
23. BUSSOLINO, F., F. BREVIARIA, C. TETTA, M. AGLIETTA, F. SANAVIO, A. MANTOVANI & E. DEJANA. 1986. Pharmacol. Res. Commun. **18** (Suppl.): 133–137.
24. SAUDER, D. N. 1984. Lymphokine Res. **3:** 145–151.
25. BAUMANN, H., G. P. JAHREIS, D. N. SAUDER & A. KOJ. 1984. J. Biol. Chem. **259:** 7331–7342.
26. BAUMANN, H., R. E. HILL, D. N. SAUDER & G. P. JAHREIS. 1986. J. Cell Biol. **102:** 370–383.
27. GAULDIE, J., C. RICHARDS, D. HARNISH, P. LANSDORP & H. BAUMANN. 1987. Proc. Natl. Acad. Sci. USA **84:** 7251–7255.
28. BARACOS, V., H. P. RODEMANN, C. A. DINARELLO & A. L. GOLDBERG. 1983. N. Engl. J. Med. **308:** 553–558.
29. SAUDER, D. N., J. SEMPLE, D. TRUSCOTT, B. GEORGE & G. H. A. CLOWES. 1986. J. Invest. Dermatol. **87:** 711–714.
30. ROSTOW, H. J. 1987. Proc. Natl. Acad. Sci. USA **84:** 1940–1944.
31. SAUDER, D. N., B. PRAEGER & B. GILCHREST. 1988. Arch. Dermatol. Res. **280:** 71–76.
32. TRON, V. A., D. ROSENTHAL & D. N. SAUDER. 1988. J. Invest. Dermatol. **90:** 378–381.
33. CAMP, R. D. R., N. J. FINCHAM & F. M. CUNNINGHAM. 1986. J. Immunol. **137:** 3469–3474.
34. KRAGBALLE, K., C. L. MARCELLO, J. J. VOORHEES & D. N. SAUDER. 1987. J. Invest. Dermatol. **88:** 8–10.
35. ANSEL, J., T. A. LUGER & I. GREEN. 1983. J. Invest. Dermatol. **81:** 519–523.
36. GAHRING, L., M. BLATZ, M. B. PEPYS & R. DAYNES. 1984. Proc. Natl. Acad. Sci. USA **81:** 1198–1202.
37. GRANSTEIN, R. D. & D. N. SAUDER. 1987. Lymphokine Res. **6:** 187–193.
38. SAUDER, D. N., F. P. NOONAN, E. C. DEFABO & S. I. KATZ. 1983. J. Invest. Dermatol. **80:** 485–489.
39. DOWER, S. K., S. M. CALL, S. GILLIS & D. L. URDAL. 1986. Proc. Natl. Acad. Sci. USA **83:** 1060–1064.
40. CHIN, J., P. M. CAMERON, E. RUPP & J. A. SCHMIDT. 1987. J. Exp. Med. **165:** 70–86.
41. DOWER, S. K. & D. L. URDAL. 1987. Immunology Today **8:** 46–51.

42. SAUDER, D. N. & P. L. KILIAN. 1987. Clin. Res. **35:** 390A.
43. KILIAN, P. L., K. L. KAFFKA, A. S. STERN, D. WOEHLE, W. R. BENJAMIN, T. M. DE CHIARA, V. GUBLER, J. J. FARRA, S. B. MIZEL & P. T. LOMEDICO. 1986. J. Immunol. **136:** 4509–4514.
44. AURON, P. E., L. J. ROSENWASSER, K. MATSUSHIMA, T. COPELAND, C. A. DIARELLO, J. J. OPPENHEIM & C. WEBB. 1985. J. Mol. Cell. Immunol. **2:** 169–177.
45. AURON, P. E., C. A. WEBB & L. J. ROSENWASSER. 1984. Proc. Natl. Acad. Sci. USA **81:** 7907–7911.
46. LOMEDICO, P. T., P. L. KILIAN, U. GUBLER, A. S. STEIN & R. CHIZZONITE. 1986. Cold Spring Harbor Symp. Quant. Biol.: 631–639. Cold Spring Harbor Laboratory. Cold Spring Harbor, NY.
47. LOMEDICO, P. T., V. GUBLE & C. P. HELLMANN, *et al.* 1984. Nature **312:** 458–462.
48. MARCH, C., B. MOSLEY, A. LARSEN, *et al.* 1985. Nature **315:** 641–648.
49. BELL, T. V., C. B. HARLEY, D. STETSKO & D. N. SAUDER. 1987. J. Invest. Dermatol. **88:** 375–379.
50. SAUDER, D. N., T. V. ARSENAULT, D. STETSKO & C. B. HARLEY. 1987. Clin. Res. **35:** 714A.
51. GUBLER, U. & B. J. HOFFMAN. 1983. Gene **25:** 263.
52. MANIATIS, T., E. F. FRITSH & J. SANBROOK. 1982. Molecular Cloning. Cold Spring Harbor, NY.
53. BOLIVAR, F. & K. BACKMAN. 1979. Methods Enzymol. **68:** 245–267.
54. SANGER, F., S. NICKLEN & A. R. COULSON. 1977. Proc. Natl. Acad. Sci. USA **69:** 2110–2114.

Keratinocyte-Derived Interleukin 3

THOMAS A. LUGER, ANDREAS KOCK,
REINHARD KIRNBAUER, THOMAS SCHWARZ,[a]
AND JOHN C. ANSEL[b]

Department of Dermatology II
University of Vienna
and
Ludwig Boltzmann Institute
of Dermato-Venerological Serodiagnosis
Alserstrasse 4
A-1090 Vienna, Austria

[a]*Department of Dermatology*
Vienna-Lainz Hospital
Walkersbergenstrasse 13
A-1130 Vienna, Austria

[b]*Dermatological Service (11C2-P)*
Veterans Administration Medical Center
3710 S.W. U.S. Veterans Hospital Road
P.O. Box 1034
Portland, Oregon 97207

Interleukin 3 (IL 3) which belongs to the family of colony stimulating factors (CSF) not only is responsible for regulating hematopoiesis but exerts a broad spectrum of biological activities.[1,2] It was first described by its ability to induce the synthesis of 20α-hydroxysteroid dehydrogenase (20α-SDH) in splenic lymphocytes of nude mice.[3] Recently, the protein has been purified, and cDNA clones encoding for murine IL 3 were isolated[4,5] facilitating the analysis of biological properties of this factor. Subsequently, it became evident that IL 3 was identical to a variety of factors previously characterized by using several distinct bioassays. Therefore, in addition to its multi-CSF activity, murine IL 3 is probably identical to lymphokines known under a variety of names such as burst promoting activity, histamine producing cell stimulating factor, P cell stimulating factor, Thy-1 inducing factor and mast cell growth factor.[1] Similar to other cytokines the murine IL 3 mediated growth signal apparently is due to the interaction of IL 3 with its specific cell surface receptor which recently has been isolated on IL 3 dependent cell lines.[6]

The broad spectrum of biological activities of IL 3 and its efficacy in stimulating hematopoiesis in sublethally irradiated mice[7] underline the importance of isolating the human homologue of murine IL 3. Although a factor with IL 3-like activity has been shown to be released by human T lymphocytes,[8] the identification of a gene encoding for IL 3 in human T cells turned out to be difficult. This may be explained by the relatively low sequence homology between IL 3 genes of different species. In spite of these difficulties a human gene that is related to the murine one for IL 3 recently has been identified and cloned.[9] This human hemopoietin factor has multi-CSF activity and therefore shows functional homology with murine IL 3. In addition the gene for human IL 3 proved to have significant structural homology with the murine IL 3 gene.[10]

Antigen and mitogen activated T lymphocytes were supposed to be the major cellular origin of IL 3.[3] However, it recently became evident that many cells other than lymphocytes as well as cell lines are capable of producing factors with IL 3 activity. Accordingly, the monomyelocytic cell line WEHI-3 has been demonstrated to produce large amounts of IL 3.[11] More recent studies have shown that nonlymphoid cells such as keratinocytes,[12,13] astrocytes[14] and corneal epithelial cells[15] also can produce a factor with an IL 3-like activity. In the present report we will briefly summarize the characteristics of synthesis and production of IL 3 by murine epidermal cells (EC) and discuss the possible evidence for a human analogue of murine EC-IL 3.

Murine Epidermal Cell Interleukin 3

Supernatants of murine EC have been demonstrated to contain a mediator which significantly stimulated the proliferative activity of IL 3 dependent cell lines (32 DCL, FDCP1, DA1). Since epidermal cells depleted of Langerhans cells as well as a murine transformed keratinocyte cell line (Pam 212) also produce considerable levels of IL 3-like activity, the keratinocyte within the epidermis appears to be the major source of this cytokine.[12] Although EC spontaneously release EC-IL 3, the addition of different stimulants which alter membrane metabolism such as the tumor promotor phorbol myristic acetate (PMA), lipopolysaccharide (LPS) and silica significantly enhanced EC-IL 3 production. Moreover, in the presence of the protein synthesis inhibitor cycloheximide spontaneous and LPS induced EC-IL 3 production was blocked completely, indicating that de novo protein synthesis was necessary rather than the release of preformed factor.[12]

Some IL 3 dependent cell lines (FDCP1, DA1) proliferate in response to both murine GM-CSF and murine IL 3.[16] In addition GM-CSF is also known to induce the proliferation of an IL 2 or IL 4 dependent murine T cell line (HT2).[17] It recently has been demonstrated that murine epidermal cells are capable of producing a factor which stimulates the proliferation of HT2 as well as FDCP cells.[18] This EC derived cytokine primarily was named keratinocyte-derived T cell growth factor (KTGF) and subsequently was identified to be GM-CSF.[19] EC supernatants induce the proliferation of both 32 DCL cells[12] which do not respond to GM-CSF[16] and HT2 cells which are not capable of growing in the presence of IL 3.[17] Therefore, it appeared to be most likely that EC produce both, factors with IL 3-like as well as GM-CSF-like activity.

Murine EC-IL 3 has a m.w. of approximately 28 kD and was found to be heat sensitive and stable between pH 4 and pH 12[12] (TABLE 1). Using HPLC chromatofocusing EC-IL 3 exhibits three isoelectric points (pI 7.8, 7.4 and 7.1).[20] This charge heterogeneity may be explained by the existence of different posttranslational glycosy-

TABLE 1. Biochemical Properties of Epidermal Cell IL3

	Murine	Human
Molecular weight	28 kD	17 kD
Heat stability		$-70°$–56°C
pH stability	4–12	—
DEAE (mM NaCl)	—	360
Hydroxylapatite (HPHT)		
(mM HPO$_4$, H$_2$PO$_4$)	230	230
Isoelectric points	7.8, 7.4, 7.1	7.8, 7.5, 5.6
Reversed phase (%CH$_3$CN)	75	10, 45, 70, 10

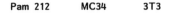

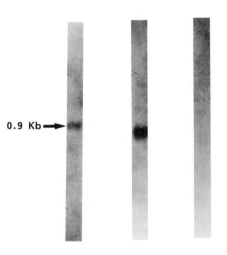

FIGURE 1. Northern blot analysis of PMA (100 ng/ml) induced IL 3 mRNA expression. Poly (A) + RNA was prepared after 12 hr incubation from Pam 212 keratinocytes, MC34 cells (mast cell line) and 3T3 fibroblasts and analyzed by Northern blotting using a ^{32}P-labeled cDNA encoding for murine IL 3. Lanes contain 7.5 μg RNA with a 24-hr autoradiographic exposure time.

0.9 Kb →

lated species which have been described for T cell IL 3.[4] Upon hydroxylapatite HPLC EC-IL 3 eluted as a single peak (225 mM sodium phosphate) and when reversed phase HPLC was applied EC-IL 3 was found to be homogenous eluting as one major peak between 70% and 80% acetonitrile.[20]

According to these biochemical characteristics EC-IL 3 appears to be indistinguishable from murine T cell IL 3. Therefore, the effect of an antiserum directed against IL 3 was tested for its effect on EC-IL 3. The antiserum blocked EC-IL 3 mediated proliferation of 32 DCL cells in a dose dependent manner,[12] indicating that both T cell IL 3 and EC-IL 3 share antigenically similar domains.

The apparent similarities between murine T cell- and EC-derived IL 3 were further analyzed by molecular criteria. Therefore, freshly isolated murine epidermal cells and Pam 212 keratinocytes were examined for IL 3 mRNA expression using a murine T cell-derived IL 3 cDNA probe by Northern blot analysis. Both murine keratinocytes and Pam 212 cells were found to contain mRNA homologous to T cell IL 3 cDNA.[19,21] The usually low constitutive IL 3 mRNA expression in the transformed keratinocyte cell line Pam 212 was markedly augmented by addition of the tumor promotor PMA. This was paralleled by the constitutive and PMA-induced IL 3 mRNA expression by the murine mast cell line MC34. However, constitutive and PMA-induced IL 3 mRNA expression by MC34 cells usually was more pronounced than in Pam 212 cells. In contrast neither constitutive nor PMA-induced IL 3 mRNA expression was noted in 3T3 fibroblasts (FIG. 1). In order to further investigate whether there is complete homology between T cell IL 3 and EC-IL 3 a S1 nuclease protection assay was performed by Kupper *et al.*[19] In case of complete homology between cDNA and mRNA a fully matched double strand cDNA/mRNA hybrid exists which is resistant to S1 digestion, while base mismatches lead to digestion of labeled cDNA by S1 nuclease. Using this approach keratinocyte mRNA apparently was insufficient to protect T cell IL 3 cDNA from S1 nuclease mediated degradation.[19] These findings indicate the existence of not completely homologous genes encoding for cytokines with IL 3 activity.

Human Epidermal Cell IL 3-like Activity

Because of the characterization of a T cell IL 3 equivalent in murine keratinocytes and the potential usage of this cytokine in treatment of myelosuppression there is considerable interest in isolating the human homologue of murine EC-IL 3. Therefore, human EC were tested for their capacity to produce a mediator which induces the proliferation of murine IL 3 dependent cell lines (32 DCL, FDCP1). Supernatants of freshly isolated human EC and keratinocyte cell lines stimulated the proliferation of both IL 3 dependent murine cell lines[13] in a dose dependent manner. However, in comparison to the murine system human EC contained low but significant levels of IL 3-like activity. Treatment of EC with PMA or LPS resulted in a significantly increased EC-IL 3 supernatant activity. Moreover, human EC-IL 3 not only stimulated the DNA synthesis of 32 DCL cells but also maintained cells proliferating over a period of several weeks.[13] Since IL 3 dependent cells in the absence of IL 3 die within 2 days, these findings indicate that human EC produce an IL 3-like mediator.

Biochemical purification of human EC-IL 3 revealed that this cytokine is indistinguishable from a human T cell-derived IL 3-like mediator.[8] Both IL 3-like factors have a m.w. of approximately 17 kD and one major isoelectric point of pI 5.6 (TABLE 1). Upon reversed phase HPLC human EC-IL 3 exhibited some heterogeneity, possibly due to either unspecific aggregation with other proteins or to differently glyosylated species. However, it was demonstrated that highly purified EC-IL 3 upon gel electrophoresis is associated with only one protein band with a m.w. of 17 kD (FIG. 2).

Since the biochemical properties of EC-IL 3 appear to be indistinguishable from a previously described EC derived natural killer (NK) cell augmenting factor (ENKAF),[22] human EC-IL 3 was tested for its effect on NK cells. NK cells are considered to be important in host immune surveillance against virus infected cells and tumors. They have been defined by their ability to lyse *in vitro* a wide variety of target cells without presensitization.[23] NK cells represent a heterogenous subpopulation of lymphocytes also termed large granular lymphocytes (LGL).[24] The activity of NK cells which is known to be highly regulated by lymphokines such as IFN-gamma and IL 2[23] was evaluated in 4-hr ^{51}Cr release assay with K562 erythroleukemia cells as targets. Preincubation of LGL with human EC-IL 3 purified to homogeneity upon SDS-PAGE resulted in a significant augmentation of NK cell activity.[25] Moreover, like a previously described EC-derived granulocyte activating mediator (EC-GRAM)[25] the highly purified EC-IL 3 was found to stimulate the release of toxic oxygen radicals by human granulocytes[26] (FIGURE 2). These findings indicate that human EC-IL 3 possibly is identical with ENKAF and EC-GRAM.

Since human EC have been demonstrated to produce a variety of different cytokines the multiple biological activities of EC-IL 3 were compared to that of other EC-derived cytokines. Human EC-IL 3 has been demonstrated to be free of IL 1, IL 2, or IFN activity.[13] Moreover, according to its biological properties and biochemical characteristics this EC cytokine has been found to be distinct from other human colony stimulating factors including G-, GM- and M-CSF[25] as well as interleukin 4 (IL 4, BSF1) and tumor necrosis factors (TNF) (TABLE 2). Although rh-G-CSF induced the proliferation of IL 3 dependent 32 DCL cells it did not affect FDCP-, NK cells or granulocytes (TABLE 2), suggesting that a relationship between EC-IL 3 and G-CSF is unlikely. Surprisingly the recently cloned human T cell IL 3 was not capable of stimulating the proliferation of IL 3 dependent cell lines and had no effect on NK cell activity,[25] indicating that human T cell IL 3 and human EC-IL 3 are distinct cytokines (TABLE 2). This is also supported by the finding that human IL 3 does not bind to the murine IL 3 receptor.[6] Using cDNA probes human EC in addition to IL 1α and IL 1β

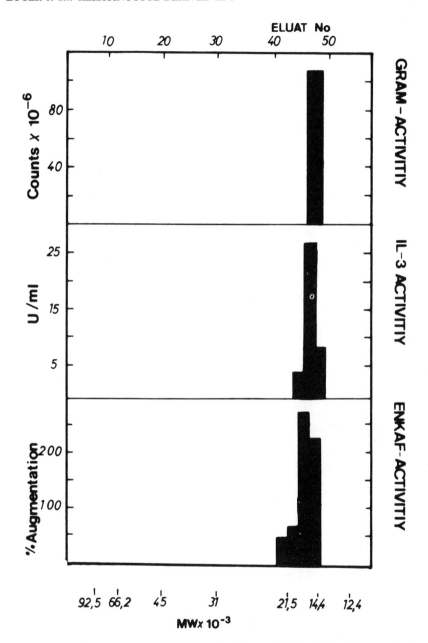

FIGURE 2. Biological activities of highly purified human EC-IL 3. Granulocyte activating mediator (GRAM) activity was tested in a Lucigenin dependent chemiluminescence assay using freshly isolated human PMNs. IL 3 activity was evaluated by the use of IL 3 dependent 32 DCL cells. Epidermal cell natural killer cell activity augmenting factor (ENKAF) activity was measured in a 4-hr Cr^{51} release assay using K562 target cells.

mRNA[27] were found to express GM-CSF mRNA[9] and IL 6 mRNA[28] but no M-CSF, IL 4 and IL 3 mRNA was detectable so far. The possibility that EC-IL 3 activity may be attributed to GM-CSF is most unlikely since rGM-CSF had no effect on IL 3 dependent cells and NK cells (TABLE 2). Moreover, an antibody directed against human GM-CSF was not capable of blocking EC-IL 3 activity.[13] The relationship between human IL 6 and EC-IL 3 is currently under investigation. Like EC-IL 3 recombinant IL 6 stimulates the proliferation of murine IL 3 dependent cell lines and enhances NK cell activity (TABLE 2). However, the question whether EC-IL 3 is IL 6 or related to other known cytokines only can be answered after gene cloning and sequencing has been performed.

TABLE 2. Biological Activities of Different Cytokines

| | IL 3 Activity[a] | | | |
	32 DCL	FDCP	NK Activity[b]	GRAM Activity[c]
Medium	152 + 80	167 ± 68	8 ± 2	9.8 ± 0.7
mu-IL 3 (25 CFUs)	5,524 ± 367	5,832 ± 235	n.d.	n.d.
rmu-GM-CSF (25 CFUs)	155 ± 77	2,823 ± 324	n.d.	n.d.
rh-IL 2 (20 U/ml)	288 ± 50	198 ± 25	72 ± 11	n.d.
rh-GM-CSF (25 CFUs)	433 ± 116	356 ± 98	10 ± 1	304.5 ± 8.0
rh-G-CSF (25 CFUs)	2,133 ± 321	410 ± 56	11 ± 2	9.2 ± 0.9
rh-CSF 1 (50 U/ml)	163 ± 27	146 ± 34	7 ± 1	n.d.
rh-IL 3 (50 U/ml)	270 ± 30	180 ± 26	9 ± 2	n.d.
rh-IL 4 (50 U/ml)	209 ± 28	157 ± 35	9 ± 2	n.d.
rh-IL 6 (50 U/ml)	1,105 ± 163	2,041 ± 124	75 ± 8	n.d.
EC-IL 3[d]	2,169 ± 225	4,930 ± 437	62 ± 5	243.2 ± 24.3

[a]IL 3 activity was evaluated by measuring (^{3}H)-thymidine incorporation of IL 3 dependent 32 DCL and FDCP cell lines. Results are expressed as cpm ± SE of 3 different experiments.

[b]EC natural killer cell activity was measured in a 4-hr Cr^{51} release assay using K562 target cells. Results are expressed as % cytotoxicity (mean ± SE) of 3 different experiments.

[c]Granulocyte activating mediator activity was tested in a Lucigenin dependent chemiluminescence assay using freshly isolated human PMNs. Results are expressed as counts × 10^{-6} (mean ± SE) of 4 different experiments.

[d]Epidermal cell-IL 3 (highly purified up to a single band on a SDS-PAGE) derived from A 431 cells.

CONCLUSIONS

In the human as well as in the murine system four major types of colony stimulating factors have been identified. They may be divided into factors exhibiting lineage specificity such as G-CSF which induces mainly colonies of neutrophilic granulocytes whereas M-CSF (CSF1) primarily induces colonies of macrophages. In contrast, colonies grown in the presence of GM-CSF or IL 3 (multi-CSF) were found to contain many different cell lineages.[29] Normal and malignant EC of human and murine origin have been shown to synthesize and release CSFs like GM-CSF[19] and IL 3[19,21] which appear to be less specific and to activate different cell types of the hematopoietic system. Similarly, the other cytokines reported to be produced by keratinocytes such as IL 1α, IL 1β[27,30] and IL 6[28] exert a rather nonspecific broad spectrum of biological activities affecting several cells involved in immune and inflammatory reactions.

It also has been observed that most of these cytokines are synthesized and released by EC in response to injurious stimuli such as UV irradiation, bacterial or viral antigen or tumor promotors.[30,31] Moreover, synthesis and release of these factors appear to be highly controlled since they may up or down regulate their own synthesis as well as interfere with the production of other cytokines by EC or other cells. Accordingly it has been found that IL 1 upregulates its own as well as GM-CSF mRNA expression in murine Pam 212 keratinocytes.[32] Similarly GM-CSF enhances GM-CSF as well as IL 1 mRNA expression in Pam 212 keratinocytes.[32,33] This hypothesis of a cytokine cascade in the epidermis is further supported by the identification of suppressor factors[34] which may represent an important down regulating pathway. Finally, nonspecific immunoregulatory EC cytokines such as IL 1, IL 3, IL 6 and GM-CSF may be involved in an early step during the initiation of host responses and thereby participate in the activation of the immune system. This is also supported by the finding that IL 1 is involved in induction of IL 2 receptors on T cells[35] leading to T cell proliferation. Moreover, it was recently reported that GM-CSF is an important mediator for inducing the maturation of epidermal Langerhans cells into potent antigen presenting dendritic cells.[36]

Although the exact role IL 3 may play in the epidermis is not yet clear, like other EC cytokines it affects a variety of cells in a nonspecific manner. Since the proliferation of mast cells and the release of mast cell-derived mediators such as histamine may be controlled by IL 3, EC through the release of IL 3 may have a regulating role in the pathogenesis of hypersensitivity reactions and allergic skin diseases. Moreover, a clinical correlation between mast cells and fibrosis has been observed in diseases such as scleroderma, scleroderma-like changes in chronic graft vs. host disease, chronic inflammation, parasitic diseases, pulmonary fibrosis and rheumatoid arthritis. It is also known that mast cell-derived histamine can affect fibroblast as well as lymphocyte functions (37). Therefore, through the release of EC-IL 3, EC may also play an important role in chronic inflammation and immunologically mediated fibrotic conditions. The active participation of EC in inflammatory reactions is further supported by the finding that EC-IL 3 directly activates granulocytes to release reactive oxygen species, known to be crucial in defense against microorganisms.[26] The activation of NK cells by EC-IL 3 appears to be of particular interest, since it recently has been shown that within the epidermis there apparently exists a new type of cell which may represent a type of natural killer cells.[38] Thus, EC through the release of EC-IL 3 may activate NK cells within the skin via an IL 2 and IFN dependent pathway. Thereby, EC may alert early nonspecific host defense mechanisms against transformed cells and various harmful microbial organisms.

The exact nature of EC-IL 3 and its relationship to T cell IL 3 is not yet fully understood and must be clarified by gene cloning and sequence analysis. However, the production of this multitargeted cytokine by EC further supports the role of the epidermis as an important immunoregulatory organ.

SUMMARY

Interleukin 3 (IL 3) initially was described as a cytokine which is produced by murine T lymphocytes and has multicolony stimulating factor (CSF) activity, activates mast cells and induces the proliferation of hematopoietic stem cell lines. In addition to T cells murine keratinocytes also produce an IL 3-like factor which according to its biological, biochemical and antigenic properties is indistinguishable from murine T cell IL 3. Moreover, by Northern blot analysis murine keratinocytes were found to express mRNA homologous to T cell IL 3 cDNA. Similarly, human keratinocytes have been shown to release an IL 3-like cytokine which also enhances the activity of natural killer cells and stimulates the release of oxygen radicals by granulocytes. However, human IL 3 mRNA could not yet be detected in human epidermal cells or epidermoid carcinoma cell lines. These findings indicate that human keratinocyte IL 3 appears to be distinct from T cell IL 3. Nevertheless, the exact nature of this cytokine remains to be clarified by sequence analysis and gene cloning. Through the production of these cytokines with IL 3-like capacity keratinocytes may participate in the regulation of the activity of different hematopoietic cells and thereby turn on early nonspecific host defense mechanisms against transformed cells and various harmful microbial organisms.

REFERENCES

1. IHLE, J. N. & Y. WEINSTEIN. 1986. Immunological regulation of hematopoietic/lymphoid stem cell differentiation by interleukin 3. Adv. Immunol. **39:** 1–50.
2. METCALF, D. 1984. The Hemopoietic Colony Stimulating Factors. Elsevier. New York, NY.
3. IHLE, J. N., L. PEPERSACK & L. REBAR. 1981. Regulation of T cell differentiation: *in vitro* induction of 20-α-hydroxysteroid dehydrogenase in splenic lymphocytes is mediated by an unique lymphokine. J. Immunol. **126:** 2184–2192.
4. FUNG, M. C., A. J. HAPEL, S. YMER, D. R. COHEN, R. N. JOHNSON, H. D. CAMPBELL & I. G. YOUNG. 1984. Molecular cloning of cDNA for mouse interleukin-3. Nature **307:** 233–237.
5. YOKOTA, T., F. LEE, D. RENNICK, C. HALL, N. ARAL, T. MOSMANN, G. NABEL, H. CANTOR, & K. ARAI. 1984. Isolation and characterization of a mouse cDNA clone that expresses mast-cell growth factor activity in monkey cells. Proc. Natl. Acad. Sci. USA **81:** 1070–1074.
6. PARK, L. W., FRIEND, S. GILLIS & D. L. URDAL. 1986. Characterization of the cell surface receptor for multi-CSF. J. Biol. Chem. **261:** 4177–4183.
7. KINDLER, V., B. THORENS, S. DEKOSSODO, B. ALLET, J. F. ELIASON, D. THATCHER, N. FARBER & P. VASSALLI. 1986. Simulation of hematopoiesis *in vivo* by recombinant bacterial murine interleukin 3. Proc. Natl. Acad. Sci. USA **83:**1001–1005.
8. STADLER, B. M., K. HIRAI, K. TADOKORO & A. L. DEWECK. 1985. Distinction of the human basophil promoting activity from a human IL 3 like factor. *In* Cellular and Molecular Biology of Lymphokines. C. Sorg, A. Schimpl & F. L. Orlando, Eds.: 479–484. Academic Press. New York, NY.
9. YANG, Y. C., A. B. CLARLETTA, P. A. TEMPLE, M. P. CHUNG, S. KOVACIC, J. S. WITEK-CLANNOTTI, A. C. LEARY, R. KRIZ, R. E. DONAHUE, G. G. WONG & S. C. CLARK. 1986. Human IL-3 (Multi-CSF): Identification by expression cloning of a novel hematopoietic growth factor related to murine IL-3. Cell **47:** 3–10.
10. YANG, Y. C. & S. CLARK, 1987. Molecular cloning and characterization of the human gene for interleukin 3 (IL 3). *In* Molecular Basis of Lymphokine Action. D. R. WELB, C. W. PRINCE & S. COHEN, Eds.: 325–337. Humana Press. Clifton, NY.
11. LEE, J. C., A. J. HAPEL, & J. N. IHLE. 1982. Constitutive production of a unique lymphokine (IL 3) by the WEHI 3 cell line. J. Immunol. **128:** 2393–2399.
12. LUGER, T. A., U. WIRTH, A. KÖCK. 1986. Epidermal cells synthesize a cytokine with interleukin 3-like properties. J. Immunol. **134:** 915–919.
13. DANNER, M. & T. A. LUGER. 1987. Human keratinocytes and epidermoid carcinoma cell lines produce a cytokine with interleukin 3-like activity. J. Invest. Dermatol. **88:** 353–361.
14. FREI, K., S. BODMER, C. SCHWERDEL & A. FONTANA. 1985. Astrocytes of the brain synthesize interleukin 3-like factors. J. Immunol. **135:** 4044–4047.
15. GRABNER, G., J. SCHREINER, T. A. LUGER, M. STUR & V. HUBER-SPITZY. 1985. Human corneal epithelial cells and a human conjunctival cell line (Chang) produce an interleukin 3 like factor. Invest. Ophthalmol. Vis. Sci. **26:** 317A.

16. HAPEL, A. J., H. S. WARREN & D. A. HUME. 1984. Different colony-stimulating factors are detected by the "interleukin-3"-dependent cell lines FDC-P1 and 32 DCL 23. Blood **64:** 786–790.
17. KUPPER, T., P. FLOOD, D. COLEMAN & M. HOROWITZ. 1987. Growth of an interleukin 2/interleukin 4-dependent T cell line induced by granulocyte-macrophage colony-stimulating factor (GM-CSF). J. Immunol. **138:** 4288–4292.
18. COLEMAN, D. L., T. S. KUPPER, P. M. FLOOD, C. C. FULTZ & M. C. HOROWITZ. 1987. Characterization of a keratinocyte-derived T cell growth factor distinct from interleukin 2 and B cell stimulatory factor 1. J. Immunol. **138:** 3314–3318.
19. KUPPER, T. S., M. HOROWITZ, F. LEE, D. COLEMAN & P. FLOOD. 1987. Molecular characterization of keratinocyte cytokines. Clin. Res. **35:** 697A.
20. KÖCK, A. & T. A. LUGER. 1985. High-performance liquid chromatographic separation of distinct epidermal cell derived cytokines. J. Chromatogr. **326:** 129–136.
21. ANSEL, J. C. 1987. Expression of IL 1 and IL 3 but not IL 4 in murine keratinocytes. Clin. Res. **35:** 247A.
22. LUGER, T. A., A. UCHIDA, A. KÖCK, M. COLOT & M. MICKSCHE. 1985. Human epidermal cells and squamous carcinoma cells synthesize a cytokine that augments natural killer cell activity. J. Immunol. **134:** 2477–2483.
23. HERBERMANN, R. B. 1985. Multiple functions of natural killer cells, including immuno-regulation as well as resistance to tumor growth. Concepts Immunopathol. **1:** 96.
24. TIMONEN, T., J. R. ORTALDO & R. B. HERBERMAN. 1981. Characteristics of human large granular lymphocytes and relationship to natural killer and K cells. J. Exp. Med. **153:** 569–575.
25. LUGER, T., A. KAPP, M. MICKSCHE & M. DANNER. 1987. Characterization of a distinct epidermal cytokine with multiple immunoregulatory properties. Clin. Res. **35:** 700A.
26. KAPP, A., M. DANNER, T. A. LUGER, C. HAUSER & E. SCHÖPF. 1987. Granulocyte-activating mediators (GRAM) II. Generation by human epidermal cells—relation to GM-CSF. Arch. Dermatol. Res. **279:** 470–477.
27. KUPPER, T. S., D. W. BALLARD, A. O. CHUA, J. S. MCGUIRE, P. M. FLOOD, M. C. HOROWITZ, R. LANGDON, L. LIGHTFOOT & U. GUBLER. 1986. Human keratinocytes contain mRNA indistinguishable from monocyte interleukin 1α and β mRNA. Keratinocyte epidermal cell-derived thymocyte-activating factor is identical to interleukin 1. J. Exp. Med. **164:** 2095–2100.
28. KÖCK, A., T. A. LUGER & J. ANSEL. 1988. Expression of IL 6 in human epidermoid carcinoma cells. Clin. Res. **36:** 377A.
29. CLARK, S. C. & R. KAME. 1987. The human hematopoietic colony stimulating factors. Science **236:** 1229–1236.
30. ANSEL, J. C., T. A. LUGER, D. LOWRY, P. PERRY, D. R. ROOP & J. D. MOUNTZ. 1988. The expression and modulation of IL 1α in murine keratinocytes. J. Immunol. **140:** 2274–2279.
31. LUGER, T. A. & J. C. ANSEL. 1988. Epidermal Cytokines. Proc. of CMD. Berlin. In press.
32. ANSEL, J. C. & P. PERRY. 1988. Modulation of GM-CSF expression in murine epidermal cells. Clin. Res. **36:** 374A.
33. ANSEL, J., M. HEINRICH, S. HEFENEIDER & P. PERRY. 1988. Regulation of IL 1 expression in murine keratinocytes by IL 1, TNFα, and GM-CSF. Clin. Res. **36:** 630A.
34. SCHWARZ, T., A. URBANSKA, F. GSCHAIT & T. A. LUGER. 1987. UV-irradiated epidermal cells produce a specific inhibitor of interleukin 1 activity. J. Immunol. **138:** 1457–1463.
35. OPPENHEIM, J. J., E. J. KOVACS, K. MATSUSHIMA & S. K. DURUM. 1986. There is more than one interleukin 1. Immunology Today **7:** 45–56.
36. WITMER-PACK, M. D., W. OLIVIER, J. VALINSKY, G. SCHULER & R. M. STEINMAN. 1987. Granulocyte/macrophage colony stimulating factor is essential for the viability and function of cultured murine epidermal Langerhans cells. J. Exp. Med. **166:** 1499–1509.
37. FERNEX, M., Ed. 1968. The Mast Cell System. Its Relationship to Atherosclerosis, Fibrosis and Eosinophils. William & Wilkins. Baltimore, MD.
38. ROMANI, N., G. STINGL, E. TSCHACHLER, M. D. WITMER, R. M. STEINMAN, R. M. SHEVACH & G. SCHULER. 1985. The Thyl bearing cell of murine epidermis. A distinctive leukocyte perhaps related to natural killer cells. J. Exp. Med. **161:** 1368–1372.

Hematopoietic, Lymphopoietic, and Proinflammatory Cytokines Produced by Human and Murine Keratinocytes

THOMAS S. KUPPER,[a] MARK HOROWITZ,[a,b]
NICHOLAS BIRCHALL,[a] HITOSHI MIZUTANI,[a]
DAVID COLEMAN,[c] JOSEPH McGUIRE,[a]
PATRICK FLOOD,[a,d] STEVEN DOWER,[e]
AND FRANK LEE[f]

*Departments of [a]Dermatology
[b]Orthopedic Surgery
[c]Medicine (Infectious Disease), and [d]Pathology
Yale University School of Medicine
500 LIC
New Haven, Connecticut 06510*

*[e]Immunex Corporation
151 University Street
Seattle, Washington 98101*

*[f]DNAX Research Institute
901 California Avenue
Palo Alto, California 94304*

INTRODUCTION

The hematopoietic colony stimulating factors (CSFs) are a class of proteins which were initially discovered and characterized by their capacity to induce the generation of leukocyte colonies from individual bone-marrow-derived precursor cells in a semisolid agar culture system (review in REF. 1). The morphology of these colonies led to a nomenclature for these cytokines as multi-CSF (IL-3), granulocyte-macrophage CSF (GM-CSF), granulocyte CSF (G-CSF), and macrophage CSF (M-CSF or CSF-1). The molecular cloning and characterization of these factors confirmed their functional identity; however, it also led to a series of unanticipated observations. First, most CSFs have significant activities on fully differentiated mature cells, and these activities appear to parallel their hematopoietic activities (for example, G-CSF activates mature neutrophils while GM-CSF is a potent activator of both mature neutrophils and macrophages). Secondly, a class of factors which by themselves have little or no hematopoietic activity synergize effectively with the above CSFs, resulting in qualitative and quantitative changes in the repertoire of cells generated. These factors include IL-1 alpha and beta (hemopoietin 1), IL-4 (BSF-1), IL-5 (eosinophil differentiation factor), and IL-6 (beta 2 interferon, BSF-2). Thirdly, inhibitors of the hematopoietic effects have been identified in several systems and include both type 1 (alpha and beta) and type 2 (gamma) interferons, transforming growth factor beta (TGF-beta), and IL-2.[2]

One of the most intriguing recent observations is that many of these hematopoietic CSFs are produced by tissues distant from the bone marrow, including mammalian epidermis. Indeed, the epidermis is separated from the bone marrow by multiple tissues, and it is difficult to imagine that cytokines produced by this tissue in modest amounts have distal effects on resident bone marrow cells. This paper will review the evidence that epidermal keratinocytes produce IL-1 alpha and beta, IL-3, IL-6, GM-CSF, G-CSF, and M-CSF, and will speculate as to the role that these factors may play in the biology of skin.

The spontaneously transformed neonatal keratinocyte cell line Pam 212 has been extremely useful in characterizing murine keratinocyte cytokine production.[3] Extrac-

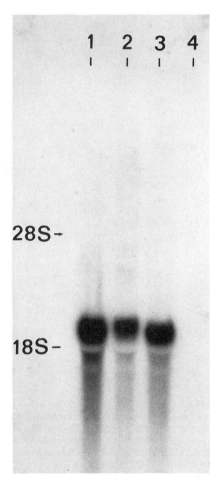

FIGURE 1. Total RNA was isolated from LPS-stimulated P388D1 murine macrophages (*lane 1*), Pam 212 murine keratinocytes (*lane 2*), LPS-stimulated Pam 212 (*lane 3*), and BW 12 murine thymomona line (*lane 4*). Northern blot analysis was performed as described previously.[15] The blot was hybridized with 32-P labelled murine IL-1 alpha cDNA (gift of Peter Lomedico, Hoffman LaRoche).

tion of RNA from this cell line and hybridization with murine cytokine cDNAs using Northern blot analysis reveals constitutive production of IL-1 alpha (FIG. 1), IL-3 (FIG. 2), and GM-CSF (FIG. 3). Using a highly sensitive S1 nuclease analysis assay, we were unable to detect either IL-4 (FIG. 4) or IL-2 (not shown) mRNA. Pam 212 production of IL-1 activity, as measured by the D10.G4.1 bioassay, also appears to be constitutive and this activity can be neutralized by a rabbit antiserum to recombinant

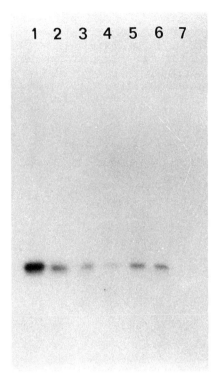

FIGURE 2. Poly A RNA was isolated by affinity chromatography from WEHI 3 murine myelomonocytic leukemia cells (*lane 1*, 2 μg; *lane 3*, 1 μg), PMA-stimulated EL4 murine T cells (*lane 3*, 2 μg), Pam 212 cells (*lane 4*, 2 μg), LPS/cyclohex-stimulated Pam 212 cells (*lane 5*, 2 μg), PMA-stimulated Pam 212 cells (*lane 6*, 2 μg), and P388D1 cells (*lane 7*, 2 μg). The blot was hybridized with 32-P labelled murine IL-3 cDNA.

FIGURE 3. Total RNA was isolated from AC5 murine bone marrow stromal cells (*lane 1*), EL4 murine thymoma cells (*lane 2*), Pam 212 cells (*lane 3*), and Pam 212 cells stimulated with PMA (*lane 4*, 10 μg/ml). The blot was hybridized with 32-P labelled murine GM-CSF cDNA.

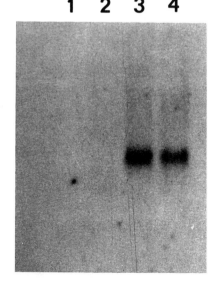

IL-1 alpha (FIG. 5). Both IL-3 and GM-CSF biological activity can be identified after RP-HPLC separation of Pam 212-conditioned medium. IL-3 activity elutes at 43% acetonitrile, and induces DA-1 but not HT-2 proliferation. GM-CSF activity elutes at 49% acetonitrile and induces both HT-2 and DA-1 proliferation. Both activities can be neutralized by antibodies to recombinant IL-3 and GM-CSF, respectively (not shown). M-CSF activity, as measured by peritoneal macrophage proliferation and the 14M1.4 (bone-marrow-derived macrophage) cell line proliferation, is also produced constitutively (Chodakewitz, J., personal communication). Primary confluent cultures of neonatal murine keratinocytes produce identical cytokine activities, also in an apparently constitutive fashion.[4,5]

Human keratinocytes have previously been shown to contain mRNAs homologous to IL-1 alpha and beta and GM-CSF by S1 nuclease analysis.[6,7] We have been unable,

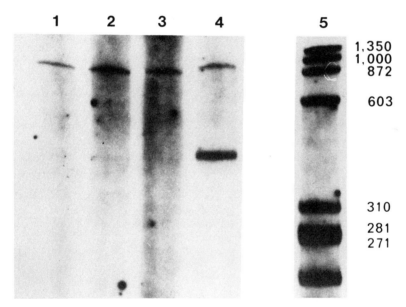

FIGURE 4. Total RNA was isolated from Pam 212 cells stimulated with PMA (*lane 1*), Pam 212 cells (*lane 2*), AC5 cells (*lane 3*), and Con-A-stimulated D9 cells (*lane 4*). The S1 nuclease analysis was performed using 32-P labelled murine IL-4 cDNA as previously described.[15]

however, to demonstrate IL-3 mRNA in cultures of human keratinocytes (not shown). IL-1 biological activity can be identified in conditioned medium of human keratinocytes and can be neutralized by antibodies to recombinant human IL-1 alpha (FIG. 6). Interestingly, all detectable IL-1 biological activity from cultured keratinocytes and from freshly obtained epidermis appears to be of the alpha form, despite the fact that IL-1 beta mRNA is readily detectable under these conditions. Similarly, GM-CSF biological activity can be identified in conditioned medium from IL-1-induced human keratinocytes as measured by proliferation of the TALL 101 cell line, which responds to both GM-CSF and IL-3 (FIG. 7). However, all of the TALL 101 stimulating activity mediated by keratinocyte-conditioned medium can be neutralized by an antibody to recombinant GM-CSF which does not cross-react with human IL-3.[8] These data also

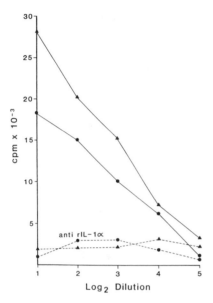

FIGURE 5. Incorporation of 3H-Thymidine into D10.G4.1 cell DNA 72 hours after stimulation with serial dilutions of either recombinant IL-1 alpha (*closed triangles*) or Pam 212 conditioned medium (*closed circles*) in the presence (*dotted line*) or absence (*solid line*) of a 1:1000 dilution of rabbit anti-rIL-1 alpha antiserum.

suggest an absence of IL-3 in these media. Studies on the G- and GM-CSF responder cell line MV 411 indicate that proliferation induced by keratinocyte-conditioned medium can be only partially neutralized by the anti-GM-CSF antibody, suggesting the presence of G-CSF in this medium; the presence of G-CSF mRNA in human keratinocytes is currently being confirmed by S1 nuclease analysis (unpublished observations). Finally, we have recently confirmed the presence of IL-6 in human keratinocytes by Northern analysis and immunoprecipitation of cells and conditioned medium, using an antibody to recombinant IL-6.[9] Interestingly, a substantial amount of IL-6 can be identified in the cell membrane fraction. We have demonstrated in the

FIGURE 6. Serial dilutions of conditioned medium from cultured human keratinocytes were tested for D10.G4.1-stimulating activity in the presence of a 1:200 dilution of rabbit antiserum to rhIL-1 alpha, rhIL-1 beta, or a rabbit antiserum to rhIL-6. 3D3 antibody was used as the costimulus for D10.G4.1.

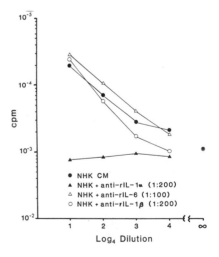

studies outlined above and elsewhere that cultured human and/or murine keratinocytes produce IL-1 alpha and beta, GM-CSF, G-CSF, M-CSF, IL-3, and IL-6. We have not found evidence for the production of IL-2 or IL-4 by cultured keratinocytes, though it now appears that dendritic epidermal T cells can produce these factors.

Gene expression for all of these cytokines appears to require an inductive signal. This inductive signal can be provided by the keratinocyte itself, since exogenous IL-1 has been shown to induce keratinocyte gene expression of these cytokines.[7] We have identified and characterized both high- and low-affinity IL-1 receptors on normal human keratinocytes (FIGS. 8 and 9). The characterization of these receptors has been published,[10] and we have found in general that stimuli which enhance IL-1 gene expression also induce markedly increased cell surface levels of IL-1 receptors.[11] Since it appears that the number of IL-1 receptors expressed on a keratinocyte reflects its responsiveness to IL-1, it may be that autocrine stimulation of keratinocytes by endogenous IL-1 is fundamental to the release of hematopoietic CSFs and other

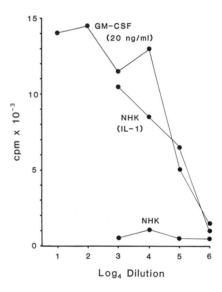

FIGURE 7. Incorporation of 3H-Thymidine into TALL 101 cells (a human T cell acute lymphocytic leukemia cell line, G. Rovera, Wistar Inst.) after 48-hour stimulation with serial dilutions of rhGM-CSF or conditioned medium from either unstimulated NHK or NHK stimulated with rIL-1 alpha for 24 hours.

cytokines. Thus, stimuli which either induce IL-1 release from ketatinocytes (active or passive) and/or enhance IL-1 gene expression (*e.g.,* UVB irradiation) might be expected to initiate a cascade of cytokines including those mentioned above with significant activities on both mature and immature leukocytes. A model for this type of activation is illustrated in FIGURES 10 and 11.

The capacity of skin to produce activators of mature granulocyte effector functions seems well suited to the protective role of the skin as interface between the host and environmental pathogens. Therefore, one interpretation of the widespread tissue distribution of such factors is that anatomical compartmentalization allows epidermal production of the hematopoietic CSFs to induce inflammation in skin by activating mature cells, and bone marrow stromal cell production of the same factors to induce hematopoiesis in the bone marrow by activating immature ones. The activities of G-CSF and GM-CSF on granulocytes and Langerhans cells, GM- and M-CSF on

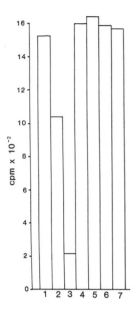

FIGURE 8. Binding of 125-I-labelled rhIL-1 alpha to mono-layers of adult human keratinocytes (10^{-10} M) grown in MCDB 153 medium (Clonetics). Radiolabelling of IL-1 alpha and binding protocols have been previously described. Unlabelled cytokines at the following concentrations were added at the initiation of the study and specific bound counts were measured: (**1**) none; (**2**) rhIL-1 alpha (10^{-10} M); (**3**) rhIL-1 alpha (10^{-8} M); (**4**) rhIL-6 (10^{-10} M); (**5**) rhIL-6 (10^{-8} M); (**6**) beta-MSH (10^{-6} M); and (**7**) rhIL-2 (10^{-8} M).

monocytes, IL-3 on mast cells, and IL-6 on B cells and plasma cells, are well suited to a peripheral tissue which frequently encounters injurious or infectious agents.

Does hematopoiesis occur in skin? Certainly pluripotent hematopoietic stem cells circulate in peripheral blood, and it is difficult to believe that such cells would not differentiate into mature leukocytes when they encounter an inflamed tissue containing high levels of hematopoietic CSFs released by resident cells. In fact, this would represent a highly efficient means of generating large numbers of granulocytes and macrophages in a specific area where their presence is required without requiring systemic release of CSFs and large scale mobilization of resident bone marrow stem cells. However, under defined conditions, hematopoietic growth factors produced by epidermis may have endocrine activities. *In vivo* UVB irradiation appears to markedly shorten the degree and duration of bone marrow suppression induced by a single large dose of 5-fluorouracil in mice,[12] presumably by inducing release of IL-1 and IL-1-inducible hematopoietic CSFs from skin.

Langerhans cells are derived from precursors which arise in the bone marrow. In mice, Langerhans cells are considered to be distinct from another population of

FIGURE 9. Scatchard analysis of 125-I-labelled rhIL-1 alpha to confluent monolayers of keratinocytes grown in MCDB 153 (KGM, Clonetics). Cells were treated with 10 ng/ml PMA (*squares*) or vehicle (*circles*) for 24 hours prior to binding assays.

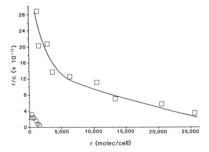

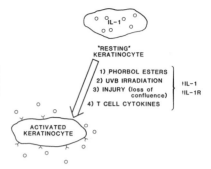

FIGURE 10. Hypothetical scheme by which a "resting," cytokine nonproducer keratinocyte is induced to become activated by releasing IL-1 and expressing IL-1 receptors.

bone-marrow-derived dendritic epidermal cells which bear both Thy-1 and a gamma-delta T cell receptors. It has been shown that GM-CSF induces the differentiation of freshly isolated Langerhans cells (which present antigen poorly to T cells) into mature highly efficient antigen-presenting cells. Whether resident Langerhans cells have matured *in situ* from immature precursors in response to ambient epidermal cytokines is, at present, unknown. Cell lines derived from such Thy-1$^+$ dendritic epidermal cells proliferate in response to GM-CSF and IL-3, and bear a striking resemblance to immature thymocytes.[13] These cells may also have matured *in situ* under the influence of hematopoietic cytokines.

What is the relationship between hematopoietic and lymphopoietic cytokines? It is clear that a common stem cell gives rise to both myeloid and lymphoid elements of blood. Thymic epithelium has long been considered to produce factors which influence the maturation of T cells from immature precursors. We have recently demonstrated that murine thymic epithelial cell lines produce the same spectrum of hematopoietic colony stimulating factors as do neonatal murine keratinocyte lines.[14] Might these factors, in conjunction with thymopoietin and other factors or cell surface markers, contribute to lymphopoiesis? Evidence that certain immature and mature T cells bear receptors for GM-CSF suggests that the divergence of T lymphoid and myeloid may occur later in differentiation than previously thought. It is therefore not unreasonable to suggest that thymocytes or dendritic epidermal T cells may respond to signals provided by hematopoietic CSFs which have been previously understood in the context of myeloid differentiation.

It is likely that the role of hematopoietic colony stimulating factors may be even more complex than the combined possibilities above. For instance, observations that certain epithelial carcinomas bear receptors for and proliferate in response to IL-1, CSF-1 and GM-CSF may suggest a role for these factors in the normal growth of

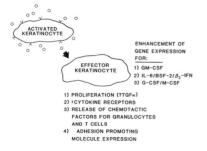

FIGURE 11. Autocrine activation of keratinocytes via IL-1/IL-1R interaction induces proliferation, cytokine release, cytokine receptor expression, and adhesion promoting molecule expression.

epithelial tissues. Hematopoiesis may be only one of many important activities mediated by the hematopoietic CSFs.

REFERENCES

1. CLARK, S. & R. KAMEN. 1987. Science **236:** 1229.
2. SIEFF, C. A. 1987. J. Clin. Invest. **79:** 1549.
3. KUPPER, T., D. COLEMAN, J. MCGUIRE *et al.* 1986. Proc. Natl. Acad. Sci. USA **83:** 4451.
4. KUPPER, T., M. HOROWITZ, F. LEE *et al.* 1987. J. Invest. Dermatol. **88:** 501.
5. GOLDMINZ, D., T. KUPPER & J. MCGUIRE. 1987. J. Invest. Dermatol. **88:** 97.
6. KUPPER, T., D. BALLARD, A. CHUA *et al.* 1986. J. Exp. Med. **164:** 2095.
7. KUPPER, T. 1988. Adv. Dermatol. **3:** 293.
8. KUPPER, T., N. BIRCHALL. Submitted.
9. KUPPER, T., L. MAY, H. MIZUTANI, N. BIRCHALL & P. SEHGAL. Ann. N.Y. Acad. Sci. In press.
10. KUPPER, T., S. DOWER, N. BIRCHALL, S. CLARK & F. LEE. 1988. J. Clin. Invest. **82.** 1287.
11. BLANTON, B., T. KUPPER, S. MCDOUGALL & S. DOWER. Proc. Natl. Acad. Sci. USA. In press.
12. BIRCHALL, N., C. GAMBA & T. KUPPER. 1988. Submitted.
13. TIGELAAR, R. 1988. Ann. N. Y. Acad. Sci. This volume.
14. BIRCHALL, N., T. POTWOROWSKI, F. LEE, C. GAMBA & T. KUPPER. Submitted.
15. KUPPER, T., M. HOROWITZ, F. LEE *et al.* 1987. J. Immunol. **138:** 420.

Effect of Keratinocyte Cytokines on Thy-1$^+$ Dendritic Epidermal Cells

R. TIGELAAR, J. NIXON-FULTON,
A. TAKASHIMA, W. KUZIEL, C. TAKIJIRI,
J. LEWIS, P. TUCKER, AND P. BERGSTRESSER

University of Texas Southwestern Medical Center
Departments of Dermatology and Microbiology
5323 Harry Hines Boulevard
Dallas, Texas 75235

INTRODUCTION

It is now well recognized that the epidermis of normal mice contains two distinct populations of dendritic, bone marrow-derived cells; Ia$^+$ Langerhans cells and Thy-1$^+$ epidermal cells (Thy-1$^+$ EC). Particularly during the past two years, knowledge about Thy-1$^+$ EC has advanced to the point that these cells are of considerable interest to a variety of investigators including cellular and molecular immunologists and cutaneous biologists. It has become apparent that Thy-1$^+$ EC are a readily accessible and perhaps unusually homogeneous population with the genotypic, phenotypic, and functional properties of a newly identified subset of T lymphocytes which express a γ/δ T cell antigen receptor (TCR), whose relationships to other T cells are just beginning to be understood, and whose relevant *in vivo* roles have yet to be defined.

Thy-1$^+$ EC normally are found distributed in a fairly regular and dendritic array throughout interfollicular epithelium as well as in a somewhat less uniform pattern within follicular epithelium, putting them in direct contact with large numbers of keratinocytes. This localization, when coupled with our rapidly advancing understanding that keratinocytes are a rich source of cytokines capable of affecting the growth, differentiation and state of activation of a variety of lymphoid cells, raises obvious questions about the effects of such cytokines on Thy-1$^+$ EC. Available evidence gives rise to the intriguing possibility that some, if not all, steps in the differentiation of bone marrow-derived, immature, CD3- TCR- precursors into CD3$^+$ TCR γ/δ^+ T cells may occur *within* the epidermis, thereby implicating keratinocytes and/or their cytokines in that process. Furthermore, data are consistent with the possibility that epidermal keratinocytes may provide accessory signals necessary for mitogen-driven proliferation of Thy-1$^+$ EC *in vitro*. Finally, some populations of activated Thy-1$^+$ EC, under appropriate circumstances, can utilize known or presumed keratinocyte cytokines as growth factors. This paper will summarize these recent data in the context of our present understanding of the phenotypic, genotypic, and functional characteristics of Thy-1$^+$ EC and their relationship(s) to other T lymphocytes.

Thy-1$^+$ EC from Normal Mice Resemble a Minor Subset of T Cells in the Thymus and Peripheral Tissues of Adult Mice and Humans

The phenotype of Thy-1$^+$ EC from normal mice [Thy-1$^+$, asialo GM1$^+$, Lyt-5$^+$, Ia$^-$, NK-1$^-$, L3T4(CD4)$^-$, and Lyt-2(CD8)$^-$] indicated that they were neither

271

Langerhans cells, natural killer (NK) cells nor typical peripheral T cells.[1-8] However, the findings that Thy-1$^+$ EC could proliferate in response to concanavalin A (Con A) as well as secrete and respond to interleukin 2 (IL-2), suggested a close relationship to cells in the T lineage.[9-12] This relationship was reinforced by the absence of Con A-responsive Thy-1$^+$ EC in *scid* mice (which lack normal T cells and T cell precursors, but have normal NK cells[13-16]), and the appearance of such Con A-reactive cells in the epidermis of *scid* recipients transplanted with normal bone marrow cells.[17]

The T cell antigen receptor (TCR) is typically a heterodimeric (usually disulfide-linked) (glyco)protein, each chain of which is the product of one of four genes (termed α, β, γ, and δ) which, during the process of differentiation, rearrange variable (V-), diversity (D-), joining (J-), and constant (C-) segments in a manner similar to that which occurs in the rearrangement of immunoglobulin genes in B cells. Most adult thymocytes and peripheral T cells express a TCR composed of an α chain and a β chain; this TCR is expressed in noncovalent association with the multichain CD3 complex.[18-24] In contrast, the CD3-associated TCR expressed on Thy-1$^+$ EC is composed of a γ/δ heterodimer.[25-31] This class of TCR is also expressed by a minor population of adult human and adult mouse thymocytes and peripheral T cells.[32-43] Furthermore, the functional capabilities of lines and/or clones of Thy-1$^+$ EC are similar to those reported for other populations of TCR γ/δ^+ cells, *i.e.*, proliferation following activation with Con A or insolubilized anti-CD3, secretion of and response to the lymphokine IL-2, and the capacity, once activated, to exhibit non-MHC-restricted, NK-like cytotoxicity.

The γ/δ TCRs Expressed by AKR/J Thy-1$^+$ EC Clones Resemble Those Expressed Preferentially by Day 14-15 Fetal Thymocytes

Virtually all TCR$^+$ thymocytes early in gestation (day 13–15) are γ/δ^+; beginning about fetal day 16, cells with other TCRs, particularly those with α/β TCRs, begin to accumulate rapidly.[38,39,44-52] The relationships between γ/δ^+ and α/β^+ cells have yet to be clarified adequately. The currently most popular theory is that they represent separate T cell lineages, γ/δ cells being the phylogenetically more "primitive" population.[24,38,42,43] An alternative and not disproved theory is that γ/δ cells represent a "precursor" population which can "switch" its TCR to α/β after appropriate signalling.[25,46]

In our analysis of TCR expression in Thy-1$^+$ EC clones established from AKR/J mice by limiting dilution microculture in the presence of Con A, IL-2, and irradiated splenic filler cells, we noted that each of five separate clones exhibited similar TCR gene expression profiles.[25,26] That is, they not only each utilized the same γ gene C region (Cγ1, which is one of three known C regions which can code for a functional protein), but they each expressed the same γ gene V region (V3, which is one of four potentially utilizable V genes in the γ1 complex). Similar preferential use of V3-Cγ1 is seen in fetal thymocytes at day 14–15 of gestation; *i.e.*, this combination is selectively expressed on the earliest thymocytes known to express a TCR. However, by day 16–17 of gestation, selective utilization of V3 declines precipitously, with increased expression of other γ chain V genes, as well as the rapid accumulation of cells expressing a variety of α/β TCRs.[44,47,48,52]

We have very recently analyzed TCR mRNA expression in these same populations using cDNA probes to specific regions within the TCR δ locus, and have found that each AKR/J clone also displays remarkably limited heterogeneity of TCR δ gene expression. That is, each clone expresses full-length (2.0 kb), potentially productive mRNA containing the same V region, named V$_{\delta 1}$,[52] and the same J region, J2 (J.

Allison et al., manuscript in preparation). Once again, preferential utilization of this same $V_{\delta 1}$ V region has also been reported in day 15 fetal thymocytes; furthermore, parallel to the situation observed for V3-Cγ1, productive expression of $V_{\delta 1}$ drops dramatically by the very next day of gestation.[52] Finally, while one explanation for the apparent lack of TCR heterogeneity among AKR/J Thy-1$^+$ EC lines and clones is strain-specific TCR gene utilization, very recently we have found that a short-term line of Con A- and IL-2-stimulated Thy-1$^+$ EC from CBA/J mice also expresses full-length mRNA for both V3-Cγ1 and $V_{\delta 1}$-J2-Cδ (J. Nixon-Fulton and W. Kuziel, unpublished observations). The reasons for this apparently restricted and similar TCR γ/δ gene utilization in Thy-1$^+$ EC from adult mice and in fetal thymocytes at a discrete and limited time in gestation are yet to be elucidated; one of the more interesting explanations is that these populations are similar because of similarities between the thymic and cutaneous microenvironments in which these cells differentiate.

Precursor Frequency Analysis Suggests That Not All Thy-1$^+$ EC from Normal Mice Can Proliferate in Response to Con A and IL-2

Recently we determined the precursor frequency of Con A-responsive Thy-1$^+$ EC freshly isolated from the epidermis of AKR/J mice.[12] After preliminary experiments to determine optimal culture conditions, limiting dilution microculture of flow cytometry-purified Thy-1$^+$ EC in the presence of Con A, IL-2, and irradiated syngeneic spleen cells (as a source of accessory cells) revealed that approximately 20% of Thy-1$^+$ EC had the capacity for sustained proliferation. The observation that approximately 80% of the Thy-1$^+$ EC failed to proliferate under the conditions employed suggested that freshly isolated Thy-1$^+$ EC are functionally heterogenous. A relatively trivial explanation for at least some of this heterogeneity would be cell damage during preparation of the epidermal cell suspensions. Alternatively, the possibilities that not all Thy-1$^+$ EC can transduce Con A binding into a mitogenic signal, and that this failure may be associated with heterogeneity of CD3-TCR complex expression, are consistent with several observations. These include: 1) evidence, using both normal T cells and mutants selected for loss of CD3-TCR complex expression, that mitogens such as Con A activate cells via binding to the CD3-TCR complex;[21,53,54] 2) phenotypic analysis of Thy-1$^+$ EC in situ indicating that while all freshly isolated Thy-1$^+$ EC appeared to be CD3$^+$, only 40–60% appeared to be Cγ1-positive;[29] 3) recent phenotypic analysis of short-term (2-week) lines of AKR/J and CBA/J Thy-1$^+$ EC stimulated with Con A and IL-2, showing that not all CD3$^+$ cells react with a monoclonal antibody directed against the TCR expressed by all long-term lines/clones of AKR/J Thy-1$^+$ EC examined to date (J. Lewis, and J. Allison, unpublished observations). The possibility that a subset of Thy-1$^+$ EC bearing a particular CD3-TCR complex responds preferentially to Con A and IL-2 under the conditions employed must be kept in mind as an explanation both for the observed precusor frequency of 20% and for the apparently limited heterogeneity of TCR expression in AKR/J Thy-1$^+$ EC clones established by stimulation with Con A.

Thy-1$^+$ EC in Nude Mice Appear to be Functionally and Phenotypically Less Differentiated Than Thy-1$^+$ EC from Normal Mice

Nude mice lack a thymus and have grossly abnormal hair growth.[55] Furthermore, Krueger et al.[56] have demonstrated that the proliferative response of nude mouse epidermis to TPA was different from either that of normal euthymic littermates or of hairless (hr/hr) mice, a strain lacking hair follicles, but with normal thymic function. The epidermis of nude mice is known to contain reduced but readily detectable numbers of Thy-1[+] EC.[5] We recently completed a detailed examination of the morphologic, functional and genotypic characteristics of nude Thy-1[+] EC, and found them to be distinct from Thy-1[+] EC in euthymic mice.[57] 1) Morphologically, nude Thy-1[+] EC are more likely to be round or angular rather than dendritic in appearance as in normal mice, and they localize predominantly in follicular epithelium rather than interfollicular epithelium as in normal mice. 2) While Thy-1[+] EC from normal mice respond modestly to IL-2 alone, nude Thy-1[+] EC give a marked IL-2 response; furthermore, this response is triggered by relatively low concentrations of IL-2 (<10 U/ml), considerably less than the concentrations found to be optimal for proliferation of true NK cells.[16] 3) Unlike Thy-1[+] EC from normal mice, nude Thy-1[+] EC fail to proliferate or secrete IL-2 in response to Con A; however, both populations proliferate and secrete IL-2 in response to phorbol ester (PMA) and calcium ionophore (Ionomycin). These functional characteristics are similar to those described for some populations of immature thymocytes,[58] and again distinguish nude Thy-1[+] EC from true NK cells, which can secrete γ-interferon following activation with PMA/Ionomycin, but do not secrete IL-2 (M. Tutt and V. Kumar, personal communication). 4) At the level of TCR gene transcription, short-term cultured nude Thy-1[+] EC express no TCR α and only nonfunctional TCR β transcripts, a pattern similar to that observed with cultured Thy-1[+] EC from normal mice. However, while such normal Thy-1[+] EC express abundant full-length, (containing a V region) potentially functional mRNA for both TCR γ and TCR δ as well as mRNA for the CD3δ chain, nude Thy-1[+] EC express only truncated, presumably germline TCR γ transcripts lacking a V region, and no TCR δ or CD3δ. This profile of gene expression not only distinguishes nude Thy-1[+] EC from normal Thy-1[+] EC but also from true NK cells, which do not transcribe substantial amounts of even truncated TCR β or γ mRNA.[15,16] These data indicate that Thy-1[+] EC derived from nude mice are not NK cells, but rather suggest that they represent cells in a T cell pathway of differentiation which are less differentiated than most Thy-1[+] EC in normal euthymic mice.

Evidence That Thy-1[+] EC Differentiation May Occur within the Cutaneous Microenvironment and Be Thymic Independent

The presence in nude epidermis of Thy-1[+] EC which are CD3[−], but express some truncated TCR gene transcripts and can be triggered to secrete IL-2 argues that at least some early steps in the differentiation of bone marrow precursors to T cells can be thymic independent. In point of fact, substantial evidence from studies of nude mice indicates that T cell differentiation (particularly of class I MHC-restricted Lyt-2[+] cells) can occur extrathymically,[59–61] albeit somewhat inefficiently and resulting in a more limited (oligoclonal) TCR α/β repertoire.[62] The possibility that epidermis can be a site of such extrathymic differentiation[63] has been reinforced by the several similarities between the epidermal and thymic microenvironments: 1) cross-reactivity between thymic epithelial and keratinocyte antigens;[64–66] 2) keratinocyte secretion of immunomodulating cytokines such as IL-1[67–68], GM-CSF[69,70], an IL-3-like molecule,[71] IL-6 (formerly known as BSF 2 or β2-interferon; T. Kupper, personal communication), α-interferon,[72] β1-interferon (see B. Nickoloff, this volume), and thymopoietin;[73] 3) presence in both epidermis and thymus of bone marrow-derived Ia[+] cells.

How then does one explain the lack of CD3[+], TCR γ/δ[+] Con A-responsive Thy-1[+] EC in nude mice? One possibility is that such cells are postthymic (e.g., bone marrow precursors traffic to the thymus, differentiate there into CD3⁺ TCR γ/δ^+ cells, and then migrate to the epidermis); thus athymic mice will lack such cells. (This scheme is illustrated in FIG. 1 by the dotted arrows going from bone marrow to thymus to skin.) For the vast majority of TCR γ/δ^+ cells in peripheral organs such as the spleen and lymph nodes, the case for their thymic dependency is strong. Young nude mice (2–3 months old) have virtually no detectable CD3⁺ T cells (either TCR α/β^+ or γ/δ^+) in their spleens; however, within weeks of receiving syngeneic thymus grafts, nude spleens contain readily detectable CD3⁺ TCR γ/δ^+ (and α/β^+) cells.[43] (See FIG. 1, solid arrows.) However, according to Shimada *et al.*,[31] nude epidermis even several months after thymus grafting still does not contain CD3⁺ Thy-1⁺ EC. In addition, we have observed that *scid* skin (devoid of dendritic Thy-1⁺ EC) becomes populated with dendritic Thy-1⁺ EC within 2 months of grafting to nude mice (J. Nixon-Fulton, unpublished observations). Taken together, these observations suggest the possibility

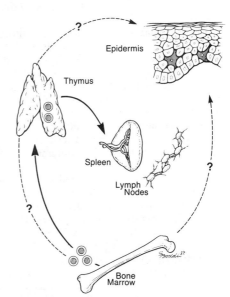

FIGURE 1. Migration pathways of TCR γ/δ^+ T cells and their precursors.

that dendritic Thy-1⁺ EC may not be thymus-derived even in mice which have a normal thymus, but rather may represent bone marrow-derived precursors which have migrated into the epidermis and differentiated *in situ* into CD3⁺ TCR γ/δ^+ cells. (See also FIG. 1.) Additional data consistent with the possibility that differentiation of TCR γ/δ^+ Thy-1⁺ EC may occur within the epidermal microenvironment comes from the studies of Romani *et al.*[5] They found very few Thy-1⁺ EC at birth, and adult densities were not reached for several weeks. These kinetics are considerably slower than that seen within the thymus for the appearance of CD3⁺ TCR γ/δ^+ cells; such cells are readily found by day 15 of gestation, and at later times comprise a relatively constant number but rapidly decreasing percentage of CD3⁺ thymocytes. If CD3⁺ TCR γ/δ^+ Thy-1⁺ EC represent cells which have migrated from the thymus into the epidermis, since such thymocytes are at maximal levels well before birth, it might be anticipated that the kinetics of appearance of Thy-1⁺ EC in the epidermis would not be as slow as they apparently are.

If the possibility is entertained that Thy-1$^+$ EC are not thymic dependent, then it is reasonable to suggest that nude mice may lack Con A-responsive, CD3$^+$ TCR γ/δ^+ Thy-1$^+$ EC because of a cutaneous microenvironmental defect rather than because they lack a thymus. The absence of reports of successful genetic dissociation in nude mice of the cutaneous defect from the defect in thymus development points to a very close linkage between the gene(s) responsible for the two abnormalities. Embryologic evidence has shown thymic dysgenesis in nude mice is linked with an abnormal third branchial cleft *ectoderm,* leading to the hypothesis that the nude locus is concerned with biosynthesis of one or more ectodermal cell proteins.[74] It is possible that the failure of nude Thy-1$^+$ EC to differentiate as they do in normal mice is related to this same biochemical defect expressed in the epidermis.

Evidence That Thy-1-Negative Epidermal Cells Can Serve as Accessory Cells in the Mitogen-Stimulated Growth of Thy-1$^+$ EC

There is substantial evidence that mitogen- and antigen-driven proliferation of highly purified, normal resting peripheral T cells (most of which are CD3$^+$ TCR α/β^+) requires exogenous accessory cells and/or their products.[75–78] Langerhans cells are known to serve such accessory cell functions for peripheral T cells.[79–84] The capacities of epidermal keratinocytes to provide critical accessory cell signals for peripheral T cells have not been studied as extensively; very recent data indicate that Ia-negative keratinocytes are unable to act as antigen-presenting cells, while Ia-positive keratinocytes have limited but demonstrable capacities for presenting peptide antigens to antigen-specific T cell clones.[85] To date there have been no studies reporting the accessory cell requirements of CD3$^+$ TCR γ/δ^+ cells such as Thy-1$^+$ EC.

Several different types of epidermal cell suspensions (containing varying proportions of Thy-1$^+$ EC, Ia$^+$ cells, and keratinocytes) can respond to Con A in the absence of added accessory cells and/or lymphokines/cytokines: 1) crude suspensions of trypsin-disaggregated epidermal sheets (typically containing 1–2% Thy-1$^+$ EC and 1–2% Ia$^+$ cells); 2) interface EC prepared by gradient centrifugation of crude EC suspensions (typically enriched to 10–30% Thy-1$^+$ EC and 5–10% Ia$^+$ cells); and 3) flow cytometry-purified populations containing >90% Thy-1$^+$ cells.[9,10,12,17] Careful inspection of the relative magnitudes of the Con A responses of such populations suggests strongly that they are indeed accessory cell dependent, and that such accessory cell functions can be provided by Thy-1$^-$ EC (*i.e.,* keratinocytes and/or Ia$^+$ cells).

TABLE 1 illustrates the results of an experiment in which the responses of AKR/J

TABLE 1. Proliferative Responses of Flow Cytometry-Enriched Thy-1$^+$ EC to Con A $+/-$ IL-2^a

Cells Cultured	% Thy-1$^+$	Added Factors			
		None	2 μg/ml Con A	10 U/ml IL-2	Con A + IL-2
Unstained, unsorted	—	52 ± 2	2,830 ± 180	180 ± 20	1,800 ± 80
Stained, unsorted	24	49 ± 4	2,500 ± 60	150 ± 4	2,680 ± 230
Sorted Thy-1$^+$ EC	96	47 ± 8	5,080 ± 710	110 ± 40	6,230 ± 160

^{a}Unsorted or sorted EC (20,000 cells/well) were cultured with the indicated factors in the absence of filler cells. Cultures were pulsed with ^{3}H-thymidine for 24 hr prior to harvest on day 5. Data shown are the mean ± SE from triplicate cultures.

TABLE 2. Proliferative Responses of Flow Cytometry-Purified Thy-1$^+$ EC to Con A and/or IL-2

	Cells Cultured	
Factor(s) Added	Stained, Unsorted IEC (14% Thy-1$^+$)[a]	Sorted Thy-1$^+$ EC (99% Thy-1$^+$)[b]
Media	90 ± 65[c]	40 ± 35
Con A (2 μg/ml)	3,250 ± 1,100	50 ± 10
IL-2 (10 U/ml)	1,930 ± 310	440 ± 210
Con A + IL-2	26,030 ± 340	24,320 ± 800

[a]2×10^4 cells/well.
[b]5×10^3 cells/well.
[c]Mean cpm ± SEM of triplicate cultures established with the indicated numbers of cells and added factors and harvested on day 6 after a 24 hr pulse with 1 μCi ^{3}H-thymidine.

EC suspensions containing different proportions of Thy-1$^+$ EC were compared. Interface EC suspensions were cultured either unstained, or after staining with a monoclonal antibody which recognizes a Thy-1-associated epitope on thymocytes and Thy-1$^+$ EC.[8,9] An aliquot of this stained EC suspension was then sorted by flow cytometry to obtain a population of "purified" Thy-1$^+$ EC; compared to unsorted cells (24% Thy-1$^+$), the sorted cells were markedly enriched (96% Thy-1$^+$). The proliferative responses to either Con A alone or to Con A and IL-2 of this flow cytometry-enriched population were approximately twice those of unsorted cells. However, since sorting had resulted in a fourfold enrichment of Thy-1$^+$ EC, the response of the sorted cells was relatively less than that expected if the only limiting factor was the number of responding Thy-1$^+$ EC per well. These results rather suggest that other factors (such as accessory cell function) may become limiting when flow cytometry-enriched cells are cultured.

The results of an experiment in which flow cytometry was used to obtain a more highly purified population of CBA/J Thy-1$^+$ EC are shown in TABLE 2. Unsorted interface EC contained 14% Thy-1$^+$ EC, while the flow cytometry-purified population was >99% Thy-1$^+$. The unsorted population responded to Con A alone; addition of exogenous IL-2 resulted in substantially augmented proliferation. In contrast, highly purified Thy-1$^+$ EC were totally incapable of proliferating in response to Con A alone; nevertheless, addition of exogenous IL-2 to this population resulted in proliferation indistinguishable from that of unsorted cells cultured with Con A and exogenous IL-2. These results suggest that rigorously purified Thy-1$^+$ EC do not respond to Con A activation by elaborating IL-2 (or IL-2-like growth factors) which can then be utilized in an autocrine manner, and rather suggest that such IL-2 production may be dependent upon accessory cells and/or signals present in unsorted EC suspensions. However, it is clear that Con A in the absence of such accessory signals does have some effect on highly purified Thy-1$^+$ EC; i.e., it apparently up-regulates expression of IL-2 receptors on Thy-1$^+$ EC so that addition of exogenous IL-2 results in proliferation indistinguishable from that of unsorted cells cultured under the same conditions. In other words, these results suggest that the Con A-induced up-regulation of IL-2 receptors is not nearly as dependent upon accessory cells/signals as is the Con A-induced secretion of IL-2. In this regard, purified resting Thy-1$^+$ EC (TCR γ/δ$^+$) resemble purified resting conventional (TCR α/β$^+$) T lymphocytes.[86–88]

The experiment shown in TABLE 3 illustrates that irradiated syngeneic EC suspensions (containing Thy-1$^+$ EC, Ia$^+$ cells, and keratinocytes) can serve an accessory role in the proliferation of activated Thy-1$^+$ EC. In this study, the responding

TABLE 3. Secondary Proliferative Responses of THY-1⁺ EC With and Without Irradiated Syngeneic EC[a]

	Media	Con A	PMA + Ionomycin
Irradiated EC	50 ± 10[b]	490 ± 210	30 ± 5
Thy-1⁺ EC	870 ± 400	520 ± 60	780 ± 380
Thy-1⁺ EC + Irradiated EC	540 ± 230	$3,560 \pm 590$	$1,570 \pm 210$

[a]CBA EC were initially stimulated with Con A (2 μg/ml) and IL 2 (10 U/ml), cultured for two weeks with IL 2, and harvested. These cultured cells were virtually all Thy-1⁺ (95% brightly stained). The cultured Thy-1⁺ EC were then placed back in culture (3000 cells/well) with or without freshly isolated γ-irradiated CBA EC (10,000/well). Proliferation in response to Con A (2 μg/ml) and PMA (3.3 ng/ml) + Ionomycin (1 μM) was assayed in triplicate cultures harvested on day 3 after a 24 hr pulse with 1 μCi ³H-thymidine.

[b]Mean cpm ± SEM.

Thy-1⁺ EC were derived from cultures established 2 weeks previously by stimulation of interface EC with Con A and IL-2. Such pre-stimulated cells were then placed in secondary cultures in the presence or absence of freshly prepared, irradiated, syngeneic interface EC and stimulated with either Con A or PMA/Ionomycin. Neither irradiated EC (10,000/well) alone nor pre-activated Thy-1⁺ EC (3000/well) alone were able to proliferate when stimulated with these agents; however, responses were observed in wells containing both populations.

These experiments all suggest that accessory cell signals are necessary for the mitogen-driven proliferation of both resting and short-term cultured Thy-1⁺ EC and that epidermal cell suspensions themselves can provide such signals. Whether these accessory signals are derived from keratinocytes and/or from the Ia⁺ cells (*e.g.*, Langerhans cells) also present in the EC suspensions utilized to date remains to be determined; studies using populations of EC depleted of Ia⁺ cells (either by flow cytometry or by antibody and complement-mediated lysis) as sources of accessory cells should resolve this issue.

Evidence That Some Thy-1⁺ EC Can Utilize Epidermal Cytokines as Growth Factors for Mitogen-Stimulated Growth

The lymphokine IL-2 is a growth factor for Thy-1⁺ EC from both normal and nude mice. Furthermore, the IL-2 responsiveness of Con A-activated Thy-1⁺ EC is increased dramatically. This increased responsiveness is most likely secondary to the up-regulation of high affinity IL-2 receptors expressed on the cell surface;[89] previous studies from our laboratory have demonstrated a dramatic increase in IL-2 receptor mRNA 3 hr after Con A stimulation of Thy-1⁺ EC clones.[25] However, it has become clear that a large number of other lymphokines and/or cytokines can affect the growth and/or differentiation of other T cells in either a positive or negative manner. It has also become clear that epidermal keratinocytes can secrete a large number of such immunomodulatory cytokines; the fact that a large portion of this volume is devoted to such factors testifies to the complexity of, as well as to the interest in this area of investigation. Given the usual location of Thy-1⁺ EC within the epidermis, it is clearly appropriate (albeit somewhat awe-inspiring!) to investigate the effects of keratinocyte cytokines on these cells.

Our preliminary data on this subject have been limited to examining the effects of two such cytokines (GM-CSF and IL-3) on the growth of selected short-term lines and long-term clones of Thy-1⁺ EC. TABLE 4 shows the results of this analysis. Each of the responding populations was established from AKR/J mice by stimulation with Con A in the presence of exogenous IL-2, and each had been maintained in continuous culture by feeding with IL-2-supplemented media and intermittent restimulation with Con A. Lines 11–17 (2 months old) and 7–17 (30 months old) had been established from cultures of interface EC, while clones 6G3 and 1D2A (25 months old) had been established by limiting dilution microculture of flow cytometry-purified Thy-1⁺ EC in the presence of irradiated syngeneic splenic filler cells.[12,25] Each population was washed thoroughly prior to culture in media containing either no added factors, or IL-2, IL-3, or GM-CSF. As expected, IL-2 was a potent growth factor for each of the populations (line 2). However, neither IL-3 nor GM-CSF alone consistently stimulated significant proliferation above background (data not shown).

Each population was also stimulated with 0.5 μg/ml Con A in the presence or absence of these same factors. This concentration of Con A had been determined in previous experiments to give a submaximal proliferative response in the absence of exogenous IL-2, presumably secondary to submaximal secretion of IL-2 by the stimulated cells. This concentration of Con A was clearly sufficient to acutely up-regulate the expression of IL-2 receptors on lines 11–17 and 7–17, as evidenced by the enhanced responses of cultures containing both Con A and IL-2 (line 4) compared to those seen in cultures containing only IL-2 (line 2). Neither IL-3 nor GM-CSF was able to augment the proliferative responses of these two lines to suboptimal Con A to the same degree as IL-2 (lines 5 and 6 *vs* line 4); proliferation in these cultures was only marginally greater than that seen with Con A alone. Very different response profiles were seen in clones 6G3 and 1D2A. While these clones responded to 0.5 μg/ml Con A alone (line 3 *vs*. line 1), their responses apparently included the acute *down-regulation* of IL-2 receptor expression (line 4 *vs* line 2). However, addition of either IL-3 or GM-CSF to these Con A-stimulated populations resulted in augmented proliferation compared to that seen with Con A alone (lines 5 and 6 *vs* line 3). Furthermore, the response of 1D2A to Con A + GM-CSF was strikingly greater than the response of that same clone to Con A and IL-2. While additional detailed studies will be necessary to accurately interpret these findings, these results suggest that at least some populations of Thy-1⁺ EC can utilize cytokines other than IL-2 for their growth, and that there is obvious heterogeneity among Thy-1⁺ EC populations in their abilities to utilize such cytokines.

TABLE 4. Responses of Thy-1⁺ EC Lines/Clones to IL-2, IL-3, and GM-CSF

Factor(s) Added	Thy-1⁺ DEC Line/Clone			
	11–17	7–17	6G3	1D2A
0	300 ± 60[b]	210 ± 25	640 ± 80	1,300 ± 190
2 U/ml IL-2	3,150 ± 250	13,420 ± 1,140	6,200 ± 2,800	62,500 ± 1,800
0.5 μg/ml Con A	1,970 ± 50	850 ± 100	1,700 ± 600	15,800 ± 9,000
0.5 μg/ml Con A + 2 U/ml IL-2	7,620 ± 500	41,700 ± 11,000	3,970 ± 1,000	33,000 ± 11,000
0.5 μg/ml Con A + 20 U/ml IL-3[a]	2,530 ± 450	1,200 ± 200	3,100 ± 1,330	43,600 ± 10,500
0.5 μg/ml Con A + 10 U/ml GM-CSF[a]	2,940 ± 710	2,000 ± 670	3,770 ± 1,460	71,600 ± 17,200

[a]Responses of each line/clone to 20 U/ml IL-3 or 10 U/ml rGM-CSF in the absence of added Con A were indistinguishable from that population's response in media alone.

[b]Mean cpm (±SD) of triplicate cultures harvested at 72 hr after a 24-hr pulse with 1 μCi ³H-thymidine.

CONCLUSIONS

Despite the substantial advances made recently in our understanding of Thy-1$^+$ EC, it is obvious that the present state of affairs is punctuated by major gaps both in data collection and in comprehension. Key unresolved issues include the precise relationship(s) of Thy-1$^+$ EC to other T cells, and the roles that the epidermis plays in the growth, activation, and differentiation of these cells. Another area of critical importance for understanding the developmental biology and physiologic function of this distinctive T cell population is the identification of the "natural" ligands for the γ/δ TCRs they express. It is conceivable that the γ/δ TCR is an evolutionary vestige of this cell type's original function as the *only* T cell population recognizing foreign antigens. Given the apparently relatively limited diversity of γ/δ TCRs, it is likely that the "natural" ligands of cells bearing such receptors are also relatively limited, perhaps to epitopes common to either certain environmental pathogens or to subtle alterations in host cells associated with malignant transformation. In any event, the facts that substantial numbers of TCR γ/δ^+ cells reside in the epidermis of normal mice and that these cells appear to resemble cells in fetal thymus at a discrete stage of ontogeny points to a special opportunity for cutaneous biologists to make substantive contributions to a more comprehensive understanding of the roles of epithelial cells in T cell ontogeny.

REFERENCES

1. BERGSTRESSER, P., R. TIGELAAR, J. DEES & J. STREILEIN. 1983. J. Invest. Dermatol. **81:** 286–288.
2. TSCHACHLER, E., G. SCHULER, J. HATTERER, H. LEIBL, K. WOLFF & G. STINGL. 1983. J. Invest. Dermatol. **81:** 282–285.
3. BERGSTRESSER, P., R. TIGELAAR & J. STREILEIN. 1983. J. Invest. Dermatol. **83:** 83–87.
4. BREATHNACH, S. & S. KATZ. 1984. J. Invest. Dermatol. **83:** 74–77.
5. ROMANI, N., G. SCHULER & P. FRITSCH. 1986. J. Invest. Dermatol. **86:** 129–133.
6. SULLIVAN, S., P. BERGSTRESSER, R. TIGELAAR & J. STREILEIN. 1985. J. Invest. Dermatol. **84:** 491–495.
7. ROMANI, N., G. STINGL, E. TSCHACHLER, M. WITMER, R. STEINMAN, E. SHEVACH & G. SCHULER. 1985. J. Exp. Med. **161:** 1368.
8. BERGSTRESSER, P., S. SULLIVAN, J. STREILIEN & R. TIGELAAR. 1985. J. Invest. Dermatol. **85:** 85–90.
9. NIXON-FULTON, J., P. BERGSTRESSER & R. TIGELAAR. 1986. J. Immunol. **136:** 2776.
10. CAUGHMAN, S., S. BREATHNACH, S. SHARROW, D. STEPHANY & S. KATZ. 1986. J. Invest. Dermatol. **96:**615–624.
11. NIXON-FULTON, J., J. HACKETT, P. BERGSTRESSER, V. KUMAR & R. TIGELAAR. 1988. J. Invest. Dermatol. **91:** 62–68.
12. TAKASHIMA, A., J. NIXON-FULTON, P. BERGSTRESSER & R. TIGELAAR. 1988. J. Invest. Dermatol. **90:** 671–678.
13. BOSMA G., P. CUSTER & J. BOSMA. 1983. Nature **301:** 527–529.
14. SCHULER, W., I. WEILER, A. SCHULER, R. PHILLIPS, N. ROSENBERG, T. MAK, J. KEARNEY, R. PERRY & M. BOSMA. 1986. Cell **46:** 963–972.
15. TUTT M., W. KUZIEL, J. HACKETT, M. BENNETT, P. TUCKER & V. KUMAR. 1986. J. Immunol. **137:** 2998–3004.
16. TUTT, M., W. SCHULER, W. KUZIEL, P. TUCKER, M. BENNETT, M. BOSMA & V. KUMAR. 1987. J. Immunol. **138:** 1.
17. NIXON-FULTON, J., P. WITTE, R. TIGELAAR, P. BERGSTRESSER & V. KUMAR. 1987. J. Immunol. **138:** 2902.
18. ACUTO, O., M. FABBI, M. BENSUSSAN, C. MILANESE, T. CAMPEN, H. ROYER & E. REINHERZ. 1985. J. Clin. Immunol. **5:** 141–157.

19. MARRACK, P. & J. KAPPLER. 1986. Adv. Immunol. **38**: 1–30.
20. MEUER, S., C. ACUTO, T. HERCEND, E. REINHERZ. 1986. Annu. Rev. Immunol. **2**: 23–50.
21. WEISS, A., J. IMBODEN, K. HARDY, B. MANGER, C. TERHORST & J. STOBO. 1986. Annu. Rev. Immunol. **4**: 593–619.
22. KRONENBERG, M., G. SIU, L. HOOD & N. SHASTRI. 1986. Annu. Rev. Immunol. **4**: 529–591.
23. MARRACK, P. & J. KAPPLER. 1987. Science **238**: 1073–1079.
24. ALLISON, J. & L. LANIER. 1987. Annu. Rev. Immunol. **5**: 503–540.
25. KUZIEL, W., A. TAKASHIMA, M. BONYHADI, P. BERGSTRESSER, J. ALLISON, R. TIGELAAR & P. TUCKER. 1987. Nature **328**: 263–266.
26. BONYHADI, M., A. WEISS, P. TUCKER, R. TIGELAAR & J. ALLISON. 1987. Nature **330**: 574–576.
27. STINGL, G., K. GUNTER, E. TSCHACHLER, H. YAMADA, W. YOKOYAMA, G. STEINER, R. GERMAIN & E. SHEVACH. 1987. Proc. Natl. Acad. Sci. USA **84**: 2430–2434.
28. KONIG, F., G. STINGL, W. YOKOYAMA, H. YAMADA, W. MALOY, E. TSCHACHLER, E. SHEVACH & J. COLIGAN. 1987. Science **236**: 834–836.
29. STINGL, G., F. KONIG, H. YAMADA, E. TSCHACHLER, J. BLUESTONE, L. SAMELSON, A. LEW, J. COLIGAN & E. SHEVACH. 1987. Proc. Natl. Acad. Sci. USA **84**: 4586–4590.
30. YOKOYAMA, W., F. KONIG, G. STINGL, J. BLUESTONE, J. COLIGAN & E. SHEVACH. 1987. J. Exp. Med. **165**: 1725–1730.
31. SHIMADA, S., S. CAUGHMAN, J. BLUESTONE, R. SCHWARTZ, S. KATZ & D. PARDOLL. 1987. Clin Res. **35**: 582A.
32. BANK, I., R. DEPINHO, M. BRENNER, J. CASIMERIS, F. ALT & L. CHESS. 1986. Nature **322**: 179–181.
33. BRENNER, M., J. MCLEAN, D. DIALYNAS J. STROMINGER, J. SMITH, F. OWEN, J. SEIDMAN, S. IP, F. ROSEN & M. KRANGEL. 1986. Nature **322**: 145–149.
34. LANIER, L. & A. WEISS. 1986. Nature **324**: 268–270.
35. LEW, A., D. PARDOLL, W. MALOY, B. FOWLES, A. KRUISBEEK, S. CHENG, R. GERMAIN, J. BLUESTONE, R. SCHWARTZ & J. COLIGAN. 1986. Science **234**: 1401–1405.
36. MOINGEON, P., S. JITSUKAWA, F. FAURE, F. TROALEN, F. TREIBEL, M. GRAZIANI, F. FORESTIER, D. BELLET, C. BOHUON & T. HERCEND. 1987. Nature **325**: 723–726.
37. NAKANISHI, N., K. MAEDA, M. HELLER & S. TONEGAWA. 1987. Nature **325**: 720–723.
38. PARDOLL, D., B. FOWLKES, J. BLUESTONE, A. KRUISBEEK, W. MALOY, J. COLIGAN & R. SCHWARTZ. 1987. Nature **326**: 79–81.
39. BLUESTONE, J., D. PARDOLL, S. SHARROW & B. FOWLKES. 1987. Nature **326**: 82–85.
40. BORST, J., R. VAN DE GRIEND, J. VAN OOSTVEEN, S.-L. ANG, C. MELIEF, J. SEIDMAN & R. BOLHUIS. 1987. Nature **325**: 683–688.
41. LANIER, L., N. FEDERSPIEL, J. RUITENBERG, J. PHILLIPS, J. ALLISON, D. LITTMAN & A. WEISS. 1987. J. Exp. Med. **165**: 1076–1094.
42. LANIER, L., A. SERAFINI, J. RUITENBERG, S. CWIRLA, N. FEDERSPIEL, J. PHILLIPS, J. ALLISON & A. WEISS. 1987. J. Clin. Immunol. **7**: 429–440.
43. PARDOLL, D., A. KRUISBEEK, J. COLIGAN & R. SCHWARTZ. 1987. FASEB J. **1**: 103–109.
44. BORN, W., J. YAGUE, E. PALMER, J. KAPPLER & P. MARRACK. 1985. Proc. Natl. Acad. Sci. USA **82**: 2925–2928.
45. SAMELSON, L., T. LINDSTEN, B. FOWLKES, P. VAN DEN ELSEN, C. TERHORST, M. DAVIS, R. GERMAIN & R. SCHWARTZ. 1985. Nature **315**: 765.
46. RAULET, D., R. GARMAN, H. SAITO & S. TONEGAWA. 1985. Nature **314**: 103–107.
47. GARMAN, R., P. DOHERTY & D. RAULET. 1986. Cell **45**: 733–742.
48. BORN, W., G. RATHBUN, P. TUCKER & J. KAPPLER. 1986. Science **234**: 479–482.
49. CHIEN, Y.-H., M. IWASHIMA, J. ELLIOT & M. DAVIS. 1987. Nature **327**: 677–682.
50. PARDOLL, D., B. FOWLKES, R. LECHLER, R. GERMAIN & R. SCHWARTZ. 1987. J. Exp. Med. **165**: 1624–1638.
51. BORN, W., C. MILES, J. WHITE, R. O'BRIEN, J. FREED, P. MARRACK, J. KAPPLER & R. KUBO. 1987. Nature **330**: 572–574.
52. CHIEN, Y.-H., M. IWASHIMI, D. WETTSTEIN, K. KAPLAN, J. ELLIOTT, W. BORN & M. DAVIS. 1987. Nature **330**: 722–727.
53. FLEISCHER, B. 1984. Eur. J. Immunol. **14**: 748–752.

54. HUBBARD, S., D. KRANZ, G. LONGMORE, M. SITKOVSKY & H. EISEN. 1986. Proc. Natl. Acad. Sci. USA **83:** 1852–1856.
55. RIGDON, R. & A. PACKCHANIAN. 1974. Tex. Rep. Biol. Med. **32:** 711–723.
56. KRUEGER, G., D. CHAMBERS & J. SHELBY. 1980. J. Exp. Med. **152:** 1329–1339.
57. NIXON-FULTON, J., W. KUZIEL, B. SANTERSE, P. BERGSTRESSER, P. TUCKER & R. TIGELAAR. 1988. J. Immunol. **141:** 1897–1903.
58. LUGO, J., S. KRISHNA, R. SAILOR & E. ROTHENBURG. 1986. Proc. Natl. Acad. Sci. USA **83:** 1862–1866.
59. MACDONALD, H., R. LEES, P. ZAECH, J. MARYANSKI & C. BRON. 1981. J. Immunol. **126:** 865.
60. KRUISBEEK, A., M. DAVIS, L. MATIS & D. LONGO. 1984. J. Exp. Med. **160:** 839–857.
61. MACDONALD, H., C. BLANC, R. LEES & B. SORDAT. 1986. J. Immunol. **136:** 4337.
62. MACDONALD, H., R. LEES, B. SORDAT & G. MIESCHER. 1987. J. Exp. Med. **166:** 195–209.
63. PATTERSON, J. & R. EDELSON. 1982. Br. J. Dermatol. **107:** 117–122.
64. SUN, T.-T., C. SHIH & H. GREEN. 1979. Proc. Natl. Acad. Sci. USA **76:** 283–286.
65. DIDIERJEAN, L. & J. SAURAT. 1980. Clin. Exp. Dermatol. **5:** 395–404.
66. LOBACH, D. & B. HAYNES. 1987. J. Clin. Immunol. **7:** 81.
67. SAUDER, D., C. CARTER, S. KATZ & J. OPPENHEIM. 1982. J. Invest. Dermatol. **79:** 34.
68. LUGER, T., M. SZTEIN & J. OPPENHEIM. 1983. Fed. Proc. **42:** 2772.
69. KUPPER, T., D. COLEMAN, J. MCGUIRE, D. GOLDMINZ & M. HOROWITZ. 1986. Proc. Natl. Acad. Sci. USA **83:** 4451–4455.
70. KUPPER, T., F. LEE, D. COLEMAN, J. CHODAKOWITZ, P. FLOOD & M. HOROWITZ. 1988. J. Invest. Dermatol. **91:** 185–188.
71. LUGER, T., U. WIRTH & A. KOCK. 1985. J. Immunol. **134:** 915–920.
72. YAAR, M., A. PALLERONI & B. GILCHREST. 1986. J. Cell Biol. **103:** 1349–1354.
73. CHU, A., J. PATTERSON, G. GOLDSTEIN, C. BERGER, S. TAKEZAKI & R. EDELSON. 1983. J. Invest. Dermatol. **81:** 194–197.
74. CORDIER, A. & J. HERMANNS. 1975. Scand. J. Immunol. **4:** 193–196.
75. DICKLER, H., C. COWING, S. SHARROW, R. HODES & A. SINGER. 1980. *In* Macrophage Regulation of Immunity. A. Rosenthal & E. Unanue, Eds.: 265–275. Academic Press. New York, NY.
76. UNANUE, E. R. 1984. Annu. Rev. Immunol. **2:** 395–435.
77. STEINMAN, R. K. INABA, G. SCHULER & M. WITMER. 1986. *In* Host Resistance Mechanisms to Infectious Agents, Tumors, and Allografts. R. Steinman & R. North, Eds.: 71–97. Rockefeller Univ. Press, New York, NY.
78. UNANUE, E. R. & P. M. ALLEN. 1987. Science **236:** 551–557.
79. STINGL, G., K. TAMAKI & S. I. KATZ. 1980. Immunol. Rev. **53:** 149–174.
80. SCHEYNIUS, A., K. GRONVIK & J. ANDERSON. 1983. Scand. J. Immunol. **17:** 283–289.
81. TSUCHID, T., M. IIJIMA, H. FUJIWARA, H. PEHAMBERGER, G. SHEARER & S. KATZ. 1984. J. Immunol. **132:** 1163–1168.
82. STEINER, G., K. WOLFF, H. PEHAMBERGER & G. STINGL. 1985. J. Immunol. **134:** 736–741.
83. INABA, K., G. SCHULER, M. WITMER, J. VALINSKY, B. ATASSI & R. STEINMAN. 1986. J. Exp. Med. **164:** 605–613.
84. WITMER-PACK, M., W. OLIVIER, J. VALINSKY, G. SCHULER & R. STEINMAN. 1987. J. Exp. Med. **166:** 1484–1498.
85. GASPARI, A. & S. I. KATZ. 1988. J. Immunol. **140:** 2956–2963.
86. HUNIG, T. 1983. Eur. J. Immunol. **13:** 596–601.
87. SCHWAB, R., M. CROW, C. RUSSO & M. WEKSLER. 1985. J. Immunol. **135:** 1714–1718.
88. GARMAN, R., K. JACOBS, S. CARK & D. RAULET. 1987. Proc. Natl. Acad. Sci. USA **84:** 7629–7633.
89. HEMLER, M., M. BRENNER, J. MCLEAN & J. STROMINGER. 1984. Proc. Natl. Acad. Sci. USA **81:** 2172–2176.

Interleukin-1 Alpha mRNA Induced by Cycloheximide, PMA, and Retinoic Acid Is Reduced by Dexamethasone in PAM-212 Keratinocytes

JOSEPH McGUIRE, ROBERT LANGDON, NICHOLAS BIRCHALL, AND THOMAS KUPPER

Department of Dermatology
Yale University School of Medicine
333 Cedar Street
P.O. Box 3333
New Haven, Connecticut 06510

INTRODUCTION

Interleukin-1 (IL-1) is a mediator of a variety of biological responses and appears to play an important role in inflammation either directly or through the activation of, or synergism with other cytokines. Although IL-1 was initially described in monocytes and macrophages, it has been found in keratinocytes both *in vivo*[1] and in culture.[2] Initially, the IL-1-like activity in keratinocytes was termed epidermal cell derived thymocyte activating factor (ETAF).[3,4] Subsequently, it was shown that human keratinocytes contain mRNA identical to monocyte IL-1 alpha and beta. Unlike monocytes, which have negligible amounts of constitutive IL-1, cultured keratinocytes express both IL-1 alpha and IL-1 beta mRNA in an unstimulated state; further, IL-1 alpha apears to be the predominant form.[2]

We have reported that hydrocortisone reduces the amount of ETAF in the conditioned media of PAM-212, a transformed murine keratinocyte cell line; A-431, a human epidermal carcinoma; and normal human cultured keratinocytes. This conclusion was based on analysis of conditioned media in a comitogenic assay using D10.G4.1, a murine helper T-cell clone, and either concanavalin A or 3D3, a clonotype-specific monoclonal antibody which activates the T-cell receptor of D10.[5]

The observation that hydrocortisone reduced ETAF raises several questions. Was the mitogenic activity measured in the D.10 assay due to IL-1 or to another ETAF mitogenic for T cells? If the decrease in mitogenic activity represented a decrease in IL-1, is the effect at the level of mRNA transcription, altered half-life of IL-1 mRNA, IL-1 translation, or posttranslational processing of IL-1?

In this report, we show that phorbol 12-myristate 13-acetate (PMA), cycloheximide, and retinoic acid increase mRNA for IL-1 alpha and that a glucocorticosteroid (GCS), dexamethasone, decreases this effect.

METHODS

The cells used in these experiments were PAM-212, a spontaneously transformed keratinocyte line from BALB/c mice, which were obtained from Dr. Pamela Hawley-

283

Nelson (National Cancer Institute).[6] Cells were grown in DMEM with 10% calf serum and the following antibiotics: penicillin (50 units/ml), streptomycin (50 μg/ml), and gentamicin (50 μg/ml). The cultures were grown in 10-cm dishes; medium was changed from DMEM/10% calf serum with antibiotics to DMEM the day before the experiment.

The general design of these experiments was to examine the amount of IL-1 alpha mRNA by Northern blot analysis of total RNA extracted by the method of Chomcyznski.[7] Cultures having been exposed to the biological modifiers of interest were rinsed twice with 4 ml PBS and drained, and then 0.5 ml of 4 M guanidinium denaturing solution D was added to the dish. Dishes were placed on a rocking platform for 5 min and the solubilized cells were then scraped from the plastic using sections of silicone stoppers sliced with a razor blade. The extraction procedure of Chomczynski[7] was adapted to 2.2-ml polypropylene tubes. After the RNA was precipitated with isopropanol, it was washed with 75% ethanol, dried, and dissolved in water. RNA was quantitated by absorbancy at 260 nm.

Total RNA 20 μg/lane was separated by electrophoresis on a 1.5% agarose/formaldehyde gel. The RNA was transferred to a nylon filter and hybridized to the insert of pIL-1-1301, which had been labelled with ^{32}P using the random primer

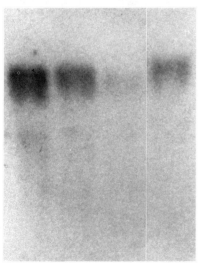

FIGURE 1. Dexamethasone reduces the amount of IL-1 alpha mRNA induced by cycloheximide in cultured PAM-212 murine keratinocytes. *Lane 1*, cycloheximide 10 mg/ml present for 4.5 hr. *Lane 2*, dexamethasone 10^{-6} M present for 5 hr; cycloheximide 10 mg/ml present for final 4 hr. *Lane 3*, dexamethasone 10^{-6} M present for 4 hr. *Lane 4*, control, no dexamethasone or cycloheximide.

FIGURE 2. Dexamethasone reduces IL-1 mRNA induced by PMA. *Lane 1*, control. *Lane 2*, dexamethasone 10^{-6} 6 hr. *Lane 3*, PMA 10 ng/ml 6 hr. *Lane 4*, PMA 10 ng/ml 6 hr preceeded by dexamethasone 10^{-6} for 2 hr. *Lane 5*, Ru 486 10^{-6} 6 hr. *Lane 6*, PMA 10 ng/ml for 6 hr preceded by dexamethasone and Ru 486 for 2 hr.

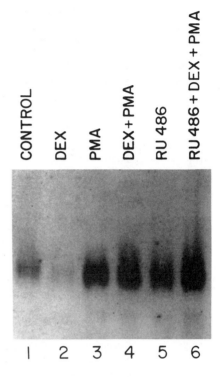

technique. pIL-1-1301[8] was a gift of Dr. Peter Lomedico (Hoffmann-LaRoche, Inc., Nutley, NJ). Autoradiographs of these northern blots were analyzed by densitometry using a Joyce-Loebl Chromoscan 3 (Vickers Instruments, Inc., 300 Commercial St., Malden, MA).

RESULTS

Cycloheximide

Cycloheximide has been shown to induce IL-1 beta in macrophages.[9] This compound, an inhibitor of protein synthesis, also induces IL-1 alpha mRNA in PAM-212 (FIG. 1). The presence of cycloheximide shown in lane 1 greatly increases the amount of mRNA, and this induction is reduced by dexamethasone (lane 2). Also demonstrated in this figure is that constitutive levels of IL-1 alpha mRNA shown in lane 4 are reduced by dexamethasone (lane 3).

PMA

PMA is known to induce IL-1 in human keratinocytes[10] and induces IL-1 alpha mRNA in PAM-212 (FIG. 2). In this experiment, dexamethasone reduced constitutive levels of IL-1 mRNA. PMA greatly increased IL-1 mRNA levels and Ru 486 by itself in some experiments had no effect and sometimes increased levels of IL-1 mRNA over constitutive levels. Ru 486 was used because it inhibits both progesterone and glucocorticosteroid binding to receptors.[11,12] PMA reduced IL-1 mRNA and dexamethasone reduced both constitutive IL-1 alpha mRNA (compare lanes 1 and 2) and also the PMA induced IL-1 mRNA (compare lanes 3 and 4).

Ru 486 either had little effect or increased IL-1 mRNA. The addition of Ru 486 to dexamethasone and PMA abrogated the inhibitory effect of dexamethasone on PMA stimulation and occasionally increased the levels of mRNA above the TPA induced levels (data not shown).

In FIGURE 3, a densitometric analysis of FIGURE 2, it is seen that dexamethasone reduces constitutive amounts of IL-1 alpha mRNA and that PMA increases it. Further, the reduction of TPA stimulation by dexamethasone is abrogated by Ru 486.

Retinoic Acid

The influence of retinoic acid on IL-1 alpha mRNA was of interest because of the report by Trechsel[13] that retinoic acid stimulates the production of IL-1 in human peripheral blood mononuclear cells and in the murine macrophage cell line P388 D. WEHI-3 cells were stimulated to release increased amounts of IL-3 by retinoic acid. IL-1 activity was measured by assay of collagenase release from rabbit articular chondrocytes. Schmitt[14] subsequently showed that the synthetic retinoid etretin increases epidermal IL-1 in hairless rats. She administered etretin intraperitoneally and analyzed lymphocyte activating factor and stimulation of PGE_2 in dermal fibroblasts by extracts of rat epidermis. Etretin effected an increase in ^3H-thymidine labelling of the epidermis as well as an increase in LAF and PGE_2. We found that retinoic acid in concentrations as low as 10^{-10} M increased IL-1 mRNA (FIG. 4). This

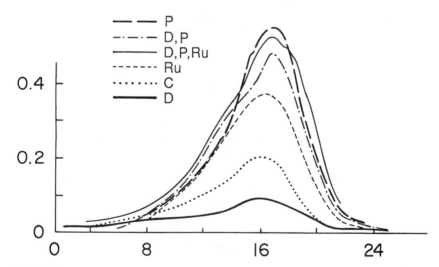

FIGURE 3. Densitometric analysis of Northern blots of IL-1 alpha, showing the influence of dexamethasone and RU 486 on PMA induction on IL-1 alpha mRNA. Conditions are as shown in FIGURE 2.

FIGURE 4. Dexamethasone reduces the induction of IL-1 mRNA by retinoic acid. *Lane 1,* control. *Lane 2,* retinoic acid 10^{-10} M 6 hr. *Lane 3,* retinoic acid 10^{-8} M 6 hr. *Lane 4,* retinoic acid 10^{-6} M 6 hr. *Lane 5,* dexamethasone 10^{-6} M 6 hr. *Lane 6,* dexamethasone 10^{-6} M and retinoic acid 10^{-6} M 6 hr. The dexamethasone was added to the cultures 1 hr before the retinoic acid.

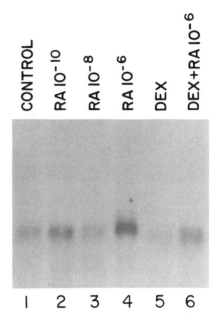

stimulation was reduced by dexamethasone (10^{-6} M) as was the constitutive level of IL-1 mRNA. Densitometric analysis of these results confirmed the stimulation of IL-1 mRNA by retinoic acid and reduction of this effect by dexamethasone 10^{-6} M (FIG. 5).

DISCUSSION

Many factors increase IL-1 alpha mRNA. We have found that dexamethasone reduces the increase induced by cycloheximide, PMA, and retinoic acid. This reduction in IL-1 alpha mRNA is probably the basis of the glucocorticoid suppression of ETAF activity we described previously in the conditioned media of cultured keratinocytes.[5] In 1970 Dillard and Bodel[15] found that cortisol reduced the amount of endogenous pyrogen released from human white blood cells in response to phagocytosis of heat-killed staphylococci. These observations probably reflect a steroid-mediated reduction of IL-1 from macrophages in the white blood cell preparation. The next important observation linking GCS and IL-1 was made by Snyder and Unanue in 1982,[16] who found that GCS reduced the amount of IL-1 released by murine peritoneal macrophages. Staruch found that serum from mice sensitized with P. acnes and then challenged with LPS contained high levels of IL-1 activity. Pretreatment of the mice with dexamethasone before LPS challenge reduced the IL-1 response.[17]

The observation reported here that IL-1 alpha mRNA is decreased by dexamethasone differs from the findings of Kern et *al.,*[18] who examined adherent human blood monocytes stimulated with LPS and found that LPS increased IL-1 beta transcription 2–4-fold, and that dexamethasone (10^{-5} M) did not influence the increase in IL-1 message. Dexamethasone did, however, inhibit the release of IL-1 into the media.

There are differences between those experiments and the ones reported here. We did not remove dexamethasone from the cells until they were harvested for analysis of IL-1 alpha mRNA. In Kern's experiment the cells were exposed to dexamethasone for two hr, washed extensively, and then recultured with LPS with and without dexamethasone for 16 hr before harvesting RNA. Kern did find by immunoprecipitation with IL-1 beta antisera that dexamethasone reduced the translation of IL-1 beta precursors and had a variable effect on the low molecular weight form. Conditioned media from these experiments showed abundant 17.5 kD IL-1 beta in the presence of LPS and none when dexamethasone was present.

In contrast, Lee et al.,[19] using a human promonocytic cell line U-937, found that dexamethasone decreased transcription of IL-1 beta, which had been induced by PMA and LPS. Further, dexamethasone caused a striking decrease in stability of IL-1 beta mRNA. In the presence of cycloheximide or actinomycin D, however, dexamethasone had no effect on the stability of IL-1 beta mRNA.

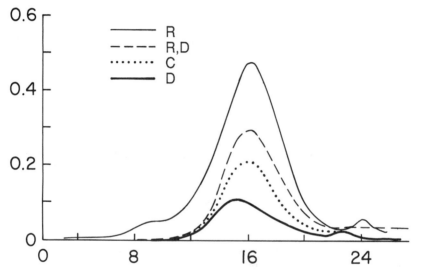

FIGURE 5. Densitometric analysis of IL-1 mRNA, showing induction by retinoic acid. *Lanes* 1, 4, 5, and 6 from FIGURE 4 are shown.

The mechanisms through which glucocorticoids influence IL-1 mRNA transcription are not known; however, RU 486, a steroid receptor antagonist, abolished or reduced the effects of dexamethasone in Lee's experiments and in those reported here. It is interesting that glucocorticosteroids, which are antiinflammatory, can regulate the production of proinflammatory cytokines, such as IL-1, by interfering with both transcription and mRNA stability. The work of Kern also indicates that IL-1 processing and delivery into the media are strongly reduced by GCS. GCS also reduce the amount of induced mRNA for IL-3[20] and IL-2.[21]

A somewhat analogous situation exists with beta-2 interferon, also termed BSF-2 or IL-6. Lipopolysaccharide (LPS) induces IL-6 mRNA in human fibroblasts, and this effect of LPS is enhanced by cycloheximide. Dexamethasone effectively reduces IL-6 mRNA induced by LPS.[22] There is an interesting relationship between IL-1 and IL-6.

IL-1 and tumor necrosis factor both induce IL-6 mRNA in human fibroblasts. Dexamethasone inhibits this increase in IL-6 mRNA.[23]

Intracellular transport and secretion of IL-1 is poorly understood and complicated by the absence of a hydrophobic leader sequence on IL-1 that would allow insertion into the cell membrane. Nonetheless, a cell membrane associated IL-1 has been described by Kurt-Jones in macrophages,[24] and we have found significant amounts of IL-1 associated with the keratinocyte membrane.[25] Efforts to fractionate human monocytes by Bakouche *et al.*[26] led to finding IL-1 in the cell membrane of fixed activated cells, cytosol, and lysosomes. There was no activity in endoplasmic reticulum. They concluded that the absence of IL-1 in the endoplasmic reticulum was consistent with nonconventional secretion of IL-1 and that IL-1 may be released via lysosomal vesicles.

GCS are used extensively to reduce inflammation associated with many diseases. The mechanism for this antiinflammatory effect is unknown but has been attributed to the induction of lipocortins, a group of phospholipase A_2 inhibitors.[27,28] These inhibitors decrease products of arachidonic acid metabolism that are proinflammatory.

An equally likely hypothesis is that GCS reduce IL-1, which would not only diminish the direct effect of IL-1 but also effects of IL-6 and GMCSF, which are induced by IL-1.[29]

IL-1 is present in normal human epidermis[1] and cultured keratinocytes.[2] The abundance of IL-1 in keratinocyte membranes[25] arms the cell with cytokine accessible to any adjacent cells and may be important in the paracrine effects of IL-1. Many biological effects have been attributed to IL-1, include fever, leucocytosis, induction of acute phase reactants, chemotaxis,[30] and growth stimulation.[31] GCS may interfere with these effects in a variety of ways: reduction of transcription of IL-1 mRNA, decrease in the stability of IL-1 mRNA, and alteration of the processing and secretion of IL-1.

SUMMARY

Keratinocytes in culture produce detectable amounts of IL-1 alpha mRNA constitutively and can be stimulated to express increased amounts of IL-1 alpha mRNA by cycloheximide, PMA, and retinoic acid. Dexamethasone decreases the amount of IL-1 mRNA induced by these agents, as well as constitutive IL-1 alpha mRNA. RU 486, which interferes with glucocorticosteroid-receptor binding, decreases inhibition of TPA stimulation of IL-1 alpha mRNA by dexamethasone, which suggests that the inhibition by dexamethasone is through a conventional ligand-receptor mechanism.

REFERENCES

1. HAUSER, C., J.-H. SAURAT, A. SCHMITT, F. JAUNIN & J.-M. DAYER. 1986. J. Immunol. **136:** 3317–3323.
2. KUPPER, T., D. W. BALLARD, A. O. CHUA, J. S. MCGUIRE, P. M. FLOOD, M. C. HOROWITZ, R. LANGDON, L. LIGHTFOOT & U. GUBLER. 1986. J. Exp. Med. **164:** 2095–2100.
3. LUGER, T. A., B. M. STADLER, B. M. LUGER, B. J. MATHIESON, M. MAGE, J. A. SCHMIDT & J. J. OPPENHEIM. J. Immunol. **128:** 2147–2152.
4. SAUDER, D. N., C. S. CARTER, S. I. KATZ & J. J. OPPENHEIM. 1982. J. Invest. Dermatol. **79:** 34–39.

5. KUPPER, T. S. & J. MCGUIRE. 1986. J. Invest. Dermatol. **87:** 570–573.
6. YUSPA, S. H., P. HAWLEY-NELSON, B. KOEHLER & J. R. STANLEY. 1980. Cancer Res. **40:** 4694–4703.
7. CHOMCZYNSKI, P. & N. SACCHI. 1987. Anal. Biochem. **162:** 156–159.
8. LOMEDICO, P. T., U. GUBLER, C. P. HELLMAN, M. DUKOVICH, J. G. GIRI, Y.-C. E. PAN, K. COLLIER, R. SEMIONOW, A. O. CHUA & S. B. MIZEL. 1984. Nature **312:** 458–462.
9. MIZEL, S. B. & D. MIZEL. 1981. J. Immunol. **126:** 834–837.
10. KUPPER, T. S., A. O. CHUA, P. FLOOD, J. MCGUIRE & U. GUBLER. 1987. J. Clin. Invest. **80:** 430–436.
11. JUNG-TESTAS, I. & E. E. BAULIEU. 1983. Exp. Cell Res. **147:** 177–182.
12. BOURGEOIS, S., M. PFAHL & E. E. BAULIEU. 1984. EMBO J. **3:** 751–755.
13. TRECHSEL, U., V. EVEQUOZ & H. FLEISCH. 1985. Biochem. J. **230:** 339–344.
14. SCHMITT, A., C. HAUSER, L. DIDIERJEAN, Y. MEROT, J.-M. DAYER & J.-H. SAURAT. 1987. Br. J. Dermatol. **116:** 615-622.
15. DILLARD, G. M. & P. BODEL. 1970. J. Clin. Invest. **49:** 2418–2426.
16. SNYDER, D. S. & E. R. UNANUE. 1982. J. Immunol. **129:** 1803–1805.
17. STARUCH, M. J. & D. D. WOOD. 1985. J. Leukocyte Biol. **37:** 193–207.
18. KERN, J. A., R. J. LAMB, J. C. REED, R. P. DANIELE & P. C. NOWELL. 1988. J. Clin. Invest. **81:** 237–244.
19. LEE, S. W., A.-P. TSOU, H. CHAN, J. THOMAS, K. PETRIE, E. M. EUGUI & A. C. ALLISON. 1988. Proc. Natl. Acad. Sci. USA **85:** 1204–1208.
20. CULPEPPER, J. A. & F. LEE. 1985. J. Immunol. **135:** 3191–3197.
21. ARA, S. K., F. WONG-STAHL & R. C. GALLO. 1984. J. Immunol. **133:** 273–276.
22. HELFGOTT, D. C., L. T. MAY, Z. STHOEGER, I. TAMM & P. B. SEHGAL. 1987. J. Exp. Med. **166:** 1300–1309.
23. KOHASE, M., D. HENRIKSEN-DESTEFANO, P. B. SEHGAL & J. VILCEK. 1987. J. Cell Physiol. **132:** 271–278.
24. KURT-JONES, E. A., D. I. BELLER, S. B. MIZEL & E. R. UNANUE. 1985. Proc. Natl. Acad. Sci. USA **82:** 1204–1208.
25. GOLDMINZ, D., T. S. KUPPER & J. MCGUIRE. 1987. J. Invest. Dermatol. **88:** 97–100.
26. BAKOUCHE, O., D. C. BROWN & L. P. LACHMAN. 1987. J. Immunol. **138:** 4249–4255.
27. FLOWER, R. J. & G. J. BLACKWELL. 1979. Nature **278:** 456–459.
28. HIRATA, F., E. SCHIFFMANN, K. VENKATASUBRAMANIAM, D. SALOMON & J. AXELROD. 1980. Proc. Natl. Acad. Sci. USA **77:** 2533–2536.
29. KUPPER, T. S. 1988. *In* Advances in Dermatology. J. P. Callen, M. V. Dahl, L. E. Golitz, L. S. Schachner & S. J. Stegman, Eds. Vol. 3:293–307. Year Book Medical Publishers. Chicago, IL.
30. SAUDER, D. N., N. L. MOUNESSA, S. I. KATZ, C. A. DINARELLO & J. I. GALLEN. 1984. J. Immunol. **132:** 828–832.
31. RISTOW, H.-J. 1987. Proc. Natl. Acad. Sci. USA **84:** 1940–1944.

Hemopoietins: Roles in Inflammation, Allergy and Neoplasia

JOHN W. SCHRADER

The Biomedical Research Centre
University of British Columbia
Vancouver, British Columbia V6T 1W5, Canada

The skin has a critical role in defense of the body against invasion by microorganisms, one facet of which is the communication between keratinocytes and the more specialized defense cells of the lymphoid and hemopoietic systems. Keratinocytes, like other epithelial cells, fibroblasts and endothelial cells, respond to the endotoxin produced by certain bacteria or the foreign nucleic acids introduced during viral infections, by releasing cytokines, a group of soluble polypeptides that regulate defense and repair responses.

Cytokines exert a diversity of effects. One function is to stimulate the production and function of the cells most directly involved in defense against microorganisms and parasites, namely, the monocytes and their derivatives and the various granulocytic cells, *i.e.*, neutrophils, eosinophils, basophils and various types of mast cell. Cytokines also stimulate the production and function of fibroblasts and epithelial cells involved in repair responses, mediate certain systemic responses to infection such as fever and the appearance of acute-phase proteins in the serum, and regulate the responses of lymphocytes.

The thymus-derived (T) and bone-marrow-derived (B) lymphocytes constitute a more specialized system for sensing intrusion by foreign organisms or substances. The clonally distributed diversity of antigen-receptors characteristic of the lymphoid system, confers the capacity to sense intrusion by innumerable foreign molecules and, through expansion of the relevant clones, to generate a specific immunological response and immunological memory. A major component of immunological responses is cytokine production and T-lymphocytes exert many of their functions through the secretion of cytokines. As discussed below, some of these cytokines are exclusive products of T-lymphocytes, while others are also released by less specialized cells such as keratinocytes.

Actions and Interactions of Cytokines

Typically, cytokines exert their effects locally and therefore act as paracrine rather than endocrine hormones. There are, however, some notable exceptions; for example, serum-borne interleukin-6 may have an important role in the acute-phase response of the liver.[1]

It is increasingly evident that the regulatory interactions involving cytokines are complex. For example, as a general rule, stimulation of keratinocytes or fibroblasts by endotoxin, or of T-lymphocytes by specific antigen, results in the release of multiple cytokines. Moreover, individual cytokines have multiple cellular targets. The action of a particular cytokine on a single target cell is typically pleotropic, with effects on growth, the expression of cell-surface and secreted proteins, and the response to other cytokines. It is not uncommon for the effects of a cytokine on a target cell to include induction of the release of yet more cytokines, leading to the generation of a chain or cascade of cytokines.[2] Finally, as discussed below, cytokines often have the potential of affecting not only neighboring cells but also the cell from which they are released.

Nomenclature

The nomenclature of the cytokines is confusing. In many instances, cytokines have been discovered and characterized by virtue of their effects on the growth of target cells and have often been termed "growth factors." However, in most cases, they modulate not only the multiplication of target cells, but also their state of differentiation and function. Moreover, it is not uncommon for a single cytokine, *e.g.,* tumor-necrosis factor (TNF), interleukin-1(IL-1), transforming growth factor β (TGFβ), or granulocyte colony-stimulating factor (G-CSF), to stimulate the growth of one target cell yet inhibit the growth of another. Thus a "growth" factor such as TGFβ may inhibit the proliferation of lymphocytes while a "cytotoxic" factor, TNF, may stimulate the growth of certain cells.

Terms based on early information on the source or action of factors can also be misleading. The name "platelet-derived growth factor" reflects the origin of the original preparations of this factor and not its exclusive source—as it is also produced by other cells, *e.g.,* fibroblasts, and macrophages. "Interferon-β_2" or interleukin-6 (IL-6) probably has no direct antiviral action. Indeed, the term "interleukin," not withstanding its original connotations, now denotes neither an exclusive origin from, nor an exclusive action upon leucocytes. This point is well exemplified by interleukin-1 (IL-1), as many nonleukocytes, including keratinocytes and fibroblasts, are both sources and targets. The "interleukin" system of nomenclature has achieved quite wide acceptance, although inclusion of a particular cytokine within this system is still quite arbitrary. A recent subcommittee of the International Union of Immunology Societies (IUIS) Nomenclature Committee recommended that any well-characterized, novel polypeptide that was involved in inflammatory responses could qualify as an "interleukin."

Hemopoietins

The hemopoietins or colony-stimulating factors are a group of cytokines primarily involved with defence against microorganisms. The term "colony-stimulating factor" arose because the original assay for these substances was based upon their capacity to stimulate hemopoietic progenitor cells in populations of bone-marrow cells that were cultured in medium jellified with agar or methyl-cellulose, to generate colonies of differentiated leukocytes. There are now four well-characterized molecules that fit in this general category; IL-3 (or multi-CSF), G-CSF, GM-CSF and CSF-1 or (M-CSF) (see TABLE 1). The function of these molecules can be summarized as the stimulation of production of blood cells from their undifferentiated progenitors and enhancement of the survival and function of fully differentiated leukocytes.

Other cytokines have also been reported to stimulate the growth of small colonies of leukocytes. These include interleukin-6 (also known as B-cell stimulating factor-2 or interferon β-2), although it is not yet clear that the IL-6 directly stimulates the hemopoietic progenitor cell rather than acting indirectly by inducing the release of other substances such as G-CSF.[3] Interleukin-5 stimulates the growth of small colonies of eosinophils, the small size of these colonies reflecting the fact that IL-5 acts mainly on the later stages of this differentiation lineage.[3] Both IL-5 and IL-6 have powerful effects on the function and growth of B-lymphocytes.

Sources

Of the cytokines listed in TABLE 1 only IL-2, interferon-γ and IL-5 appear to be exclusively products of activated T-lymphocytes. IL-1 and IL-6 are also made by macrophages, fibroblasts and many other cell-types including keratinocytes. GM-CSF is released not only by activated T-cells, but also fibroblasts, macrophages, endothelial cells and keratinocytes in response to stimulation by endotoxin or by other cytokines.

TABLE 1. Cytokines: Sources and Targets

Cytokines	Sources	Targets
IL-2	T-lymphocytes	Lymphohemopoietic
IL-3	T-lymphocytes	Hemopoietic
IL-5	T-lymphocytes	Lymphohemopoietic
GM-CSF	T-lymphocytes, macrophages, fibroblasts, endothelial cells	Hemopoietic
IL-4	T-lymphocytes, mast cells	Lymphohemopoietic plus other cell types
IFN-γ	T-lymphocytes	Lymphohemopoietic plus other cell types
IL-6	T-lymphocytes, macrophages, fibroblasts, epithelial cells	Lymphohemopoietic plus other cell types
TNFα,β	T-lymphocytes, macrophages, fibroblasts, epithelial cells	Lymphohemopoietic plus other cell types
IL-1α,β	T-lymphocytes, lymphohemopoietic cells, fibroblasts, endothelial cells, epithelial cells	Lymphohemopoietic plus other cell types
TFGβ	T-lymphocytes, B-lymphocytes, megakaryocytes and others	Lymphohemopoietic plus other cell types
G-CSF	Macrophages, endothelial cells, epithelial cells	Neutrophil lineage
CSF-1	Macrophages, fibroblasts, endometrium	Macrophage lineage, Trophoblast

Interleukin-4 is released by activated T-cells and by at least some mast cell lines, although the physiological significance of the latter observation is not clear.

Interleukin-3 has been considered to be solely a product of activated T-lymphocytes. The data presented by Dr. Luger and Dr. Kupper in this volume, however, raise the possibility that IL-3 is also produced by keratinocytes. Confirmation of these observations and investigation of the regulation of IL-3 production in keratinocytes and its physiological and pathological significance are important tasks for the future.

Interferon-γ is exclusively a product of activated T-lymphocytes. Although it does not act as a growth factor, except perhaps to enhance the growth of certain T-lymphocytes, it has very important effects on the state of differentiation of many of the cells involved in defence and repair, regulating the expression of important cell-surface molecules such as major histocompatibility antigens and the receptors for the Fc fragment of immunoglobulin. Interferon-γ also has strong synergistic or antagonistic influences on the effects of many cytokines on their target cells.[4]

Several of the cytokines mentioned in TABLE 1 are not produced at all by lymphocytes. These include G-CSF, the major sources of which appear to be endothelial cells, epithelial cells (including keratinocytes, as reported by Sauder in this volume), fibroblasts and macrophages. Available evidence suggests that the secretion of G-CSF is not constitutive, but must be induced by stimulation with endotoxin or with other cytokines.

Another molecule not released by activated T-cells is CSF-1. Fibroblasts and monocytes produce CSF-1, and again stimulation by endotoxin or by other cytokines seems to be necessary for its release.[3]

Molecular Nature of Cytokines

Although the cytokines outlined in TABLE 1 share broad molecular similarities, all being polypeptides of around 12–14 Kd, with their activity depending on the integrity of at least one disulfide bond, there is no compelling evidence that they are members of a gene family. The exceptions are the family of three TGFβ which have a distinct structure and are related to inhibin, and the two unrelated families of IL-1α and β and TNFα and β, respectively. Comparison of amino-acid sequences has not revealed any other strong similarities among the various polypeptides, although there are suggestions of distant relationships between some cytokines. For example, GM-CSF and IL-2 share some regions of sequence similarity and the position of the critical disulfide bond appears to be conserved.[5] Likewise, G-CSF and IL-6 have a similar arrangement of cysteine residues and there are amino acid similarities in the third exon of both.[3] Both the GM-CSF and IL-3 genes can be found on a 9 Kb fragment of the human genome, this close proximity suggesting that they may have shared a common origin.[3] However, at present these relationships remain conjectural and have not yet led to any clues about structure-function relationships in these molecules, for example, concerning potential receptor-binding domains.

We noted that a group of cytokines share a motif of 5 or 6 similar amino acids at the N-terminus.[5] In the human, GM-CSF, IL-2, IL-3, IL-6, erythropoietin and IL-1β all have alanine followed by proline as the N-terminal amino acids. The functional significance of this motif has yet to be determined, but probably reflects a selective pressure for these molecules to interact with common enzyme or receptor rather than origin from a common ancestral gene.

Both IL-3[6] and GM-CSF[7] have now been produced by total chemical synthesis, a procedure which has allowed rapid accumulation of information about which parts of the polypeptide chain can be omitted without great loss of biological activity and which are thus unlikely to be directly involved in receptor binds. In the case of IL-3, Clark-Lewis and colleagues, have shown that N-terminal half of the molecule has detectable biological activity, suggesting that the important residues that interact with the receptor are found in the N-terminal half of the molecule.[6] Similar studies with GM-CSF have defined a region approximately 20 amino acids from the N-terminus as critical for biological activity.[7]

Targets of Cytokines

Some of the cytokines discussed in TABLE 1 are restricted in their activity to cells of the lymphohemopoietic system, i.e., those derived from the pluripotential lymphohemopoietic stem cell. Interleukin-3 has the broadest spectrum of activity of any of the hemopoietins and affects the progenitors of all hemopoietic lineages which have yet been investigated.[8] The sole exceptions appear to be cells of the T- and B-lymphocyte series and there is as yet no clear evidence that IL-3 affects any cell already committed to the T- and B-lymphocyte lineage.[8] IL-3 also affects the very early cells in lymphohemopoietic differentiation, including a pluripotential hemopoietic stem cell that can give rise to both hemopoietic and T- and B-lymphocytes when injected into an irradiated animal.[8] Like other hemopoietins, interleukin-3 also affects well-differentiated cells of certain lineages, including the mast cell, the macrophage and the megakaryocyte.[8]

The *in vivo* administration of chemically synthesized or recombinant IL-3 has been shown to stimulate multiple components of the hemopoietic system. The spleens of mice treated for 3 days with chemically synthesized IL-3 showed increased numbers of stem cells, various hemopoietic progenitor cells, immature granulocytes, mast cells and megakaryocytes.[9]

Interleukin-3 has a special relationship with a particular subclass of mast cell, namely, that associated with mucous membranes, the skin and lymphoid tissue. These mast cells appear to be absolutely dependent upon IL-3 for their generation from progenitors and also for their subsequent survival.[10] The subcutaneous injection of synthetic interleukin-3 to mice resulted in a cellular infiltrate at the injection site that included neutrophils, eosinophils and immature mast cells.[9]

Granulocyte-macrophage colony-stimulating factor affects a narrower range of hemopoietic cells, principally macrophages and neutrophilic and eosinophilic granulocytes and their progenitors.[3] GM-CSF not only enhances production of these cells but also can stimulate the function of the mature cells, for example, increasing the oxidative burst in response to bacterial products.

Granulocyte colony-stimulating factor is more restricted in its action, primarily stimulating the multiplication and differentiation of committed neutrophil progenitor cells and function of mature neutrophils.[3]

CSF-1 or macrophage CSF is again more restricted in its target specificity, principally stimulating the differentiation and function of macrophages. CSF-1 synergizes with factors such as IL-3 that act earlier in the differentiation pathway. Thus, at least in tissue culture, the combination of IL-3 and CSF-1 (although not either alone) stimulates the growth of large colonies of macrophages from human macrophage progenitors.[3]

CSF-1 is particularly interesting because there is some evidence that its activity is not restricted to cells of the hemopoietic system. Thus, trophoblast cells appear to respond to CSF-1, and it is produced in relatively large quantities in the pregnant uterus. CSF-1 is also interesting in that there is good evidence for the existence of a membrane-bound form of the molecule.[3] In theory, membrane-bound cytokines should be very effective mediators of cell-cell interactions.

Interleukin-5 appears to be restricted in its target range to the lymphohemopoietic system. It acts as a B-cell growth factor and also influences T-lymphocytes. *In vivo* experiments suggest that its major function may be to enhance the terminal differentiation of eosinophils.

Interleukin-4 was initially described as a factor which stimulated the growth of B-lymphocytes (B-cell stimulating factor-1). It acts as a growth factor for T-lymphocytes and may be important in expression of T-cell functions such as the generation of cytotoxic T-cells. It also has as yet ill-defined effects upon the growth of hemopoietic cells in tissue culture. Receptors for interleukin-4 have been found outside the lymphohemopoietic system, *e.g.,* occurring on cells from neural tissue,[3] and its target range may be broad and analogous to that of IL-1 and IL-6.

IL-1 and IL-6 not only share a similar range of target cells, but in many instances appear to exert similar effects. It is not clear in which cases this apparent similarity in action may reflect the stimulation of the release of one by the other—as has been shown for the "hepatocyte stimulating activity" of IL-1 which results from the release of IL-6.[1]

Interleukin-2 was originally described as a T-cell growth factor There is now evidence, however, that not only T-lymphocytes, but also B-lymphocytes and nonlymphoid cells such as macrophages can respond to IL-2.

Transforming growth factor β, which is produced by activated T-cells and by many other cell types, is an important regulator of defense and repair responses. As is the case for IL-1, many cells have receptors for TGFβ and its effects vary with the target cell. As noted by Moses in this volume, in some cases, e.g., fibroblasts, TGFβ can stimulate growth, whereas in others, e.g., epithelial cells, it stimulates differentiation and inhibits growth.

Physiology and Pathology of Cytokines

In general terms the cytokines discussed above are only released in response to some pathological stress. As discussed above, one set of stimuli are relatively nonspecific, e.g., the endotoxin of certain bacteria, the foreign nucleic acids introduced by viral infections, or other cytokines. Specific activation of T-lymphocytes, with resultant release of cytokines is also obviously a response to an abnormal situation. Tissue injury leading to blood coagulation will result in the release from platelets of cytokines such as PDGF and TGFβ. These will then induce the secretion of IL-1 from macrophages, etc., and thus initiate "cascades" of cytokine production.[2]

Relatively little is known about how the cellular composition of an inflammatory response is regulated, i.e., why neutrophils or macrophages or eosinophils, for example, predominate in a particular instance. Presumably the balance of those cytokines that primarily affect specific lineages is important, e.g., IL-5 tending to promote increases in eosinophils, but there is little direct evidence. IL-3, however, probably does have a unique relationship with the subclass of mast cells that increase in number at epithelial surfaces during T-cell activation and are likely to be the key cells in many nonallergic responses.

There is very little evidence that any of the cytokines discussed are produced constitutively in the absence of pathological stimuli, although CSF-1 has been reported to be detectable in normal human serum. There is also some preliminary evidence from studies involving antibodies against G-CSF, that G-CSF may be responsible for the normal levels of neutrophils in the peripheral blood. Nevertheless, available data suggest that the cytokines discussed here are only released following stimulation by microorganisms or their products, or as a result of injury or immunological responses.

Roles of Cytokines in Normal Hemopoiesis

One consequence of the fact that cytokines appear to be released only during pathological situations, is that they are unlikely to play a role in the normal, steady state of production of lymphoid and other cells of hemopoietic origin. A careful search for evidence of the production of factors such as IL-3, GM-CSF, G-CSF or M-CSF in normal bone-marrow has failed to find evidence for the production of these proteins or of the respective mRNAs.[3] There are a number of reasons, however, why these data do not completely exclude a role for these factors in hemopoiesis.

One relates to the fact that methods for detecting small amounts of proteins and mRNA are not sufficiently sensitive to exclude the production of biologically significant amounts of these factors. This problem is intensified by the possibility that if presented on the surface of cells within bone-marrow, very small amounts of these hemopoietins could in theory be regulating steady-state hemopoiesis. One possibility is that the relevant cytokines could be present as trans-membrane proteins, as has been demonstrated for forms of CSF-1, TNF and TGFα. Alternatively, cytokines could be bound to the extra-cellular matrix associated with cell membranes, as Dexter's group has demonstrated for substances like IL-3 and GM-CSF.[3] There is some evidence from culture systems that generate hemopoietic cells *in vitro* for long periods in the absence of soluble growth factors, that cell surface-bound molecules could be critical in regulating normal hemopoiesis,[3] but whether these represent small amounts of known cytokines in cell-bound form, or, perhaps more likely, as yet undescribed molecules, is unknown.

Paracrine and Autocrine Mechanisms

One theme that has emerged recently is that of the potential physiological importance of paracrine or autocrine mechanisms involving cytokines. There is now a significant number of examples where cells that are known to be responsive to a particular cytokine can be shown to be capable of producing it. For example, fibroblasts produce and respond to platelet-derived growth factor and IL-1, macrophages produce and respond to GM-CSF and CSF-1, and as shown in this volume keratinocytes produce and respond to TGFα and IL-1. In these instances the production of autostimulatory factors is tightly regulated, and when the initiating stimuli are extinguished, factor production ceases.

We have been interested in the role of the pathological production of autostimulatory factors in oncogenesis. Some five years ago we found a variant of an immortalized but nontumorigenic, IL-3-dependent hemopoietic cell line that had spontaneously gained the capacity to grow autonomously in the absence of exogenous IL-3.[11] We observed that this variant had simultaneously begun to produce IL-3 and acquired the capacity to grow as a disseminated leukemia when transferred into a syngeneic mouse. There is now very good evidence from a number of systems that the pathological onset of the constitutive production of an autostimulatory growth factor by an immortalized cell can result in transformation into a tumorigenic cell.[11]

Analysis of the significance of autocrine or paracrine mechanisms of cytokine action in response to injury or invasion and in oncogenesis can at present be best approached using antibodies that neutralize cytokine action. It is possible that the action on a cell of a cytokine produced by that cell may be intrinsically difficult to inhibit with antibody. This is obviously the case if the cytokine interacts with its receptor and delivers an effective signal within the cell during the synthetic and secretory process. A more interesting possibility is that receptor-cytokine complexes could be formed during synthesis within the same cell, but that these complexes might not transmit effective signals until they had reached the cell-surface and were able to associate with critical signal transducing proteins, *e.g.,* G proteins.

SUMMARY

Cytokines link keratinocytes with lymphocytes and the specialized phagocytic and granulocytic cells of the hemopoietic system and are critical elements in the response of the skin to invasion or injury. The cytokine network is complex and includes potential autoregulatory circuits. A better understanding of these networks and of the function and structure of cytokines may lead to a more rational approach to the design of drugs and to new treatments.

REFERENCES

1. GAULDIE, J., C. RICHARDS, D. HARNISH, P. LANDSDORP & H. BAUMANN. 1987. Proc. Natl. Acad. Sci. USA **84:** 7251–7255.
2. SCHRADER, J. W. 1988. Immunol. Cell Biol. **66:** 111–122.
3. SEILER, F. R. Ed. 1988. The Colony Stimulating Factors. Behring Institute. Mitteilungen.
4. WONG, G. H. W., I. CLARK-LEWIS, J. A. HAMILTON & J. W. SCHRADER. 1984. J. Immunol. **133:** 2043.
5. SCHRADER, J. W., H. J. ZILTENER & K. B. LESLIE. 1986. Proc. Natl. Acad. Sci. USA **83:** 2458–2462.
6. CLARK-LEWIS, I., R. AEBERSOLD, H. J. ZILTENER, J. W. SCHRADER, L. E. HOOD & S. B. H. KENT. 1986. Science **231:** 134–139.
7. CLARK-LEWIS, I., A. LOPEZ, L. B. TO, M. A. VADAS, J. W. SCHRADER, L. E. HOOD & S. B. H. KENT. 1988. J. Immunol. **141:** 881–889.
8. SCHRADER, J. W. 1986. Annu. Rev. Immunol. **4:** 205–230.
9. SCHRADER, J. W., I. CLARK-LEWIS, H. J. ZILTENER, L. E. HOOD & S. B. H. KENT. 1987. *In* Immune Regulation by Characterized Polypeptides. G. Goldstein, J.-F. Bach & H. Wigzell, Eds.: 475–484. A. R. Liss. New York, NY.
10. CRAPPER, R. M., W. R. THOMAS & J. W. SCHRADER. 1984. J. Immunol. **133:** 2174–2179.
11. SCHRADER, J. W. & R. M. CRAPPER. 1983. Proc. Soc. Natl. Acad. Sci. USA **80:** 6892–6896.

Normal Human Keratinocytes Contain an Interferon-like Protein That May Modulate Their Growth and Differentiation

M. YAAR,[a] A. V. PALLERONI,[b] AND B. A. GILCHREST[a]

[a]United States Department of Agriculture
Human Nutrition Research Center on Aging
at Tufts University
711 Washington Street
Boston, Massachusetts, 02111
and

[b]Department of Experimental Oncology and Virology
Hoffman-La Roche, Inc.
340 Kingsland Street
Nutley, New Jersey 07110

INTRODUCTION

Epidermal growth and differentiation is a complex process *in vivo* integral to the normal function and appearance of human skin and to the pathophysiology of many distressing dermatologic diseases. Our understanding of the epidermis has been greatly augmented over the past decade by the advent of keratinocyte culture systems suitable for quantitative short-term and long-term studies of these cells. Rheinwald and Green[1] developed the first system for growth of disaggregated cells into stratified colonies, relying on serum mitogens and a 3T3 fibroblast "feeder" layer. This system and minor modifications of it have allowed elucidation of probable physiologic roles for hydrocortisone,[1] epidermal growth factor,[1] vitamin A,[2] calcium,[3] and the cAMP pathway[4] in the epidermis, as well as identification of useful histologic and biochemical markers for keratinocyte differentiation.[5] Other systems, relying on growth factor supplements and reduced calcium ion concentration, have contributed further to this body of information.[6–9] The system developed in our laboratory, the first permitting keratinocyte cultivation in the absence of serum,[10] has allowed identification of a potent, previously unknown growth factor in hypothalamic extracts,[11] and of autocrine and paracrine growth modulation of keratinocytes by other skin-derived cells *in vitro*.[12]

Despite these rapid and important increases in our understanding of *positive* modulators of keratinocyte growth *in vitro*, to date there are very little data regarding possible *negative* modulators, classically termed chalones. Since chalones were first conceptualized by Bullough in 1962,[13] more than 600 papers concerning chalones have been published, but only recently have chalones been isolated or characterized in a pure state.

Interferons (IFN) are a family of glycoproteins presently classified as alpha, beta or gamma on the basis of physiochemical characteristics.[14] IFN were first identified by their antiviral activity during studies of viral interference,[15] but have since been widely recognized to inhibit proliferation of both normal and malignant cells *in vitro*[16–18] and are under investigation as antitumor drugs in several clinical trials.[19]

Since we had shown that keratinocytes are capable of producing IFN under condition of viral infection,[34] it was interesting to see whether IFN may function as a physiological regulator of keratinocyte growth *in vivo* with properties of negative growth factor or chalone.[49,50]

MATERIALS AND METHODS

Antibodies

Polyclonal antibodies to recombinant human IFN-alpha were prepared by subcutaneous injections of New Zealand white rabbits at four different sites with a total of 1 mg of recombinant human IFN-alpha (monomer form) in Freund's complete adjuvant. The animals were given booster injections of 500 μg in incomplete Freund's adjuvant weekly for a total of 4 weeks. One week after the 4th injection, blood was obtained from each rabbit. The animals were bled before immunization and their pre- and postimmunization serum samples were screened for the presence of antibodies to recombinant human IFN-alpha by Ouchterlony agarose double diffusion analysis and antibody neutralizing bioassay.[20] Antibodies ($0.48-1.26 \times 10^6$ IFN neutralizing units/ml) were detected in the immune sera only.

Purification of Antibodies

Purification of antirecombinant human IFN-alpha antibodies from two rabbit sera was performed by conventional affinity chromatography using a column of sepharose 4B coupled to recombinant human IFN-alpha.[21] Either IFN-alpha antiserum or preimmune serum (0.3–0.7 ml) was added to the column. The column was washed with 7 column vol. of phosphate-buffered saline (PBS). The specific bound antibody was then eluted. Pooled peak fractions were concentrated by ultrafiltration on a YM10 Amicon filter and used for immunofluorescence and immunoblotting.

Tissue and Cells Culture

Newborn foreskin or cutaneous biopsies of healthy adult volunteers were embedded and frozen in tissue Tek O.C.T. Primary keratinocyte cultures were prepared from newborn foreskin as described.[22] At confluence dishes were washed twice with 0.02% EDTA, incubated in 0.25% trypsin at 37°C, and disaggregated to form a single cell suspension. Keratinocytes were inoculated either on glass coverslips or on dishes coated with human fibronectin 10 μg/cm^2.[23] Cultures were maintained at 37°C in 8% CO_2 and provided with serum-free hormone supplemented medium (SFHSM).[24]

Experimental Procedures

One or more of the following parameters was examined for each cell strain:

1. Presence of IFN in Cultured Keratinocytes. At confluence either cells on cover slips were fixed with acetone at $-20°C$ for 30 sec[25] and used for indirect immunofluorescence, or confluent keratinocyte sheets were detached from dishes with medium containing 1.2 μg/ml Dispase II,[26] then embedded and frozen in tissue Tek II O.C.T.

and processed for indirect immunofluorescence. Other confluent keratinocyte cultures were prepared in a small volume of 0.5% Nonidet P-40 in tris buffered saline containing 1 mM phenylmethyl sulphonylfluoride.[27] The extracts were used either for immunoblotting or for analyzing IFN activity by reduction of cytopathic effect of vesicular stomatitis virus.

2. IFN-alpha Effect on Cell Growth. At the time of seeding before cells had attached and with each subsequent feeding (three times weekly) IFN-alpha was added to paired cultures at final concentrations of 0, 25, 50, 250, 500, 1250, 2500 and in some experiments 5000 units/ml. After one week, 3 paired plates at each IFN concentration were (a) fixed and stained to permit comparison of colony size;[28] (b) rinsed and then trypsinized to yield a single cell suspension and counted in a hemacytometer chamber; or (c) rinsed and exposed to 1 N NaOH, and processed for protein determination according to the colorimetric BioRad Protein Assay.[29] Three additional paired plates at each IFN concentration were continued in culture and subsequently maintained in SFHSM lacking IFN. After 7 additional days' growth, the plates were processed as above in order to examine the reversibility of the IFN effect.

3. Effect of IFN on Cell Attachment. Keratinocytes were plated in SFHSM with or without 2500 units/ml IFN-alpha and examined using a Zeiss phase microscope (160X) 2 hr, 6 hr, and 24 hr after initial seeding, without reference to culture identity, for gross differences in cell attachment, spreading and morphology. At 24 hr the medium was collected from each plate and centrifuged at 1200 rpm for 5 min. The supernatant was aspirated, the pellet (unattached cells) was resuspended in 100 μl PBS, and 2 separate aliquots of the suspension from each dish were then counted using a hemacytometer. The dishes containing adherent colonies were then incubated for 10 min at 37°C in 1 ml of 0.5% trypsin, vigorously sprayed with a pasteur pipet and examined under phase microscopy to assure that no adherent cells remained. Three separate aliquots of the cell suspension from each dish were then counted using a hemacytometer.

4. Effects of IFN on Keratinocyte Differentiation. Keratinocyte cultures were maintained in SFHSM supplemented with IFN 0 or 2500 units/ml for 7 days, then processed to examine markers of terminal differentiation, or placed in SFHSM lacking IFN for an additional 7-day period prior to such processing, in order to examine the reversibility of the IFN effect on differentiation.

One set of cultures was harvested to yield separately attached cells and cells shed into the medium during the preceding 48–72-hr feeding cycle. Medium from each plate was centrifuged for 5 min at 1200 rpm, the supernatant aspirated, and the pellet of desquamated cells resuspended in 1 ml of PBS. This single cell suspension was divided equally into 2 tubes and recentrifuged. One pellet was resuspended in 0.2 ml of PBS and counted in a hemacytometer chamber to determine total cell number. The second pellet was resuspended in 10 ml of 1% SDS containing 1% 2-mercaptoethanol (2ME), incubated 10 min at room temperature, and centrifuged at 2400 rpm for 10 min, aspirated, resuspended and counted as above to determine the proportion of cells with cornified (SDS & 2ME resistant) envelopes.[30] Cells in colonies attached to the plate, not removed with the medium, were trypsinized to yield a single cell suspension, centrifuged, resuspended, divided and processed in the same way as cells recovered from the medium.

Separate cultures maintained under identical conditions were subjected to keratin extraction.[31] A 30-μg aliquot of reduced protein from each dish was run on an 8.5% PAGE with appropriate molecular weight standards. The gel was stained overnight with 0.025% Coomassie brilliant blue G-250 solution, destained with 10% isopropanol followed by 10% acetic acid solution for 3–4 hr, and fixed in 10% acetic acid.

A final set of cultures was processed as previously described[33] for transmission electron micrographs of vertically sectioned keratinocyte colonies.

Immunofluorescence

Six-μm sections of fresh frozen tissues or cells were incubated with immune rabbit serum, preimmune rabbit serum, column-purified antibodies from immune serum, and column-purified antibodies from preimmune serum. The second antibody used was fluorescein-tagged goat antirabbit IgG.

Immunoblotting

Extracted keratinocyte proteins and pure recombinant IFN-alpha (2.5 μg of a standard preparation) were reduced and separated by 12% SDS PAGE, then electrophoretically transferred to nitrocellulose paper using a BioRad Trans-Blot apparatus overnight at 4°C, 60 V in tris-glycine buffer with 20% methanol.[27] Antigens on the nitrocellulose paper were incubated with either IFN-alpha antiserum or preimmune serum at 1:250–1:500 dilution, column-purified IFN-alpha antiserum, or column-purified preimmune serum at 1:50–1:500 dilution. Specific binding of antibodies was identified by immunoperoxidase staining of the nitrocellulose paper strips.

IFN Activity of Keratinocyte Extracts

IFN activity of keratinocyte extracts was determined quantitatively by reduction of cytopathic effect of vesicular stomatitis virus using a bovine kidney cell line of epithelial origin.[34] The extract samples used to measure IFN activity were serially diluted and added to infected cells either immediately after viral innoculation or 24 hr later. Cultures were incubated at 37°C until the virally infected control cells displayed a 100% cytopathic effect. The IFN titer was then read as the reciprocal of the extract dilution that protected 50% of the cell monolayer. A laboratory standard of IFN was included in all assays.

RESULTS

Antibody Binding

Two sera obtained from rabbits immunized against recombinant human IFN-alpha produced staining of the epidermal basal layer in tissue cross sections with occasional staining in the first few suprabasilar layers (FIG. 1a), while control specimens incubated with preimmune serum were negative (FIG. 1b).

To determine whether the IFN staining was retained in cultured epidermis, second passage human keratinocyte cultures grown under serum-free conditions[4] were either detached from the dish by the use of Dispase II[26] and frozen or grown on cover slips and fixed in -20°C acetone for 30 sec.[26] Immunofluorescent staining was present in the basal layer of stratified colonies in a pattern analogous to that observed *in vivo* (FIG. 2a), and individual cells displayed bright cytoplasmic fluorescence (FIG. 2b). Stratified colonies incubated with preimmune serum were negative (FIG. 2c and d).

To exclude the possibility that the staining pattern was due to antibodies nonspecifically induced during the immunization procedure, the reactive rabbit antisera were applied to an IFN affinity column of sepharose 4B coupled to recombinant human IFN-alpha.[21] Material eluted from the IFN affinity column

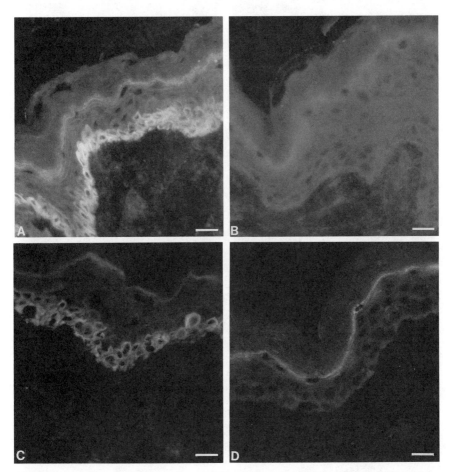

FIGURE 1. Binding of anti-IFN anti-serum (**A**) and column-purified anti-IFN antibodies (**C**) to normal human skin. Bright fluorescence is present in the basal layer of the epidermis. The remainder of the viable epidermis and the stratum corneum are negative. Faint nonspecific linear fluorescence is frequently present in both these sections and control sections within the stratum granulosum (**B** and **D**). Representative sections from one of two sera and one of 8 tissue donors are shown. *Bar* 20 μm. (From Yaar *et al.*[49] Reprinted by permission from the *Journal of Cell Biology*.)

(specifically bound) and material in the column washes (nonbound) were separately concentrated and used in the above procedure. Skin sections reacted with the purified antibodies present in the concentrated eluant revealed the same fluorescent staining pattern observed with whole sera (FIG. 1c), while the comparably concentrated washes gave no staining whatsoever (FIG. 1d). Similarly, cultured epidermis reacted with the

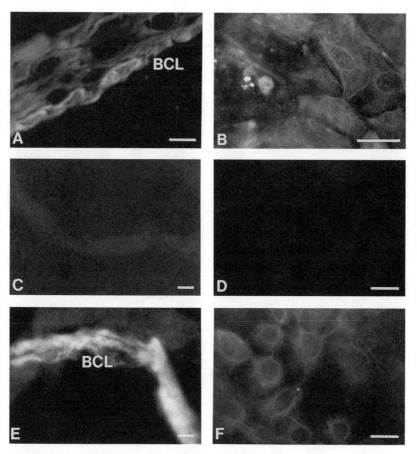

FIGURE 2. Binding of anti-IFN antiserum to stratified keratinocyte cultures. Antibodies from immune serum (**A**) and column-purified anti-IFN antibodies (**E**) bound to the basal cell layer (BCL) of cultured keratinocyte colonies sectioned vertically and displayed bright cytoplasmic fluorescence in the respective *en face* preparations (**B** and **F**). Antibodies from preimmune serum did not bind to vertically sectioned keratinocyte colonies (**C**) or to keratinocytes grown on cover slips (**D**). *Bar* 5 μm. (From Yaar *et al.*[49] Reprinted by permission from the *Journal of Cell Biology*.)

purified antibodies present in the concentrated eluant revealed the same fluorescent staining pattern observed with whole sera (FIGS. 2e and f), while the comparably concentrated washes again gave no staining.

IFN Activity in Keratinocyte Cell Extracts

To determine whether the positive staining pattern was associated with the presence of biologically active IFN, cell extracts and conditioned medium from confluent keratinocyte cell cultures, nonconditioned medium, and extracting solution alone prepared as described[25] were analyzed for IFN activity in a cytopathic effect assay that could detect as little as 2 neutralizing units/ml of IFN activity.[33] IFN activity was found exclusively in the cell extracts and was titered to 38 neutralizing units/ml in triplicate samples, 0.4% of the peak values measured previously by a different methodology in the medium of comparably confluent keratinocyte cultures infected with herpes simplex virus.[34]

Characteristics of IFN Present in Cultured Keratinocytes

To determine the molecular weight of IFN present in cultured keratinocytes, confluent cultures were extracted as above, the proteins separated by SDS PAGE and transferred to nitrocellulose paper,[35] and the lanes reacted separately with either immune sera, column-purified anti-IFN antibodies from these sera, appropriate control sera, or column-purified preimmune sera. Antigen binding on nitrocellulose paper was identified by immunoperoxidase staining.[27] Immune sera and the affinity column adherent antibodies from these sera bound principally to a ~20-kD recombinant human IFN-alpha standard and, to a lesser degree, to its ~40-kD dimer (FIG. 3) as expected.[36] In lanes containing cell extracts, antibodies derived from immune sera

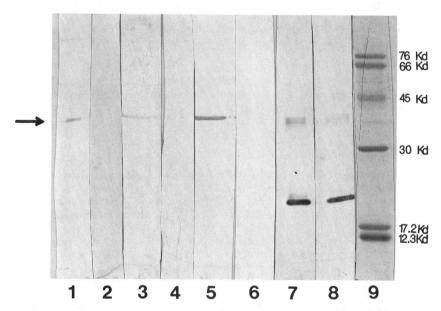

FIGURE 3. Immunoblot analysis of anti-IFN antibody binding to keratinocyte cell extracts. Antibodies used are either rabbit serum containing antibodies against recombinant human IFN-alpha (*lane 1*) or column-purified antibodies from immune rabbit serum (*lanes 3 and 5*). One of 3 positive sera is shown in this immunoblot. A ~40-kD band is recognized by immune serum and purified antibodies but not by preimmune serum (*lane 2*) or column-purified antibodies from preimmune sera (*lanes 4 and 6*). Column-purified antibodies which recognized the ~40-kD band (*lanes 1, 3 and 5*) also bind to the ~20-kD band of pure recombinant human IFN-alpha and to its dimer at 40 kD (*lanes 7 and 8*). Molecular weight standards are shown in *lane 9*. (From Yaar *et al.*[49] Reprinted by permission from the *Journal of Cell Biology*.)

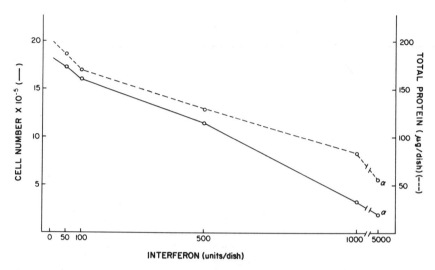

FIGURE 4. Keratinocyte growth inhibition by IFN-alpha. Paired keratinocyte cultures derived from a newborn donor were inoculated at 2×10^5 cells/35 mm dish and maintained for 7 days in serum-free medium supplemented at each feeding with IFN-alpha 0–2500 units/ml. One representative experiment is shown. Each point is the mean of 2 separate cell counts or 2 separate protein measurements.

recognized only a protein of ~40 kD. Preimmune sera and affinity column eluants prepared from these sera did not bind to either recombinant human IFN-alpha or the ~40-kD protein in the cell extract (FIG. 3).

IFN Effect on Cell Growth

IFN-alpha consistently inhibited keratinocyte growth, although degree of inhibition varied moderately among donor strains and with inoculation density. After 7 days, control cultures contained 6.5 ± 0.33 (mean ± SEM) times the inoculated cell number, while cultures supplemented with 2500 units/ml of IFN-alpha contained only 2.1 ± 0.5 times the initial inoculum, a highly significant ($p < 0.001$ sign test) mean growth inhibition of 65 ± 8% for 8 experiments. Mean inhibition of keratinocyte growth by 2500 units/ml of IFN-alpha as determined by total protein per dish in 4 experiments was 48 ± 14%. Over the dose range tested there was an approximately linear decline in cell yield at 7 days (FIG. 4), and paired dishes stained with Rhodanile blue revealed a proportionate reduction in keratinocyte colony size (FIG. 5). Statistically significant ($p < 0.001$) keratinocyte growth inhibition was obtained with >250 units/ml of IFN-alpha in these experiments, but progressively greater inhibition was observed throughout the dose range examined.

Paired IFN-supplemented and control keratinocyte cultures not processed after 7 days but maintained an additional 7 days in medium lacking IFN, revealed that the IFN-induced inhibition was reversible (FIG. 5). The impression, gained from daily phase microscopic examination of cultures in the second week and from comparison of stained dishes on day 7 and 14, was confirmed by cell counts: the ratio of cell number/dish on day 14 vs day 7 ranged from 1.1 to 4.0 among experiments and was consistently greater for initially IFN-inhibited cultures than for controls, perhaps because the latter were approaching confluence during the second week. In contrast, cultures maintained 14 days in IFN supplemented medium revealed continued profound growth inhibition.

IFN Effect on Cell Attachment

There was no effect of IFN-alpha on keratinocyte attachment rate after 24 hr. Absolute attachment rates for both control and IFN treated inocula varied moderately, depending on handling procedure, but never differed from each other by more than 5% within a single experiment, as calculated by counting either attached or floating cells.

IFN Effect on Keratinocyte Differentiation

Routine phase microscopic observations of cultures grown in the presence of IFN revealed only very subtle morphologic changes, other than reduction in average colony diameter. Possible effects of IFN on keratinocyte differentiation were nevertheless examined quantitatively using biochemical techniques and in greater qualitative detail by transmission electron microscopy (EM).

IFN consistently promoted cornified envelope formation and cell shedding in cultures maintained 7 days in the presence of 2500 units/ml IFN (TABLE 1),

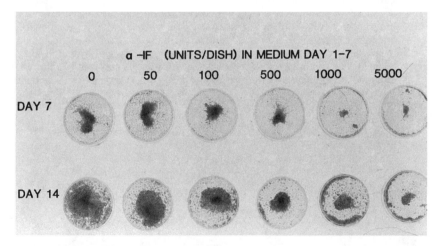

FIGURE 5. Reversal of IFN-induced inhibition of human keratinocyte growth. Keratinocyte cultures derived from a newborn donor were plated at 2×10^5 cells/35 mm dish and maintained for 7 days in serum-free medium supplemented at each feeding with IFN 0–2500 units/ml. After 7 days one plate at each IFN dose was fixed and stained with Rhodanile blue and paired plates were maintained an additional 7 days in medium lacking IFN, then processed as before. One representative experiment is shown. Cell counts (mean cell number × 10 on day 7 vs day 14, for each IFN dose) were 19.3 vs 25.4 for 0 units/ml, 17.5 vs 19.2 for 25 units/ml, 12.6 vs 18.0 for 50 units/ml, 11.2 vs 12.1 for 250 units/ml, 1.1 vs 1.63 for 500 units/ml, 0.6 vs 1.0 for 2500 units/ml. Total protein determinations in paired dishes also correlated well.

TABLE 1. Effect of Interferon on Keratinocyte Differentiation

Experiment No.	Percent Terminally Differentiated[a]		Percent Detached[b]	
	Control	IFN-Treated	Control	IFN-Treated
1	3.5	5.5	0.3	2.9
2	2.0	6.4	1.0	4.2
3	4.5	7.2	2.1	11.9
4	9.6	13.3	2.4	4.8
5	4.5	18.0	2.0	6.8

[a]Number of SDS/2ME-resistant cells divided by total number of cells on day 7 × 100.
[b]Cells counted in the cell layer on day 7 divided by cells counted in a 3-day medium collection, day 4–7, × 100.

approximately doubling the ratio of SDS/2ME-resistant (cornified) cells to total cells, compared to untreated controls, a statistically significant shift ($p < 0.04$ sign test). Exact ratios varied from experiment to experiment, depending on inoculation density and degree of confluence after the one-week growth period.

Paired control cultures processed after 7 days for EM revealed stratified colonies consisting of 6–10 cell layers, including one or two adherent surface layers with loss of intercellular desmosomes and thickened plasma membranes, representing cross-linked envelopes[31] (FIG. 6a). After 14 days, appearance was similar, but colonies contained up to 13 cell layers, including 3–8 with cross-linked envelopes, and one or two cell layers at the base of these differentiated zone resembling a stratum granulosum. In contrast, IFN-treated cultures processed after 7 days revealed colonies with only 4–5 cell layers; cells with thickened plasma membranes were either absent from the colony surface or very loosely adherent (FIG. 6b). IFN-treated cultures processed after an additional 7 days in medium lacking IFN ("reversed" cultures) contained 5–7 cell layers, including 0–3 with cross-linked envelopes and an interrupted single cell layer with keratohyaline granules, an appearance intermediate between the 7-day and 14-day control cultures.

No quantitative or qualitative differences in keratin protein profiles were apparent

FIGURE 6. Electron micrographs of cross sections through keratinocyte colonies 7 days after inoculation. (**a**) Control culture after 7 days. Note 5 cell layers and absence of cornified envelopes. (**b**) Culture supplemented with IFN 2500 units/ml after 7 days. Note 4 cell layers with detaching cells at colony surface. Thickened cell membranes represent cornified envelopes (*).

between controls and IFN-treated cultures of 7 days or 14 days. All demonstrated bands corresponding to: 58 kD, 56 kD, 52 kD, and 50 kD keratins.

DISCUSSION

We have shown that a protein cross reacting with recombinant human IFN-alpha is present in the proliferative compartment of normal human epidermis *in vivo* and *in vitro* in the absence of viral infection and that extracts of cultured keratinocytes contained IFN-like antiviral activity. The use of multiple newborn and adult skin donors effectively eliminates the possibility that clinically undetectable viral infection was responsible for these findings. The precise nature of the IFN-like substance in the basal layer of human epidermis remains unclear, however. If the substance is indeed IFN-alpha the absence of its usually predominant ~20-kD species must be explained, even though dimers of recombinant human IFN-alpha of ~40 kD are known to exist *in vivo* and *in vitro* and to persist under reducing conditions as in SDS PAGE.[37] Present data do not allow us to distinguish between a modified monomeric form of IFN-alpha, a persistent dimer of the otherwise predominant monomer, and a previously unrecognized class of IFN.

This study also demonstrates that IFN profoundly and reversibly inhibits the growth of human keratinocytes. After 1 week, optimal concentrations of IFN-alpha decreased cell yield by an average of approximately 70% and by more than 90% in some experiments, while cell attachment is not altered. The inhibition is sustained for at least 2 weeks during continuous exposure to IFN and is completely reversible, with treated cultures resuming growth at a rate equal to or greater than that of controls almost immediately after removal of IFN from the medium.

In addition to inhibiting growth, IFN promotes keratinocyte terminal differentiation as assessed by desquamation of cells from the colony surface, by proportion of cornified cells; and by the electron microscopic criteria of thickened plasma membranes; lack of desmosomes, and absent or pycnotic nucleus in upper cell layers of the colonies.

The epidermis constantly renews itself. In normal skin, however, the majority of the cells in the germinative basal layer compartment are blocked either in G_1, G_2, or G_0, and do not cycle unless stimulated.[37] The precise controls for epidermal cell proliferation *in vivo* are virtually unknown, although circumstances such as wounding, ultraviolet irradiation, certain disease states, and chronological aging have undeniable influences.[38,39] Substances reported to influence epidermal population dynamics *in vitro* include epidermal growth factor,[1] calcium,[3] cyclic nucleotides,[4] prostaglandins,[40] vitamin A,[2] and various tissue extracts,[41-43] but their physiological roles in normal or diseased skin are speculative. The present data suggest that IFN may have a physiological or therapeutic role in disorders such as psoriasis, that are characterized by reversible epidermal hyperplasia, accelerated epidermal turnover rate, and compromised keratinocyte differentiation. Chalones were first conceptualized by Bullough in 1962 as tissue specific, species-nonspecific substances that can inhibit cell division.[12] Although tentatively identified in the epidermis[12,44] and avidly investigated over many years, chalones have proven elusive.[44,45] Although IFN do not satisfy all the original criteria for epidermal chalones,[13] their demonstrated ability to inhibit growth of cultured keratinocyte profoundly and reversibly[46] make them excellent candidates for assuming such a function *in vivo*. IFN-alpha is known to behave as a negative feedback inhibitor for bone marrow cells at least *in vitro*,[47] and the presence of IFN in the amniotic fluid of pregnant women during the second and third trimester in the absence of detectable viral infection[48] suggests that IFN may participate in the regulation of fetal development. The present report demonstrates that an IFN-like protein is constitutively present in the proliferative layer of the epidermis. Intuitively, one might expect a chalone to be present in the suprabasilar nonproliferative compartment of the epidermis. However, the existence of such a regulatory factor in the basal layer, within potentially dividing cells that under normal conditions are nevertheless noncycling, is also plausible. We suggest that the IFN-like protein observed in those studies may therefore serve as a chalone in this precisely regulated tissue.

SUMMARY

Epidermal growth and differentiation is a complex process which depends upon a balance between positive and negative growth signals, and in normal skin the majority of the cells in the germinative basal layer do not proliferate unless stimulated. Using the indirect immunofluorescent method, it can be demonstrated that purified polyclonal antibodies to interferon-alpha specifically bind to the basal layer of human epidermis in cross sections of normal skin and to the basal layer of cultured keratinocyte colonies. Furthermore, extracts of keratinocyte cultures contain interferon bioactivity. With Western blot analysis, antibodies to interferon recognize a band of ~40 kD both in keratinocyte lanes and in recombinant interferon lanes that give in addition a band of ~20 kD. Addition of interferon to rapidly growing keratinocytes inhibits their growth by as much as 90% and promotes their terminal differentiation. The growth inhibitory effect of interferon is completely reversible. These data demonstrate that interferon or a closely related protein is present in human epidermis and suggest that this protein may act as a physiologic modulator of keratinocyte growth and differentiation.

REFERENCES

1. RHEINWALD, J. G. & H. GREEN. 1977. Nature (London) 265: 421–424.
2. CHOPRA, D. P. & B. A. FLAXMAN. 1975. J. Invest. Dermatol. 64: 19–22.
3. BOYCE, S. T. & R. G. HAM. 1983. J. Invest. Dermatol. 81: 33s–40s.
4. GREEN, H. 1978. Cell 15: 801–811.
5. GREEN, H. 1979. Harvey Lect. 74: 101.
6. HENNINGS, H. & K. A. HOLBROOK. 1983. Exp. Cell Res. 143: 127.
7. HENNINGS, H., K. A. HOLBROOK & S. H. YUSPA. 1983. J. Cell. Physiol. 116: 265.
8. TSAO, M. C., B. J. WAITHAL & R. G. HAM. 1982. J. Cell. Physiol. 110: 219.
9. BOYCE, S. T. & R. G. HAM. 1983. J. Invest. Dermatol. 81: 335.
10. MACIAG, T., R. E. NEMORE, R. WEINSTEIN & B. A. GILCHREST. 1981. Science 211: 1452–1454.
11. GILCHREST, B. A., W. L. MARSHALL, R. L. KARASSIK, R. WEINSTEIN & T. MACIAG. 1984. J. Cell. Physiol. 120(3): 377.
12. GILCHREST, B. A., R. L. KARASSIK, L. M. WILKINS, M. A. VRABEL & T. MACIAG. 1983. J. Cell. Physiol. 117: 235.
13. BULLOUGH, W. S. 1962. Biol. Rev. 37: 307–342.
14. STEWART, W. E. 1980. Nature (London) 286: 110.
15. ISSACS, A. & J. LINDENMAN. 1957. Proc. R. Soc. London Ser. B. 147: 258–267.
16. BROUTY-BOYE, D. 1980. Lymphokine Res. 1: 99–110.
17. GRESSER, I., M. T. THOMAS & D. BROUTY-BOYE. 1971. Nature (London) 231: 20–21.
18. HICKS, N. J., A. G. MORRIS & D. C. BURKE. 1981. J. Cell Sci. 49: 225–230.
19. BORDEN, E. C. 1984. Cancer. 54: 2770–2776.
20. IWATSUKI, K., J. VIAC, A. REANO, M. J. STAQUET & J. THIVOLET. 1985. Clin. Res. 33: 649 (A).

21. SIMON, M. & H. GREEN. 1984. Cell **36**: 827–834.
22. GILCHREST, B. A. 1979. J. Invest. Dermatol. **72**: 219–223.
23. GILCHREST, B. A., R. E. NEMORE & T. MACIAG. 1980. Cell Biol. Int. Rep. **11**: 1009–1016.
24. MACIAG, T., J. CERUNDOLO, S. ISLEY, P. R. KELLELY & R. FORAND. 1979. Proc. Natl. Acad. Sci. **76**: 5674–5678.
25. STANLEY, J. R., P. HAWLEY-NELSON, M. YAAR, G. R. MARTIN & S. I. KATZ. 1982. J. Invest. Dermatol. **78**: 456–459.
26. GREEN, H., O. KEHINDE & J. THOMAS. 1979. Proc. Natl. Acad. Sci. **76**: 5665–5668.
27. STANLEY, J. R., D. T. WOODLEY & S. I. KATZ. 1984. J. Invest. Dermatol. **82**: 108–111.
28. RHEINWALD, J. G. & H. GREEN. 1975. Cell **6**: 331–334.
29. REISNER, A. N., P. NEMES & C. BUCHOLTS. 1975. Anal. Biochem. **64**: 509–516.
30. GREEN, H. 1977. Cell **11**: 405–416.
31. FUCHS, E. & H. GREEN. 1980. Cell **19**: 1033–1042.
32. GILCHREST, B. A., M. A. VRABEL, E. FLYNN & G. SZABO. 1984. J. Invest. Dermatol. **83**: 370–376.
33. RUBINSTEIN, S. P., P. C. FAMILLETTI & S. PESTKA. 1981. J. Virol. **37**: 755–758.
34. SCHNIPPER, L. E., M. LEVINE, S. CRUMPACKER & B. A. GILCHREST. 1984. J. Invest. Dermatol. **82**: 94–96.
35. TOWBIN, H., T. STAEHELIN & J. GORDON. 1979. Proc. Natl. Acad. Sci. **76**: 4350–4354.
36. PESTKA, S., B. KELDER, D. K. TARNOWSKI & S. J. TARNOWSKI. 1983. Anal. Biochem. **132**: 328–333.
37. GROVE, G. L., R. L. ANDERTON & J. G. SMITH, JR. 1976. J. Invest. Dermatol. **66**: 236–238.
38. BADEN, H. P. 1984. *In* Pathophysiology of Dermatologic Diseases. N. A. Soter & H. P. Baden, Eds.: 101–126. McGraw-Hill Book Co. New York, NY.
39. GILCHREST, B. A. 1984. *In* Pathophysiology of Dermatologic Diseases. N. A. Soter & H. P. Baden, Eds.: 44–52. McGraw-Hill Book Co. New York, NY.
40. THOMAS, D. R., G. W. PHILPOTT & B. M. JAFFE. 1974. Exp. Cell Res. **84**: 40–46.
41. GILCHREST, B. A., W. L. MARSHALL, R. L. KARASSIK, R. WEINSTEIN & T. MACIAG. 1984. J. Cell Physiol. **120**: 377–383.
42. O'KEEFE, E. J., R. E. PAYNE & N. RUSSELL. 1985. J. Cell Physiol. **124**: 439–445.
43. PEEHL, D. M. & R. G. HAM. 1980. In Vitro **16**: 516–525.
44. MARKS, F. & K. H. RICHTER. 1984. Br. J. Dermatol. **111s**: 58–63.
45. LEHMAN, W., H. GRAETZ, H. SCHUNCK, M. SCHUTT & P. LANGEN. 1983. Acta Histochem. **27**(S): 63–71.
46. NICKOLOFF, B. J., T. Y. BASHAM, T. C. MERIGAN & V. B. MORHENN. 1984. Lab. Invest. **51**: 697–701.
47. MOORE, R. N., H. S. LARSEN, D. H. HOROHOV & B. T. ROUSE. 1984. Science **223**: 178–180.
48. LEBON, P., S. GIRARD, F. THEPOT & C. CHANY. 1982. J. Gen. Virol. **59**: 393–396.
49. YAAR, M., A. V. PALLERONI & B. A. GILCHREST. 1986. J. Cell Biol. **103**: 1349–1354.
50. YAAR, M., R. L. KARASSIK, L. E. SCHNIPPER, G. SZABO & B. A. GILCHREST. 1985. J. Invest. Dermatol. **85**: 70–74.

Keratinocytes Produce a Lymphocyte Inhibitory Factor Which Is Partially Reversible by an Antibody to Transforming Growth Factor-Beta[a]

BRIAN J. NICKOLOFF

Department of Pathology and Dermatology
University of Michigan
M4232 Medical Science I
1301 Catherine Road
Ann Arbor, Michigan 48109-0602

INTRODUCTION

A major advance in the understanding of the biology of the skin has been the recognition that the skin is more than a simple protective coat, that is, only a passive target for a variety of injurious stimuli, which alter its barrier function. The concept that keratinocytes (KCs) are metabolically active and dynamically responsive has emerged, in large part, based on immunogically related research demonstrating that KCs produce IL-1 (initially termed epidermal thymocyte activating factor, ETAF).[1,2] When KCs were found to produce thymopoietin,[3] which like IL-1 could also activate T-lymphocytes, many investigators began emphasizing the proinflammatory capacity of KCs. Since the initial assay of ETAF used partially stimulated mouse thymocytes, and conditioned media (CM) from cultured human KCs required overnight dialysis, we were interested in asking what would happen when human KCs and human peripheral blood mononuclear leukocytes (PBML) were placed directly in contact together.[4]

To our surprise, when either gamma interferon pretreated (to induce HLA-DR,5) or control, nontreated KCs were mixed with allogeneic PBML, only minimal activation by the PBML occurred.[4] In fact, instead of promoting lymphocyte proliferation, when CM from normal cultured KCs was added to 2-way mixed lymphocyte reactions (MLRs) or lectin-driven lymphocyte reactions, there was inhibition of the PBML proliferation. Part of the inhibition in the CM was due to PGE_2, but complete reversal of the inhibition could not be achieved in the presence of indomethacin, which greatly reduced PGE_2 levels.[4] We named this additional inhibitory factor, which could be transferred by cell-free supernatants, KC-derived lymphocyte inhibitory factor (KLIF). KLIF was presumed to have a molecular weight >3500 because it was nondialyzable; it did not appear to be related to any interferon because anti-α,β,γ interferon antibodies did not alter its activity. KLIF did not inhibit early T cell activation because IFN-γ production and lymphocyte enlargement occurred, but it did seem to be related to an IL 2-dependent pathway, because addition of large amounts of IL-2 partially restored allogeneic T cell proliferation in response to KCs.[6] When we

[a]This work was partially supported by National Institutes of Health Grant AM35390.

reviewed the literature for other examples of cells which produced IL-1 and inhibition of T cell proliferation involving IL-2, we found that glioblastoma cells also had these characteristics.[7] Not only did cultured glioblastoma cells produce a T cell suppressive factor, but clinically, patients with glioblastoma manifested impaired T-cell mediated immunity. Recently,[8] this T cell suppressor factor from human glioblastoma cells has been found to have an amino acid sequence very similar to transforming growth factor-beta (TGF-β).

TGF-β is a cytokine which has recently been found to be produced by cultured KCs.[9] Because TGF-β is also capable of inhibiting T cell proliferation,[10] we asked whether TGF-β or a TGF-β-like molecule could be responsible for the inhibitory effect of KC CM on MLRs and thus, be equivalent to, or contributing to KLIF. Indirect evidence to support such a hypothesis is the report that TGF-β, when added exogenously to human T cells, specifically blocks the IL-2-dependent step in T cell proliferation.[10] To more directly test our hypothesis that TGF-β was an important contributing cytokine to KLIF, KC CM was preincubated with neutralizing antibody to TGF-β and found to partially reverse the inhibitory effect on the MLR.

METHODS AND MATERIALS

Keratinocyte Culture

Normal human KCs were obtained from either foreskins or following plastic surgery and single cell suspensions were prepared as previously described.[11,12] The KCs were grown on plastic petri dishes (Lux, Flow Laboratories Inc., McLean, VA) using a low-calcium, serum-free medium (KGM, Clonetics Corp., San Diego, CA), and cells between passage number 3–8 were utilized. The KCs were maintained in a humidified incubator at 37°C with 95% air/5% CO_2. To prepare KC CM, a 10-cm-diameter plate containing subconfluent KCs approximately $2–3 \times 10^6$ cells) was incubated with RPMI (Gibco, Grand Island, NY) containing 10% FCS (Hyclone Laboratories, Inc., Logan, UT) and indomethacin (10 μg/ml in DMSO, Sigma Chemical Co., St. Louis, MO) for 2–3 days. The KC CM was then centrifuged at 1500 rpm for 5 min to obtain a cell-free supernatant which was used in the MLR.

Mixed Lymphocyte Reaction

The MLR was performed by obtaining the defibrinized venous blood from two healthy unrelated volunteers followed by Ficoll-Hypaque (Pharmacia, Piscataway, NJ) gradient centrifugation as previously described.[4] The interface cells containing PBML were washed twice and resuspended in RPMI and indomethacin. 6×10^5 PBML from each donor were added to round-bottom, 96-well plates (Costar, Cambridge, MA) in volume of 30 μl. 200 μl of either fresh RPMI + 10% FCS or KC CM was added and the PBML were maintained for 6 days at 37°C with 95% air/5% CO_2. 1 μCi/well of ^{3}H-thymidine (specific activity = 32 Ci/mmol; New England Nuclear, Boston, MA) was added for the last 12 hours and the cells collected on a PHD cell harvester (Cambridge Technology, Cambridge, MA). Cell-associated radioactivity was measured by liquid scintillation counting, and the results were expressed as the mean of triplicate wells.

TGF-β and Neutralizing Antibody to TGF-β

TGF-β was purchased from Collaborative Research Inc. (Bedford, MA) and reconstituted using 5 mm HCl in silanized vials and stored at −20°C until use. Neutralizing rabbit antibody to TGF-β was purchased from R & D Systems, Inc. (Minneapolis, MN), diluted with sterile water, and used immediately. The control nonimmune IgG purified rabbit-antisera was obtained from rabbit sera placed over a protein A sepharose column (BioRad, Richmond, CA).

RESULTS

Keratinocyte Conditioned Media Inhibits the MLRs

In control MLR, the degree of PBML proliferation was variable between different experiments, either using the same unrelated donors or different donors. Based on 4 different MLRs, the control values using 200 μl of fresh RPMI + 10% FCS for ³H-thymidine incorporation ranged between 23,116 and 38,714 cpm. The mean value for the control MLRs was 33,344 and the mean value for the MLRs performed substituting 200 μl of KC CM was 10,850 (FIG. 1). The difference between using fresh RPMI + 10% FCS and substituting KC CM was statistically significant ($p < 0.01$) and this degree of inhibition was similar to our previous report.[4] By phase contrast microscopy, the control MLR contained numerous small and large clusters of enlarged lymphocytes (FIG. 2A), whereas the MLR containing KC CM had no large clusters and only a few small clusters of enlarged lymphocytes (FIG. 2C). Thus, there was a

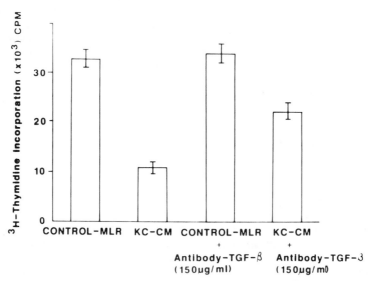

FIGURE 1. KC CM inhibits the MLR. By substituting RPMI + 10% FCS which has been in contact with cultured KCs for fresh RPMI + 10% FCS, the mean ³H-thymidine incorporation decreased from 33,344 to 10,850 cpm. When the KC CM was preincubated with neutralizing antibody to TGF-β, there was an increase from 10,850 to 22,315 cpm.

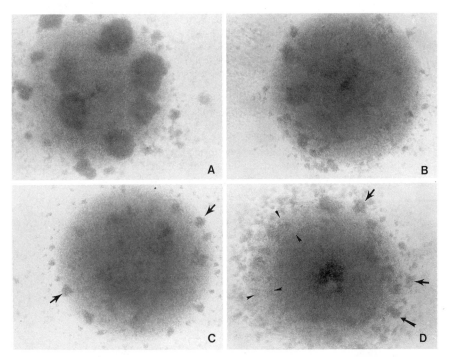

FIGURE 2. Phase contrast microscopic appearance of the MLR. (**A**) Control MLR using fresh RPMI + 10% FCS reveals numerous small and large clusters of enlarged lymphocytes. (**B**) When 100 ng/ml of TGF-β was added to the MLR containing fresh RPMI + 10% FCS, there was reduction in the size and number of clusters of enlarged lymphocytes. (**C**) When KC CM was used in the MLR there were no large clusters and only a few small clusters (*arrows*) of enlarged lymphocytes. (**D**) When KC CM was preincubated with antibody to TGF-β, many more clusters of enlarged lymphocytes (*arrows*) as well as a rim of enlarged lymphocytes (*arrow heads*) were present.

good correlation between the degree of lymphocyte activation and proliferation as assessed visually compared to the amount of ³H-thymidine incorporation.

TGF-β Inhibits the MLR Which Can Be Reversed Using Neutralizing Antibody to TGF-β

Before determining whether the neutralizing antibody to TGF-β could influence the inhibitory effect of KC CM on the MLR, the direct effect of exogenously added TGF-β was measured. FIGURE 3 reveals a dose-dependent inhibition of the MLR by TGF-β between 0.4 ng/ml and 100 ng/ml. There was approximately 50% inhibition of the control MLR by 100 ng/ml TGF-β. The inhibitory effect of 100 ng/ml TGF-β on the MLR was confirmed by phase contrast microscopy (FIG. 2B). When 1 ng/ml of TGF-β was preincubated with the neutralizing antibody to TGF-β (100 μl/ml), the inhibitory effect of the TGF-β on the MLR was reversed (FIG. 3). Since we were unable to measure the amount of active TGF-β in the KC CM, we estimated that based on the degree of inhibition, there was approximately 1–10 ng/ml TGF-β. Therefore, we also used the neutralizing antibody to TGF-β at a concentration of 150 μg/ml so as to block as much TGF-β activity in the KC CM as possible. (Based on the package insert provided by R & D Systems, 50–100 μg/ml of antibody should neutralize approximately 1 ng/ml of TGF-β using mink lung epithelial cells.)

Preincubating Keratinocyte Conditioned Media With Neutralizing Antibody to TGF-β Partially Reverses the Inhibition of the MLR

When the control MLR reaction was performed in the presence of the neutralizing antibody to TGF-β, there was no significant effect either in the phase microscopic

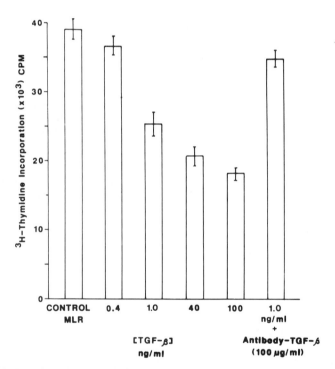

FIGURE 3. Dose-dependent inhibition of the MLR by TGF-β. When exogenous purified TGF-β was added (0.1–100 ng/ml) to the MLR, there was a decrease in the degree of lymphocyte proliferation as measured using ^{3}H-thymidine incorporation. When neutralizing antibody to TGF-β was combined with 1 ng/ml of TGF-β, the inhibition was reversed.

appearance of the lymphocytes or in their degree of ^{3}H-thymidine incorporation (FIG. 1). However, when the KC CM was preincubated with the same amount of neutralizing antibody to TGF-β, there was a significant reversal of the inhibitory effect. The ^{3}H-thymidine incorporation (FIG. 1) went from 10,850 cpm up to 22,315 cpm ($p <$ 0.01). Also, the phase contrast microscopic appearance of the reaction revealed more clusters of enlarged lymphocytes in the wells treated with the neutralizing antibody

(FIG. 2D). However, in none of the 4 experiments using 150 $\mu g/ml$ of antibody to TGF-β was the reversal of the inhibition by the KC CM on the MLR complete. When a nonimmune IgG-purified fraction of normal rabbit antisera was used at the same protein concentration, no significant enhancement or inhibition on the MLR was observed.

DISCUSSION

When cultured KCs are grown in the presence of RPMI + 10% FCS containing indomethacin, the resultant CM inhibits the MLR (FIGS. 1 and 2). However, when the KC CM was preincubated with neutralizing antibody to TGF-β, part of this inhibition was reversed (FIG. 1). Thus, the KC CM does not inhibit the MLR simply by utilizing up growth factors, but actually results from the production of TGF-β or a TGF-β-like molecule. Since the reversal of the inhibition was not complete by adding the neutralizing antibody to TGF-β, this could imply three different possibilities. First, there may be consumption of important growth-requiring factors by the KCs which limit the degree of lymphocyte proliferation in the subsequent MLR, and hence incomplete or submaximal lymphocyte response occurs despite the neutralization of the TGF-β. Second, there may be more active TGF-β present in the KC CM than can be completely neutralized by the added antibody. Also contributing to this problem are the incomplete details regarding the activation or conversion of inactive TGF-β to active TGF-β. When better discriminating reagents and assays are available to permit distinction between inactive and active forms of TGF-β, this problem can be addressed. Third, there may be other non-TGF-β inhibitory factors produced by KCs which continue to inhibit the MLR despite the presence of neutralizing antibody to TGF-β and indomethacin.

Possible KC-derived inhibitory factors which must be considered include PGE_2 since the indomethacin does not completely eliminate all PGE_2 production but only reduces it below 3 nM. Other possible inhibitory factors include IFN-β_1,[13] epidermal-cell-derived lymphocyte-differentiating factor (ELDIF),[14] and urocanic acid.[15] Seen in this context, it is possible to draw an analogy between ETAF and KLIF. When the lymphocyte-enhancing capacity of the cultured KC, ETAF, was first discovered, it was felt that this activity resided only in a single molecule (i.e., IL-1). More recently, it has become clear that several different cytokines possess this ability to stimulate partially activated thymocytes including IL-3[16] and IL-6 (IFN-β_2).[17,18] Thus, in a similar fashion, it is possible to envision that multiple inhibiting factors contribute to KLIF activity. With the rapid advance in molecular and cellular biology, a more complete list of different, defined cytokines will refine our previous observation of "factors" or "activities."

In summary, the delicate balance of cutaneous immunohomeostasis and inflammation, depends on the relative amounts and potency of both proinflammatory and/or antiinflammatory mediators as depicted in FIGURE 4. Not only can KCs produce these different mediators, but other cells in the skin, particularly the Langerhans cell (LC), contribute to this immunohomeostatic balance. LCs are dendritic, HLA-DR-positive, potent antigen-presenting cells[19] which can produce many cytokines including IL-1 and tumor necrosis factor (TNF).[20] With respect to TGF-β, TNF may be important, because TNF is capable of reversing the inhibitory effects of TGF-β.[21] Given the similar biological effects of TGF-β and cyclosporin A,[22] together with the therapeutic efficacy of cyclosporin A in inflammatory dermatoses such as psoriasis,[23] additional studies on the modulation of TGF-β which may serve as an "endogenous cyclosporin-A-like" molecule seem indicated. Despite the rapidly growing complex array of inflammatory cytokines and the largely unexplored synergistic/antagonist interaction between various cutaneous cellular constituents, there can be no doubt that the KC will be increasingly appreciated as an important, dynamic participant in inflammatory reactions, rather than simply as a passive target. Finally, it is clear that KC-lymphocyte interactions are bidirectional[4] and that a full understanding of any individual cytokine must be taken within the context of an intricate network of tissue and cell-specific interactions.[24,25]

SUMMARY

The active participation by keratinocytes (KCs) during cutaneous inflammation involves production of various immunologically active molecules. While interleukin-1

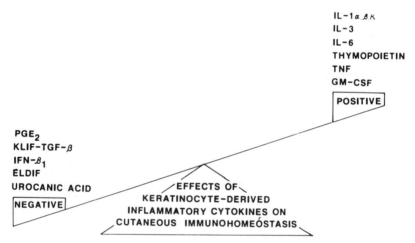

FIGURE 4. The balance in the skin immune network between lymphocyte-stimulating versus lymphocyte-inhibiting factors produced by keratinocyte influences on cutaneous immunohomeostasis.

(IL-1) is a well-known KC-derived activator of lymphocytes, less attention has been directed towards characterization of non-PGE KC-derived lymphocyte inhibitory factors. To determine whether transforming growth factor-β (TGF-β), which has recently been found to be produced by KCs, may be a biologically important constituent of KC-conditioned medium, we measured the ability of neutralizing antibody of TGF-β to reverse the inhibitory effect of KC-conditioned medium on mixed lymphocyte reactions (MLR). Not only did exogenously added TGF-β inhibit the MLR, but KC-conditioned media also inhibited the MLR, and this inhibition was partially reversed using the neutralizing antibody as detected by ^{3}H-thymidine incorporation and phase contrast microscopy. Thus, KC-derived TGF-β may serve as an important inhibitor of lymphocyte proliferation. These results suggest that the balance of cutaneous immunohomeostasis may involve several different KC-derived factors which may be either lymphocyte activating such as IL-1, or lymphocyte inhibitory such as PGE$_2$ and TGF-β.

REFERENCES

1. SAUDER, D. N., C. S. CARTER, S. I. KATZ & J. J. OPPENHEIM. 1982. Epidermal cell production of thymocyte activating factor (ETAF). J. Invest. Dermatol. **79:** 34–39.

2. LUGER, T., B. M. STADLER, S. I. KATZ & J. J. OPPENHEIM. 1982. Murine epidermal cell derived thymocyte activating factor resembles murine interleukin 1. J. Immunol. **128:** 2147–2152.

3. CHU, A. C., J. A. K. PATTERSON, G. GOLDSTEIN, C. L. BERGER, S. TAKEZAKI & R. L. EDELSON. 1983. Thymopoietin-like substance in human skin. J. Invest. Dermatol. **81:** 194–197.

4. NICKOLOFF, B. J., T. Y. BASHAM, J. TORSETH, T. C. MERIGAN & V. B. MORHENN. 1986. Human keratinocyte lymphocyte reactions *in vitro.* J. Invest. Dermatol. **87:** 11–18.

5. BASHAM, T. Y., B. J. NICKOLOFF, T. C. MERIGAN & V. B. MORHENN. 1984. Recombinant gamma interferon induces HLA-DR expression on cultured human keratinocytes. J. Invest. Dermatol. **83:** 88–92.

6. MORHENN, V. B. & B. J. NICKOLOFF. 1987. Interleukin-2 stimulated resting T lymphocyte response to allogeneic gamma interferon treated keratinocytes. J. Invest. Dermatol. **89:** 464–468.

7. FONTANA, A., H. HENGARTNER, N. TRIBOLET & E. WEBER. 1984. Glioblastoma cells release interleukin 1 and factors inhibiting interleukin 2-mediated effects. J. Immunol. **132:** 1837–1844.

8. WRANN, M., S. BODNER, R. MARTIN, C. SIEPL, K. FREI, E. HOFER & A. FONTANA. 1987. T cell suppressor factor from human glioblastoma cells is a 12.5 Kd protein closely related to transforming growth factor-β. EMBO J. **6:** 1633–1636.

9. SHIPLEY, G. D., M. R. PITTELKOW, J. J. WILLE, R. E. SCOTT, & H. L. MOSES. 1986. Reversible inhibition of normal human prokeratinocyte proliferation of type β transforming growth factor-growth inhibitor in serum free medium. Cancer Res. **46:** 2068–2071.

10. KEHRL, J. H., L. M. WAKEFIELD, A. B. ROBERTS, S. JAKOWLEW, M. ALUARCZ-MON, R. DERYNCK, M. B. SPORN & A. S. FAUCI. 1986. Production of transforming growth factor β by human T lymphocytes and its potential role in the regulation of T cell growth. J. Exp. Med. **163:** 1037–1050.

11. LIU, S. C. & M. A. KARASEK. 1978. Isolation and growth of adult human epidermal keratinocytes in cell culture. J. Invest. Dermatol. **71:** 157–162.

12. NICKOLOFF, B. J., T. Y. BASHAM, T. C. MERIGAN & V. B. MORHENN. 1984. Antiproliferative effects of recombinant alpha and gamma interferons on cultured human keratinocytes. Lab. Invest. **51:** 697–701.

13. TORSETH, J., B. J. NICKOLOFF, T. Y. BASHAM & T. C. MERIGAN. 1987. Beta interferon produced by keratinocytes in human cutaneous herpes simplex infection. J. Infect. Dis. **155:** 641–648.

14. NICOLAS, J. F., D. KAISERLIAN, M. DARDENNE, M. FAURE, & J. THIVOLET. 1987. Epidermal cell derived lymphocyte differentiating factor (ELDIF) inhibits *in vitro* lymphoproliferative responses and interleukin 2 production. J. Invest. Dermatol. **88:** 161–166.

15. DEFABO, E. C. & F. P. NOONAN. 1983. Mechanism of immune suppression by UV-irradiation *in vivo.* I. Evidence for the existence of a unique photoreceptor in skin and its role in photoimmunology. J. Exp. Med. **157:** 84–98.

16. LUGER, T. A., U. WIRTH & A. KOCK. 1984. Epidermal cells synthesize a cytokine with interleukin 3-like properties. J. Immunol. **134:** 915–919.

17. NICKOLOFF, B. J. 1988. Keratinocytes produce beta interferon and a lymphocyte inhibitory factor which is partially reversible by an antibody to transforming growth factor-beta. *In* the collected abstracts of this conference: no. 34.

18. NIJSTEN, M. W. N., E. R. GROOT, H. J. DUIS, H. J. KLASEN, C. E. HACK & L. A. AARDEN. 1987. Serum levels of interleukin-6 and acute phase responses. Lancet **ii:** 921.

19. STINGL, G., S. I. KATZ, E. M. SHEVACH, A. S. ROSENTHAL & I. GREEN. 1978. Analogous functions of macrophages and Langerhans cells in the initiation of immune response. J. Invest. Dermatol. **71:** 59–64.

20. SHI, T., J. LARRICK, Y. CHIANG & V. MORHENN. 1988. Tumor necrosis factor is released from activated Langerhans cells. Clin. Res. **36:** 253A.

21. RANGES, G. E., I. S. FIGARI, T. ESPEVIK & M. A. PALLADINO. 1987. Inhibition of cytotoxic T cell development by transforming growth factor β and reversal by recombinant tumor necrosis factor α. J. Exp. Med. **166:** 991–998.

22. ESPEVIK, T., I. S. FIGARI, M. R. SHALABY, G. A. LACKIDES, G. D. LEWIS, H. M. SHEPARD & M. A. PALLADINO. 1987. Inhibition of cytokine production by cyclosporin A and transforming growth factor β. J. Exp. Med. **166:** 571–576.

23. ELLIS, C. N., D. C. GORSULOWSKY, T. A. HAMILTON, J. K. BILLINGS, M. D. BROWN, J. T. HEADINGTON, K. D. COOPER, O. BAADSGAARD, E. A. DUELL, T. M. ANNESLY, J. G. TURCOTTE & J. J. VOORHEES. 1986. Cyclosporine improves psoriasis in a double blind study. J. Am. Med. Assoc. **256:** 3110–3116.

24. SAUDER, D. N., D. WONG, R. MCKENZIE, D. STETSKO, D. HARNISH, V. TRON, B. NICKOLOFF, T. ARSENAULT & C. B. HARLEY. 1988. The pluripotent keratinocyte: Molecular characterization of epidermal cytokines (abstract). J. Invest. Dermatol. **90:** 605.

25. KOHASE, M., L. T. MAY, I. TAMM, J. VILCEK & P. B. SEHGAL. 1987. A cytokine network in human diploid fibroblasts: Interactions of β-interferons, tumor necrosis factor, platelet-derived growth factor, and interleukin-1. Mol. Cell. Biol. **7:** 273–280.

Gamma Interferon-Induced Expression of Class II Major Histocompatibility Complex Antigens by Human Keratinocytes

Effects of Conditions of Culture

VERA B. MORHENN[a] AND GARY S. WOOD[a,b,c]

Departments of [a]Dermatology and [b]Pathology
Stanford University Medical Center
Stanford, California 94305
and
[c]Department of Dermatology
Palo Alto Veterans Hospital
Palo Alto, California 94304

INTRODUCTION

In a number of skin diseases including lichen planus, contact dermatitis, and mycosis fungoides (MF), including the poikilodermatous form of this disease, keratinocytes express class II antigens of the major histocompatibility complex (MHC).[1-3] In MF in particular, not only HLA-DR but also HLA-DQ and HLA-DP antigens are expressed by keratinocytes in lesional skin.[4] We and others have reported recently that recombinant gamma interferon (rIFN-γ) induces the expression of HLA-DR antigen on human keratinocytes but little if any HLA-DQ.[5,6] With the demonstration of DQ antigen expression on diseased skin *in situ*, we asked whether this expression could also be due to the effects of local secretion of IFN-γ by T cells and whether this expression was time dependent. Moreover, since another class II antigen of the MHC, named HLA-DP, has recently been described, we ascertained whether this antigen also could be induced by rIFN-γ.[7]

Changes in the calcium ion concentration cause alterations in the growth pattern of human as well as murine keratinocytes; therefore, we determined whether changing the calcium ion concentration in human keratinocyte cultures resulted in changes of class II MHC antigen expression after rIFN-γ treatment.[8,9] Also, we determined the effects of the presence and absence of serum on this induction.

The rIFN-γ not only induces expression of DR antigen, but also inhibits the growth of keratinocytes. In order to determine whether DR expression would occur without significant reduction in cell numbers, we performed a concentration curve comparing the effects of this lymphokine on DR expression and cell number in one and the same set of culture plates.

The rIFN-γ induces DR antigen on squamous cell carcinoma cell line (SCL-1) cells, a transformed human epithelial cell line.[10] Since the regulation of class II antigens in neoplastic epithelial cells may be different than in normal keratinocytes, we also examined the induction of HLA-DQ and DP by rIFN-γ in SCL-1 cells under various conditions of culture.

MATERIALS AND METHODS

Lymphokine, Monoclonal Antibodies (mAbs) and Cell Staining

The monoclonal antibodies anti-HLA-DR, DP, and DQ were obtained from Becton Dickinson, Mountain View, CA. The isotope control mAbs, anti-Leu-2b and anti-Leu-3a were also purchased from Becton Dickinson. The mAbs, VM-1 and VM-2, directed against the basal cell layer of normal human skin, were developed in this laboratory and have been described previously.[11,12] The rIFN-γ was a generous gift of Genentech Corp., South San Francisco, CA, and was used at 100 U/ml unless otherwise stated.

One million epidermal cells were stained for 30 min with the mAb diluted in 5% heat-inactivated, fetal calf serum (FCS) in phosphate buffered saline (PBS) containing 0.02% sodium azide (AZ) as described previously.[13] The cells were washed with 5%

TABLE 1. Expression of Class II Antigens by rIFN-γ-Treated Normal Keratinocytes as a Function of the Medium Used[a]

Days after rIFN-γ Treatment	KGM with Low Ca (0.1 mM)			KGM with High Ca (1.8 mM)			DMEM Plus Serum		
	DR	DP	DQ	DR	DP	DQ	DR	DP	DQ
2	81[b]	36	4	90	44	10	N.T.[c]	N.T.	N.T.
4	86	71	25	75	84	21	40	18	11
6	69	77	29	65	52	18	N.T.	41	0
7	N.T.						45	24	0
8	91	77	26	72	50	14	51	17	1
10	76 (492)	49 (327)	44 (89)	N.T.	N.T.	N.T.	42	19	6
12	N.T.	N.T.	N.T.	N.T.	N.T.	N.T.	30	34	14
13	N.T.	N.T.	N.T.	N.T.	N.T.	N.T.	67	34	5
14	78 (530)	83 (330)	74 (246)	N.T.	N.T.	N.T.			
18	64 (376)	65 (339)	63 (158)	N.T.	N.T.	N.T.			

[a]Composite of several experiments.

[b]Number of positive cells expressed as % of total; mean fluorescence/cell (in linear units) is given in parentheses. The isotype controls have been subtracted for both values.

[c]N.T. = not tested. In prior experiments, DR expression was low on day 2 (see REF. 5).

FCS/PBS/AZ, stained wtih fluorescein isothiocyanate conjugated rabbit antimouse IgG (R/M-FITC) (ICN Immunobiologicals, Lisle, IL) for 30 min, and washed with and then resuspended in 5% FCS/PBS/AZ. The number of fluorescent cells was determined by fluorescence microscopy or fluorescence activated cell sorter (FACS) analysis.[13] On the FACS, the mean fluorescence/cell was determined using a linear scale.

Cell Culture Conditions

Skin obtained from meloplasty specimens was processed as described previously.[14] Single cell suspensions, consisting mainly of keratinocytes, were prepared as described.[14] Other keratinocyte cultures, derived from breast skin, were obtained from Clonetics, San Diego, CA. Keratinocytes were grown using 3 different media: complete medium, keratinocyte growth medium (KGM), and keratinocyte defined medium (KDM). In the first, dispersed cells were suspended in complete medium, which consists of Dulbecco's modified Eagle's medium (DMEM) supplemented with 10% heat-inactivated FCS, 50 μg/ml gentamicin and 2 mM L-glutamine, and were seeded at $1.8–2.2 \times 10^6$ cells/3.5 cm collagen-coated Petri dish (Lux, Miles Scientific, Naperville, IL).[14] (The calcium ion concentration for DMEM is 1.8 mM.) For culture is serum-free medium, the method described by Ham and Boyce was used.[9,15] The cells were trypsinized and seeded at 3×10^4 cells/3.5 cm Petri dish in KGM (Clonetics), which contains bovine pituitary extract (BPE). The unadjusted calcium ion concentration in KGM is 0.1 mM. For some experiments, we increased the calcium ion concentration to 1.8 mM by adding appropriate amounts of $CaCl_2$ (high calcium medium). Using the third method, adult keratinocytes were grown in KDM (Clonetics), which is a fully defined medium without BPE. Parenthetically, both KGM and KDM have a short shelf life and should not be used after prolonged storage.

For some experiments, the cells were seeded on glass coverslips inserted into a plastic Petri dish and grown in KGM. At the end of the incubation period the medium was removed, and the coverslip was left in the dish where it was washed with PBS, removed to fix in cold acetone and stained. The coverslip was then inverted on a glass slide over Tris-buffered glycerol and examined with fluorescence microscopy.

Cell Harvesting and Cell Counts

At the times indicated, cultures were washed once with PBS, 1 ml of 0.3% trypsin/0.1% EDTA in GNK (150 mM NaCl, 0.04% KCl, 0.1% glucose, 0.084% $NaHCO_3$, pH 7.3) was added, and the plates incubated for 10 min at 37 degrees. The detached cells were transferred to tubes, the plates rinsed $\times$ 1 with complete medium to remove the residual cells and this rinse combined with the 1 ml aliquots already harvested. The cells were diluted with trypan blue and total and viable cell numbers determined using a hemocytometer.

RESULTS

Effect of Culture Medium on rIFN-γ-Induction of Class II Antigen Expression by Keratinocytes

Keratinocytes grown in KGM at the high calcium ion concentration (1.8 mM) showed HLA-DR expression by 90% of the cells on day 2 after addition of rIFN-γ (100 U/ml) (TABLE 1). On this day, significant, albeit less, expression of DP and minimal expression of DQ occurred. In contrast, keratinocytes grown in complete medium showed a different pattern of class II MHC antigen expression in the presence of rIFN-γ (TABLE 1). As previously reported, HLA-DR expression was virtually maximal by day 4, a time point when DP and DQ expression was still low.[5] Thus, on day 4, only small numbers (11%) of cells expressed HLA-DQ. From that time on, expression of HLA-DP increased, whereas expression of HLA-DQ stayed very low (TABLE 1). Cells grown in complete growth medium plus rIFN-γ were morphologically attenuated after 13 days in culture and, therefore, these experiments were not continued beyond this point.

In order to determine the effect of a low calcium ion concentration (0.1 mM) on the expression of class II MHC antigens by keratinocytes treated with rIFN-γ, similar experiments were repeated using KGM without additional calcium. The results were similar to those found in high-calcium KGM. In the low-calcium medium, DR expression was high by day 2 and continued at least until day 18. On day 2, HLA-DP also was expressed. By day 10, the mean expression/cell of DP was 327 linear units/cell compared to 492 linear units/cell for DR and only 49% of these cells expressed DP compared to 76% for DR (TABLE 1). Significant HLA-DQ expression appeared on day 4, albeit the expression of this class II MHC antigen again was not as strong as DR. Only about 25% of the cells expressed DQ antigen on this day, but this expression increased with time and eventually reached comparable levels with DP by day 10.

To determine whether the total absence of serum proteins or BPE affected the expression of HLA-DR antigen, keratinocytes were grown in KDM without added proteins and 5, 10 and 13 days after the addition of the lymphokine HLA-DR antigen expression was assessed (TABLE 2). In contrast to cells growth in KGM, the cells grown in KDM expressed very little DR antigen on day 5 of culture. The induction of DR by rIFN-γ in KDM did not reach the values observed on day 2 for KGM until about day 13 of culture. Thus, serum factors appear to play a major role in the induction of DR antigen expression by rIFN-γ.

Effect of Culture Medium on rIFN-γ-Induced Growth Inhibition of Keratinocytes

Since keratinocytes without the growth-promoting factors found in BPE do not grow rapidly, we predicted that growth inhibition by rIFN-γ in KDM would be less striking. The growth of keratinocytes was in fact influenced less by rIFN-γ in KDM than in any of the other media used. By day 10 of culture, only a small difference in total cell number was seen in control (0.96×10^5 cells/plate) vs rIFN-γ-treated (0.72×10^5 cells/plate) cultures. By day 13, however, a threefold difference in total number of cells/plate between control (1.38×10^5 cells/plate) and rIFN-γ-treated (0.45×10^5 cells/plate) cultures occurred.

Dose of rIFN-γ Required for Induction of Keratinocyte Class II MHC Antigen Expression vs Growth Inhibition in KGM

In order to determine whether the concentration of rIFN-γ needed to induce DR expression was equivalent to that which resulted in growth inhibition of normal keratinocytes, we performed a dose response curve for the two parameters determining percent of total cells expressing DR as well as viable cell numbers in one and the same Petri dish (FIG. 1). Recombinant IFN-γ at 3 U/ml, which by day 8 had not significantly inhibited cell proliferation, did induce DR expression. Similarly, 10 U/ml of the lymphokine, which on day 4 had not resulted in a significant decrease in cell number, had induced DR expression on 48% of the cells on this day. Thus, DR

TABLE 2. Expression of HLA-DR by Keratinocytes in KDM Plus rIFN-γ (100 U/ml)

Days in Culture	Number of Positive Cells (% of total)[a]	Mean Fluorescence per Cell[a]
5	10	11
10	31	211
13	76	663

[a]The isotype control has been subtracted.

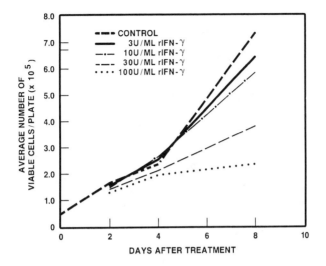

Time after rIFN-γ Treatment (Days)	Concentration of rIFN-γ (U/ml)	Number of Cells Expressing HLA-DR (%)[a]
2	3	3
	10	48
	30	77
	100	87
4	3	1
	10	48
	30	89
	100	96
8	3	13
	10	75
	30	93
	100	91

[a]Isotype control has been subtracted.

FIGURE 1. Keratinocytes grown in KGM were treated with various concentrations of rIFN-γ. On days 2, 4 and 8 the cells were harvested and the number of viable cells/plate as well as the percent of these cells expressing HLA-DR determined. For the cell counts the SEM was less than 25% except for the control on day 8, where the SEM was 30%. The *table* below the figure indicates the percent of total cells which express HLA-DR for each concentration of rIFN-γ and for each day of culture.

expression was manifested on keratinocytes before inhibition of growth could be documented. These data may explain the existence of skin diseases (*e.g.,* contact dermatitis) in which keratinocytes express DR antigen, yet no atrophy of the epidermis is apparent either clinically or histologically.

Effect of Trypsin on Class II MHC Antigen Expression by Keratinocytes in KGM

To determine whether the delay in HLA-DQ antigen expression and the smaller numbers of cells demonstrating this antigen were due to increased sensitivity of the

HLA-DQ antigen to the effects of trypsin treatment, we performed two types of experiments using KGM. For the first, keratinocytes were scraped off the culture plates and then stained for DR, DQ, and DP. As a control, replicate plates were trypsinized and stained in the conventional manner. After scraping, most of the cells were either totally fragmented or dead making quantitation of staining by FACS analysis impossible on these cell preparations. However, on examination using fluorescence microscopy, most of the few, viable keratinocytes that could be visualized were stained with mAb against DR, whereas no cells stained for the DP or DQ antigens. To further explore this question, a second type of experiment was performed. Keratinocytes were seeded on coverslips and 4 days after rIFN-γ treatment, the cells were stained for DR, DQ, and DP while still attached. Whereas most cells expressed HLA-DR at this time, only about 50% demonstrated DP and only a very few cells expressed DQ staining and this only very weakly. Thus, the delayed expression of DQ is not an artifact of trypsinization.

The Effect of the Calcium Ion Concentration on Expression of VM-1 and VM-2 Antigens by Keratinocytes Grown in KGM

To determine whether the switch in calcium ions was causing the human keratinocytes to become more differentiated as has been described in the mouse system,[8] the cells were labeled with both the mAbs VM-1 and VM-2, antibodies which stain only basal cells of the epidermis *in situ*.[11,12] On day 4, after changing the calcium ion concentration from 0.1 mM to 1.8 mM, virtually every cell expressed the cell surface antigens which bound the mAbs VM-1 and VM-2 in both types of cultures (data not shown). Thus, changing the calcium ion concentration does not change the proliferating phenotype of the attached cells. That the calcium switch does not induce differentiation of human keratinocytes has been reported recently using other markers of differentiation.[16]

Effect of Culture Medium on rIFN-γ-Induced Class II MHC Antigen Expression by SCL-1 Cells in KGM and Complete Medium

To determine whether SCL-1 cells, which also can be induced to express HLA-DR, showed a similar pattern of expression of the other two class II MHC antigens as did

TABLE 3. Class II Antigen Expression by rIFN-γ-Treated SCL-1 Cells in Complete Medium

	Class II Antigen Staining[a]		
Day	DR	DP	DQ
2	50	1	1
4	69	1	3
6	N.T.[b]	6	7
7	94	5	5
10	74	2	5
13	81	12	8
14	77	30	22

[a]Number of positive cells (% of total); the value for the isotype controls has been subtracted.
[b]N.T. = not tested.

TABLE 4. Expression of Class II MHC Antigens by rIFN-γ-Treated SCL-1 Cells and Normal Keratinocytes in Low Calcium Medium (KGM)

Cells	Days after rINF-γ	Class II Antigen Expression[a]		
		DR	DP	DQ
SCL-1	2	74	2	2
	4	56	1	3
	6	71	1	3
	8	75	1	0
	11	82	6	2
	14	88	11	25
Normal Keratinocytes	2	N.T.[b]	N.T.	N.T.
	4	86	71	25
	6	69	77	29
	8	91	77	26
	10	76	49	44
	14	78	83	74
	18	64	65	63

[a]Number of positive cells (% of total). The value for the isotype controls has been subtracted.
[b]N.T. = not tested.

normal keratinocytes, we examined the DP and DQ antigen expression of this cell type grown in complete medium. In this medium, the SCL-1 cells showed little induction of HLA-DP and DQ until about day 13, when low levels of these antigens could be documented (TABLE 3). Next we compared the induction of class II MHC antigen expression by the SCL-1 cells and normal keratinocytes, both grown in the presence of low calcium KGM. The expression of HLA-DP and DQ by SCL-1 cells was not improved significantly (TABLE 4). Thus, SCL-1 cells do not express DP and DQ antigens after rIFN-γ treatment as readily as untransformed keratinocytes, regardless of whether low or high calcium culture conditions are used.

DISCUSSION

Gamma interferon induces keratinocytes to express not only HLA-DR but also HLA-DP and DQ antigens. In all the media tested, the appearance of these class II MHC antigens appears to be time dependent with DR expression occurring first, followed by DP and then DQ. In KGM, with either low or high calcium ion concentrations, DR and DP expression can be detected within 48 hrs of rIFN-γ treatment, whereas the expression of DQ is delayed. Thus, the induction of expression of class II MHC antigens on keratinocytes by rIFN-γ appears to be related to the duration of exposure to the lymphokine as well as serum factors but not to the relative calcium ion concentration. A low level of DQ induction on keratinocytes has been reported previously.[6]

The induction of expression of HLA-DP and DQ antigens by rIFN-γ on transformed cells is not as rapid as in normal keratinocytes. Furthermore, the total number of cells expressing DP and DQ is less than in untransformed keratinocytes. Expression of HLA-DR by cells of skin melanomas but not by basal cell carcinomas has been documented.[17,18] In one of four squamous cell carcinomas, DR expression by keratinocytes was observed.[19] Based on the results in this paper, it would be interesting to determine whether DP and DQ antigens are expressed by squamous cell carcinomas *in situ*.

That cells grown in KGM expressed DR antigen at an early point in time and on virtually every cell was surprising. All of these cells also expressed the surface antigens recognized by the mAbs VM-1 and VM-2, antigens expressed only by basal cells *in situ*.[11,12] Thus, it is the highly proliferative basal cell which is growing under these conditions of culture, and these cells express HLA-DR antigen almost immediately after rIFN-γ exposure. In contact dermatitis, the lower layers of the keratinocytes in the epidermis, presumably the cells in the proliferating pool, are most intensely stained with mAb against HLA-DR.[2] Thus, the *in vitro* results are consistent with the *in situ* observations and suggest that the basal cells may be the keratinocyte subpopulation which augments antigen presentation by Langerhans cells in contact dermatitis.

Keratinocytes grown in KDM do not express class II antigens as rapidly as those grown in KGM, which contains BPE, or in complete medium, which contains serum. It is conceivable that in the intact human epidermis, *in vivo,* the protein composition is not identical to that found in normal human serum. Possibly, the basement membrane at the dermoepidermal junction acts as a type of filter allowing only those proteins and other substances of lower molecular weight to penetrate to the epidermis. Thus, the defined media used for some of these experiments may more accurately reflect the *in vivo* situation in the epidermis. Alternatively, serum factors may bind rIFN-γ resulting in less active lymphokine being available for the cell surface receptors for this protein when keratinocytes are cultured in serum-containing medium.

Keratinocytes express HLA-DR antigen in a wide variety of skin disorders ranging from contact dermatitis which shows no atrophy of the epidermis to the poikiloderma-tous form of mycosis fungoides where epidermal atrophy can be pronounced. Our findings that short incubations of keratinocytes with low doses of rIFN-γ induce HLA-DR antigen expression but do not result in pronounced reduction in cell numbers, whereas longer incubations with higher concentrations of the lymphokine induce DR as well as DP and DQ expression and reduce cell proliferation, provide further evidence that these two skin disorders may be mediated by IFN-γ.

We have shown recently that rIFN-γ-treated keratinocytes are capable of stimu-lating allogeneic resting T cells in the presence of recombinant interleukin-2 (IL-2), *in vitro.*[20] Surprisingly, this stimulation was not inhibited by the addition of a monoclonal antibody against HLA-DR antigen. Thus, the stimulation of allogeneic lymphocytes by rIFN-γ-treated keratinocytes appears to be due to expression of an antigen other than DR. The observation that significant numbers of keratinocytes express HLA-DP antigen on day 4 after the addition of rIFN-γ may provide an explanation for rIFN-γ-treated keratinocytes' capacity to stimulate foreign T lymphocytes.

SUMMARY

Normal human keratinocytes grown in MCDB 153 plus bovine pituitary extract and treated with recombinant gamma interferon (rIFN-γ) express HLA-DR, DP and DQ antigens. The expression of these class II MHC antigens is time dependent: DR and DP appear before DQ. The delay in HLA-DQ expression is not due to the effects of trypsinization of cultures prior to analysis. Increasing the calcium ion concentration from 0.1 to 1.8 mM does not alter the expression of these antigens. Keratinocytes grown without serum proteins or bovine pituitary extract exhibited markedly delayed expression of DR. By contrast, keratinocytes grown in Dulbecco's modified Eagle's medium (DMEM) plus 10% fetal calf serum express DR and DP but only very small amounts of DQ after treatment with rIFN-γ. Expression of HLA-DR occurs at doses of rIFN-γ that are too low to cause growth inhibition. The cells of the squamous cell carcinoma cell line SCL-1, whether grown in MCDB 153 plus bovine pituitary extract or DMEM plus 10% fetal calf serum, express HLA-DQ and DP on only small numbers of cells after treatment with the lymphokine. Thus, the conditions of culture, possibly the presence of a serum factor(s), influence the expression of class II antigens in normal keratinocytes. Furthermore, rIFN-γ does not induce DP and DQ antigens readily in transformed squamous cells cultured in either serum-containing or serum-free medium.

ACKNOWLEDGMENT

We thank Mrs. E. A. Pfendt for her expert technical assistance and Dr. A. B. Cua for help with the cell counts.

REFERENCES

1. TJERNLUND, U. M. 1980. Ia-like antigens in lichen planus. Acta Dermatovener. (Stockholm) **60:** 309–315.
2. MACKIE, R. M. & M. L. TURBITT. 1983. Quantification of dendritic cells in normal and abnormal human epidermis using monoclonal antibodies directed against Ia and HTA antigens. J. Invest. Dermatol. **81:** 216–220.
3. TJERNLUND, U. M. 1978. Epidermal expression of HLA-DR antigens in mycosis fungoides. Arch. Dermatol. Res. **261:** 81–86.
4. WOOD, G. S. Unpublished data.
5. BASHAM, T. Y., B. J. NICKOLOFF, T. C. MERIGAN & V. B. MORHENN. 1984. Recombinant gamma interferon induces HLA-DR expression on cultured human keratinocytes. J Invest. Dermatol. **83:** 88–91.
6. VOLC-PLATZER, B., H. LEIBL, T. LUGER, G. ZAHN & G. STINGL. 1985. Human epidermal cells systhesize HLA-DR alloantigens *in vitro* upon stimulation with γ-interferon. J. Invest. Dermatol. **85:** 16–19.
7. SHAW, S., A. H. JOHNSON & G. M. SHEARER. 1980. Evidence for a new segregant series of B cell antigens that are uncoded in the HLA-D region and that stimulate secondary allogeneic proliferative and cytoxic responses. J Exp. Med. **152:** 565–580.
8. HENNINGS, H., K. A. HOLBROOK & S. H. YUSPA. 1983. Potassium mediation of calcium induced terminal differentiation of epidermal cells in culture. J. Invest. Dermatol. **81:** 50s–55s.
9. BOYCE, S. T. & R. G. HAM. 1983. Calcium-regulated differentiation of normal human epidermal keratinocytes in chemically defined clonal culture and serum-free serial culture. J. Invest. Dermatol. 81(Suppl.): 33s–40s.
10. NICKOLOFF, B. J., T. Y. BASHAM, T. C. MERIGAN & V. B. MORHENN. 1985. Immunomodulatory and antiproliferative effect of recombinant alpha, beta and gamma interferons on cultured human malignant squamous cell lines: SCL-1 and SW-1271. J. Invest. Dermatol. **84:** 487–490.
11. OSEROFF, A. R., L. DICICCO, E. A. PFENDT & V. B. MORHENN. 1985. A murine monoclonal antibody (VM-1) against human basal cells inhibits the growth of human keratinocytes in culture. J Invest. Dermatol. **84:** 257–262.
12. MORHENN, V. B., A. B. SCHREIBER, O. SORIERO, W. MACMILLAN & A. C. ALLISON. 1985. A monoclonal antibody specific for proliferating skin squamous cells: Potential use in the diagnosis of cervical neoplasia. J. Clin. Invest. **76:** 1978–1983.
13. MORHENN, V. B., C. J. BENIKE, D. J. CHARRON, A. J. COX, G. MAHRLE, G. S. WOOD & E. G. ENGLEMAN. 1982. Use of the fluorescence-activated cell sorter to quantitate and enrich for subpopulations of human skin cells. J. Invest. Dermatol. **79:** 277–282.
14. LIU, S.-C. & M. A. KARASEK. 1978. Isolation and growth of adult human epidermal keratinocytes in cell culture. J. Invest. Dermatol. **71:** 157–162.

15. BOYCE, S. T. & R. G. HAM. 1986. Normal human epidermal keratinocytes. *In In Vitro Models for Cancer Research.* M. M. Weber & L. Sekely, Eds. Vol. **3:** 245–274. CRC Press Inc. Boca Raton, FL.

16. SHIPLEY, G. D. & M. R. PITTELKOW. 1987. Control of growth and differentiation *in vitro* of human keratinocytes cultured in serum-free medium. Arch. Dermatol. **123:** 1541a–1544a.

17. BRÖCKER, E.-B., L. SUTER & C. SORG. 1984. HLA-DR antigen expression in primary melanomas of the skin. J. Invest. Dermatol. **82:** 244–247.

18. GUILLEN, F. J., C. L. DAY & G. F. MURPHY. 1985. Expression of activation antigens by T cells infiltrating basal cell carcinomas. J. Invest. Dermatol. **85:** 203–206.

19. KAMEYAMA, K., T. TONE, H. ETO, S. TAKEZAKI, T. KANZAKI & S. NISHIYAMA. 1987. Recombinant gamma interferon induces HLA-DR expression on squamous cell carcinoma, trichilemmoma, adenocarcinoma cell lines, and cultured human keratinocytes. Arch. Dermatol. Res. **279:** 161–166.

20. MORHENN, V. B. & B. J. NICKOLOFF. 1987. Interleukin-2 stimulates resting human T lymphocytes' response to allogeneic, cultured gamma interferon treated keratinocytes. J. Invest. Dermatol. **89:** 464–469.

Expression of THY-1 Protein by Murine Keratinocytes

D. A. CHAMBERS, R. L. COHEN, S. F. MARSCHALL,
P. S. JACOBSON, AND M. S. OSTREGA

*Department of Biological Chemistry
and
Center for Research in Periodontal Diseases
and Oral Molecular Biology
University of Illinois at Chicago
P.O. Box 6998
Chicago, Illinois 60680*

The presence of murine Thy-1$^+$ epidermal cells was reported by this laboratory in October, 1982[1,2] and soon afterwards by other investigators. Such cells constitute a heterogeneous array of bone-marrow-derived (Thy-1$^+$/vimentin$^+$) and keratin-containing cells (Thy-1$^+$/keratin$^+$), of different morphotypes, most often dendritic. These experiments examine the expression of Thy-1 protein in keratinocytes and the effect of biologic response modifiers on them.

Neonatal BALB/c mouse epidermal cells were prepared and cultured as described.[3] Epidermal cell cultures were depleted of Thy-1$^+$ cells by treatment with

TABLE 1. Expression of Thy-1 Protein in Epidermal Cultures Initially Depleted of Thy-1$^+$ Cells[a]

	Days in Culture	% Thy-1$^+$ Cells		% Thy-1$^+$/Keratin$^+$
		Control	Depleted	Depleted
Exp. 1	2–4	1.0	0.3	100
	5–7	2.8	1.4	100
	8–10	5.0	1.6	100
Exp. 2	2–4	1.8	1.3	100
	5–7	3.6	2.2	100
	8–10	5.8	2.3	100

[a]Cells obtained from neonatal BALB/c mouse epidermal culture as described in the text were released from the culture dishes and double-labeled with anti-Thy-1.2, antivimentin or antikeratin antibodies and visualized by immunofluorescence microscopy. Data presented is from 2 of 4 representative experiments. Each value represents in excess of 100 cells counted in each experiment.

Thy-1.2 antibody (Becton Dickinson, Mountain View, CA) and complement (Accurate Scientific, Westbury, NY). Thy-1$^+$/keratin$^+$ cells and Thy-1$^+$/vimentin$^+$ cells were revealed by antikeratin antibody (Dr. Dennis Roop, National Institutes of Health, Bethesda, MD) or antivimentin antibody (Miles Scientific, Naperville, IL).

Thy-1-depleted cultures showed no alteration in cell kinetics over a period of 12 days, suggesting that the Thy-1$^+$ cell does not directly regulate epidermal cell culture kinetics. Two days after culture initiation keratinocytes began to express Thy-1 protein of comparable staining intensity to the bone-marrow-derived Thy-1[+] cell seen in control cultures (TABLE 1). Expression of Thy-1 protein plateaued by day 12. Only Thy-1[+]/keratin[+] cells were found in Thy-1-depleted cultures, suggesting that no resident bone marrow (vimentin[+]) cell expressed Thy-1 protein under conditions allowing keratinocyte expression of Thy-1.

Addition of dibutyryl cAMP or cholera toxin to undepleted cultures enhanced DNA synthesis, but Thy-1[+] cells were markedly reduced (TABLE 2). Preliminary experiments suggest that Thy-1[+] lymphocytes cultured in the presence of cholera toxin also show diminished levels of cell surface Thy-1 protein. When epidermal cells were cultured under conditions of restricted calcium, Thy-1 expression was less than 10% of control cultures (TABLE 2).

These experiments reveal that keratinocytes in cultures depleted of Thy-1[+] cells can activate and express the Thy-1 gene. In epidermal cultures, such activation

TABLE 2. Effect of Biological Response Modifiers on Thy-1 Expression in Epidermal Cell Culture[a]

	% of Control Values[b]		
Days in Culture	cAMP N = 3	Cholera Toxin N = 4	Low Calcium N = 2
2–4	49 ± 8	47 ± 21	38 ± 6
5–7	54 ± 7	57 ± 20	9 ± 5
8–10	57 ± 11	45 ± 9	8 ± 2

[a]Cells obtained from neonatal BALB/c mouse epidermal cultures[3] were released from the culture dish and reacted with FITC-labeled anti-Thy-1.2 antibody. Data represents in excess of 500 cells counted in each treatment group in each experiment.

[b]Percentage is equal to the ratio of Thy-1[+] cell in treated vs untreated cultures × 100.

appeared to be inhibited under conditions which stimulated cell proliferation (*e.g.,* cAMP or diminished calcium), inferring that keratinocyte Thy-1 expression is a component of the keratinocyte's differentiation program.

REFERENCES

1. CHAMBERS, D. A., R. L. COHEN & M. HEISS. 1983. Heterogeneity of epidermal cells detected by the presence of Thy-1 antigen in athymic (nude) and normal BALB/c mice. *In* Proceedings of the 4th International Workshop on Immune-Deficient Animals in Experimental Research, Chexbres, Switzerland, Oct. 1982. Exp. Cell Biol. **52:** 129–132.
2. CHAMBERS, D. A. 1985. The Thy-1[+] epidermal cell: perspective and prospective. Br. J. Dermatol. **113** (Suppl. 28): 24–33.
3. COHEN, R. L., M. E. A. F. ALVES, V. C. WEISS, D. P. WEST & D. A. CHAMBERS. 1984. Direct effects of minoxidil on epidermal cells in culture. J. Invest. Dermatol. **82**(1): 90–93.

Role of TPA-Inducible Keratinocyte Collagenase in Extracellular Collagen Breakdown

H. Y. LIN, B. BIRKEDAL-HANSEN,
AND H. BIRKEDAL-HANSEN

University of Alabama School of Dentistry
University of Alabama at Birmingham
Birmingham, Alabama 35294

Previous studies by us[1] and by others[2] have shown that cultured skin and mucosal keratinocytes express collagenase activity against interstitial collagens when properly stimulated. The enzyme is recognized by antibody to human fibroblast collagenase and appears to be either identical or highly homologous to the fibroblast enzyme. The purpose of this study was to determine whether induced normal human keratinocytes dissolve reconstituted type I collagen fibrils by a collagenase-dependent pathway.

Subconfluent and early confluent human infant foreskin keratinocytes secreted low but detectable levels of procollagenase when supplemented with bovine pituitary extract[3] (FIG. 1). Addition of 10^{-6}–10^{-8} M TPA resulted in a 40–50-fold stimulation of collagenase secretion during the ensuing 24-hr period (FIG. 1) to a yield maximum of

TABLE 1. Induction of Procollagenase by TPA[a]

Incubation Period (hr)	Collagenase Activity (U/10^6 Cells) TPA Concentration (M)				
	10^{-6}	10^{-7}	10^{-8}	10^{-9}	Control
24	1.44	0.46	0.11	0.08	0
48	1.50	0.51	0.11	0.10	0.004
72	1.55	0.57	0.13	0.10	0.014
96	1.64	0.64	0.13	0.11	0.022

[a]Normal human keratinocytes were seeded overnight at 100,000 cells per cm^2, then supplemented with 10^{-9} to 10^{-6} M TPA. Aliquots were removed at 24-hr intervals and the total collagenase activity (active and latent) measured by radiofibril assay.[4]

1.4 U/10^6 cells (TABLE 1). The effect of TPA, however, was only transient, and very little enzyme (<0.1 U/10^6 cells/day) was secreted after the first 24-hr period. The cells also secreted low levels of collagenase inhibitor (TIMP) (0.0–0.2 U/10^6 cells/ day) but the expression of inhibitor did not respond to TPA. In order to examine the extent to which human foreskin keratinocytes utilize procollagenase in the degradation of collagen fibrils, the cells were seeded directly on a film of trypsin-resistant, reconstituted [^{3}H]-type I collagen fibrils and the release of radioactivity was monitored. TPA-induced cells secreted high levels of procollagenase, but the enzyme remained latent throughout the incubation period and the cells failed to degrade the collagen fibrils unless the activation of the proenzyme was assisted by addition of exogenous protease activity (trypsin). When supplemented with 0.1–1.0 μg/ml trypsin,

TPA-induced cells dissolved the fibrils in 2–3 days. Uninduced cells which secreted low levels of procollagenase failed to degrade the collagen fibrils under any circumstance. The dissolution of the collagen fibrils by TPA-induced cells was blocked by inhibiting monoclonal antibodies to collagenase, whereas control antibodies which react with the enzyme but do not inhibit the catalytic activity of the enzyme failed to block the

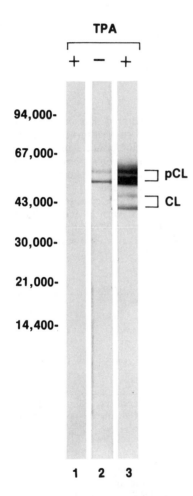

FIGURE 1. Induction of procollagenase by TPA. Conditioned keratinocyte culture medium was subjected to SDS/PAGE, transferred to nitrocellulose paper, and stained with the immunoalkaline phosphatase procedure using rabbit antibody to human fibroblast procollagenase (7 μg/ml). *Lane 1:* control stained with preimmune IgG. *Lane 2:* medium from cells grown in the absence of TPA. *Lane 3:* medium from cell cultures supplemented with 10^{-7} M TPA. pCL, 57,000/52,000 procollagenase; CL Mr 46,000/42,000 activated collagenase.

process. Likewise, affinity-purified polyclonal antibodies raised against human fibroblast procollagenase blocked the cell-mediated collagen breakdown, whereas irrelevant or preimmune IgG did not. These findings show that human foreskin keratinocytes can degrade reconstituted collagen fibrils by a collagenase-dependent pathway, but the reaction, at least in this model system, proceeds slowly, if at all, unless the activation of the secreted proenzyme is assisted by addition of exogenous protease activity.

REFERENCES

1. LIN, H. Y., B. R. WELLS, R. E. TAYLOR & H. BIRKEDAL-HANSEN. 1987. J. Biol. Chem. **262:** 6823–6831.
2. PETERSEN, M. J., D. T. WOODLEY, G. P. STRICKLIN & E. J. O'KEEFE. 1987. J. Biol. Chem. **262:** 835–840.
3. WILLE, J. I., M. R. PITTELKOW, G. D. SHIPLEY & R. E. SCOTT. 1984. J. Cell Physiol. **121:** 31–44.
4. BIRKEDAL-HANSEN, H. 1987. Methods Enzymol. **144:** 140–171.

An Effect of Sunscreens on Cutaneous Vitamin D_3 Synthesis

L. Y. MATSUOKA,[a] J. WORTSMAN,[b] J. MacLAUGHLIN,[c]
AND M. HOLICK[c]

[a]*Jefferson Medical College*
Department of Dermatology
Thomas Jefferson University
Philadelphia, Pennsylvania 19107

[b]*Southern Illinois School of Medicine*
Department of Medicine
Springfield, Illinois 62703

[c]*Boston University School of Medicine M-1013*
Vitamin D, Skin and Bone Research Laboratory
80 East Concord Street
Boston, Massachusetts 02118

Sunlight triggers a variety of photochemical reactions in the skin. Acute exposure to the wavelengths 290 to 315 nm (ultraviolet light B, UVB) causes erythema, whereas chronic exposure produces increased incidence of skin cancer. Solar irradiation also mediates the synthesis of vitamin D_3. The adverse effects of UVB can be prevented by topical sunscreens such as para-aminobenzoic acid (PABA). The present study determined: 1) the effect of PABA on the production of vitamin D precursors in human skin specimens *in vitro,* 2) the effect of PABA on vitamin D synthesis *in vivo,* and 3) the long-term effect of chronic PABA use on vitamin D stores.

MATERIALS AND METHODS

Effect of PABA on Vitamin D Production in Vivo

Surgically obtained human skin was divided into nine pieces. Three pieces received no treatment (sham); three were covered with 0.1 ml ethanol; and three were covered with a 5% PABA-ethanol solution. Specimens treated with ethanol or PABA-ethanol were then exposed to 1 minimal erythemic dose of simulated solar ultraviolet radiation (UVR). Following UVR or sham exposure, the basal epidermal layer was separated and the lipid fraction extracted. HPLC was used to quantify 7-dehydrocholesterol and previtamin D content of the lipid extract.

Effect of a Single Application of PABA on Cutaneous Vitamin D Synthesis In Vivo

Eight subjects were randomly assigned to one of the two study groups. Four subjects were treated with a whole body application of 5% PABA in 55% ethanol (SPF 8), and the other four individuals were untreated (controls). One hour later all the subjects received total body exposure to 1 MED in a phototherapy unit. Blood samples were obtained 24 and 2 hours before total body exposure to UVR and 1, 2, 3, 7 and 14 days after UVR. Serum D_3 was separated by a modified two-stage HPLC procedure and quantitated by UV absorbance at 254 nm.

Effect of Chronic Use of PABA on Vitamin D Nutritional Status

Twenty patients applying sunscreens to prevent skin cancer recurrences were recruited. Nineteen age-matched individuals served as controls. Vitamin D nutritional status was assessed by serum 25-(OH)D concentrations.

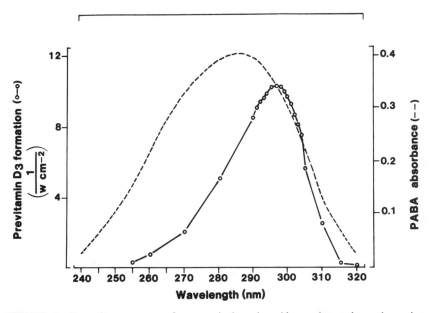

FIGURE 1. Absorption spectrum of para-aminobenzoic acid superimposed on the action spectrum of previtamin D_3 (pre-D_3) formation. Pre-D_3 formation spectrum was obtained by plotting the reciprocal of photoenergy as a function of wavelength. At any wavelength, no more then 5% of the 7-dehydrocholesterol was converted to pre-D_3. (From Matsuoka *et al.*[1] Reprinted by permission from the *Journal of Clinical Endocrinology and Metabolism*.)

RESULTS

The absorption spectrum of PABA overlapped the action spectrum for previtamin D photosynthesis in human epidermis (FIG. 1). PABA, which absorbed radiation in the 260–400-nm range, comprised most of the spectrum responsible for the photosynthesis of vitamin D_3 (260–315 nm).

Effect of PABA on Vitamin D Production In Vitro

Skin samples treated with ethanol and exposed to simulated solar radiation showed appreciable conversion (approximately 15%) of 7-dehydrocholesterol to previtamin D_3

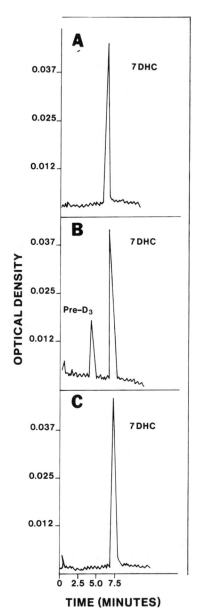

FIGURE 2. Chromatograms of vitamin D_3 precursors in lipid human skin extracts. Specimen (**A**) is untreated skin; specimen (**B**) received a topical application of vehicle (ethanol) solution prior to UVB irradiation; specimen (**C**) received a topical application of para-aminobenzoic acid (PABA) in ethanol prior to UVB irradiation. PABA blocked the photoisomerization of 7-dehydrocholesterol (7-DHC) to previtamin D_3 (pre-D_3). (From Matsuoka et al.[1] Reprinted by permission from the *Journal of Clinical Endocrinology and Metabolism*.)

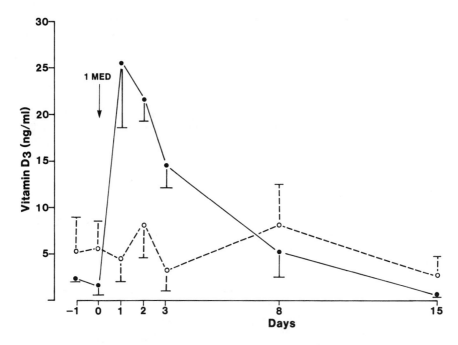

FIGURE 3. Serum vitamin D_3 concentration (mean ± SEM) in normal subjects. PABA (*open circles*) or vehicle (*closed circles*) was applied to the entire skin before one MED of UVB. PABA suppressed UVB-stimulated cutaneous vitamin D formation. (From Matsuoka *et al.*[1] Reprinted by permission from the *Journal of Clinical Endocrinology and Metabolism*.)

FIGURE 4. Serum concentration of 25-hydroxyvitamin D in chronic sunscreen users and in age- and sex-matched controls. Vitamin D stores were significantly lower in chronic PABA users. (From Matsuoka *et al.*[2] Reprinted by permission from *Archives of Dermatology*.)

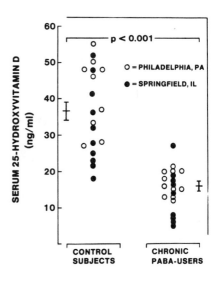

(FIG. 2B). In contrast, previtamin D_3 was absent in skin specimens treated with PABA (FIG. 2C) and was also undetectable in the stratum basale of sham skin specimens not exposed to simulated solar radiation (FIG. 2A).

Effect of a Single Application of PABA on Cutaneous Vitamin D Synthesis In Vivo

Normal volunteers who received whole body application of PABA before UVB did not exhibit a rise in D_3 levels (FIG. 3). In subjects applying vehicle, serum D_3 concentrations increased from 1.5 ± 1.0 (mean $\pm$ SEM) to a peak of 25.6 ± 6.7 ng/ml ($p < 0.01$).

Effect of Chronic Use of PABA on Vitamin D Nutritional Status

Vitamin D stores were decreased in patients using sunscreens chronically. The 25-(OH)D concentration of chronic PABA users was 16.1 ± 1.3 was compared to 36.6 ± 2.5 ng/ml in age-matched control subjects ($p < 0.001$) (FIG. 4).

CONCLUSIONS

Sunscreens suppress the cutaneous synthesis of vitamin D_3. Chronic use of these agents can lead to vitamin D deficiency.

REFERENCES

1. MATSUOKA, L. Y., L. IDE, J. WORTSMAN, J. A. MACLAUGHLIN & M. F. HOLICK. 1987. Sunscreens suppress vitamin D_3 synthesis. J. Clin. Endocrinol. Metab. **16:** 1165–1168.
2. MATSUOKA, L. Y., J. WORTSMAN, N. HANIFAN & M. F. HOLICK. Lower body stores of vitamin D among sunscreen users: a preliminary study. Arch. Dermatol. In press.

Differential Effects of 1,25-Dihydroxyvitamin D_3 on Proliferation and Biochemical Differentiation of Cultured Human Epidermal Keratinocytes Grown in Different Media

JOHN A. McLANE AND MARION KATZ

Hoffmann-La Roche, Inc.
340 Kingsland Street
Nutley, New Jersey 07110

Neonatal foreskin keratinocytes were grown in culture by a variety of techniques to assess the potential modulation of proliferation and differentiation by 1 alpha,25-dihydroxyvitamin D_3 (1,25-$(OH)_2D_3$). Reports in the literature indicate that 1,25-$(OH)_2D_3$ is a potent inhibitor of cellular proliferation.[1] This report extends that study to include cellular responses observed in other culture conditions. Keratinocytes were grown to 90% of confluency in either modified MCDB 153 (KGM, Clonetics) supplemented with calcium to 1.1 mM, DMEM with growth factor supplements (no serum), or DMEM plus 1.5 mM $CaCl_2$. Some cultures were also grown to confluency prior to the addition of 1,25-$(OH)_2D_3$. After additional growth for 7 to 14 days in the presence of 1,25-$(OH)_2D_3$ at micromolar or lower concentrations, cultures were assayed for cell proliferation, cornified envelope formation, and presence of specific keratins and involucrin. The responses of keratinocytes to 1,25-$(OH)_2D_3$ in each medium differed significantly. Therefore, although it appears that 1,25-$(OH)_2D_3$ may be an inhibitor of proliferation and stimulator of differentiation, this effect can be obliterated or even reversed by different culture media.

Human neonatal foreskins were collected by circumcision and placed into tubes containing DMEM medium with 10% serum. On arrival at the laboratory they were mechanically trimmed of excess dermis, and treated with a solution of trypsin/EDTA (0.05%/0.02%) at 4°C overnight. The epidermis was stripped from the dermis and was agitated in buffered saline to remove basal keratinocytes, and the stratum corneum layer was removed. The separated cells were centrifuged, resuspended in medium, counted and plated onto mitomycin-C-treated 3T3 cells as appropriate.[2] The keratinocytes were plated at a density of approximately 20,000 cells/cm^2 in dishes or wells of assorted size depending on the experiment. Cells were cultured in Dulbecco's modified Eagle's medium (DMEM) without serum, supplemented with the following growth factors: epidermal growth factor (25 ng/ml), hydrocortisone (203 ng/ml), insulin (5 μg/ml), transferrin (5 μg/ml), prostaglandin E1 (50 ng/ml), cholera toxin (0.1 μg/ml), and selenous acid (2 ng/ml).[1] During the experimentation with compounds, cholera toxin and hydrocortisone were not added to the medium used to feed the cells. All cultures were incubated in humidified atmosphere of 5% CO_2 at 37°C with media changed three times per week.

For each experiment every culture dish or well received the same number of cells from the same culture source. At the termination of the experiment the number of cells per dish/well was determined by counting on an electronic particle counter (Coulter Counter). Each dish was counted at least three times and all treatments including controls were done in at least triplicate.

After an aliquot of cells had been removed for counting, a solution of SDS/DTT was added to the cells to a final concentration of 1% SDS/5 mM DTT. The cells were solubilized for one hour at 37°C and an aliquot removed for enumeration. Aliquots were counted either with a hemacytometer or placed into isotonic buffered saline and counted with a Coulter Counter.

Keratinocytes grown in KGM (1.1 mM Ca) are still proliferating rapidly with a doubling time of approximately 41 hours for the cells in the population. In DMEM without serum, first passage keratinocytes had a doubling time of approximately 105 hours between 5 and 10 days. Cells grown in KGM are only inhibited about 10% in the total number of cells counted at the highest doses of 1,25-(OH)$_2$D$_3$. At lower doses of 1,25-(OH)$_2$D$_3$, cells were actually stimulated to grow. In the KGM medium, even with

TABLE 1. Effect of 1,25-(OH)$_2$-Vitamin D$_3$ on Proliferation and Envelope Formation in Human Keratinocytes in Serum-Free Medium[a]

Dose (M)	Treatment (Medium) KGM	DMEM	DMEM (1.5 mM Ca)	DMEM Preconfluent	DMEM Confluent
Cells/well					
10^{-10}	108.2	94.3	107.1	99.67	121.8
10^{-8}	94.2	66.9	99.1	80.75	109.3
10^{-6}	86.8	55.3	65.4	63.92	103.4
Envelopes/cell					
10^{-10}	78.1	127.2	111.0	79.41	180.5
10^{-8}	82.6	184.5	122.6	124.83	230.3
10^{-6}	187.8	219.9	150.1	219.51	306.4

[a]Keratinocytes from first passage were grown for seven days in the presence of 1,25-(OH)$_2$D$_3$ at the doses indicated. Control values were from ethanol-treated cultures treated under the same conditions in each set of experiments. Medium treatment is as indicated in abstract. Experimental results, shown as percent of controls, were averages from triplicate wells; experiments were repeated at least two times with similar results (SE averages 21%).

calcium supplement, envelope formation was inhibited when compared to control cultures at all but the highest doses (TABLE 1).

Several reports in the literature demonstrate that increasing extracellular calcium results in decreases in the proliferation and increases in the differentiation of human keratinocytes.[3] When 1,25-(OH)$_2$D$_3$ is added along with extra calcium, the percent decrease in cellular proliferation and increase in envelope formation is not as great as in cultures without excess calcium. The biological changes induced by extracellular calcium are not greatly enhanced by the presence of 1,25-(OH)$_2$D$_3$. The active form of vitamin D$_3$, along with other factors, is thought to mobilize the intracellular Ca in keratinocytes. The intracellular Ca may be close to the maximum amount that can be mobilized when the cells are grown with excess extracellular Ca, and the 1,25-(OH)$_2$D$_3$ is not able to increase this amount further.

It has been shown in reports that there are numerous differences in the biochemistry and biological responses of keratinocytes that are preconfluent versus those that are confluent. The biological effects of 1,25-(OH)$_2$D$_3$ on cultures differ greatly depending on whether the cultures are confluent or not. In confluent cultures 1,25-(OH)$_2$D$_3$

actually stimulates the growth of the cells at all doses tested. The confluent cultures grown in the presence of 1,25-(OH)$_2$D$_3$ had the greatest percentage of cells forming envelopes (even more then when extracellular calcium is added). In the preconfluent cultures 1,25-(OH)$_2$D$_3$ had its greatest effect on inhibiting cellular proliferation but a poor effect in stimulating envelope formation (TABLE 1).

CONCLUSIONS

1. The optimum culture conditions to study the biological effects of the active form of vitamin D$_3$ depend on whether cellular proliferation or differentiation is to be investigated. There is not a reciprocal response between inhibiting proliferation and stimulating envelope formation when keratinocyte cultures are treated with 1,25-(OH)$_2$D$_3$.

2. Preconfluent keratinocyte cultures that are not proliferating rapidly (as in the KGM medium) are adequate for investigating effects of agents on proliferation.

3. The effect of 1,25-(OH)$_2$D$_3$ on cultures of keratinocytes may be added to the growing list of differences in responses between confluent and preconfluent cultures.

REFERENCES

1. SMITH, E. L. et al. 1986. J. Invest. Dermatol. **86:** 709–712.
2. RHEINWALD, J. G. & H. GREEN. 1975. Cell **6:** 331–343.
3. SHIPLEY, G. D. & M. R. PITTELKOW. 1987. Arch. Dermatol. **123:** 1541a–1544a.

Characterization of Keratinocyte Chemotaxis and Adhesion

A. SANK, N. MARTINET, AND G. MARTIN

National Institutes of Health
National Institute of Dental Research
Building 30, Room 414
Bethesda, Maryland 20892

Wound repair requires the coordinated migration of epidermal and fibroblast cells. These cells move *en bloc,* perhaps as a result of proliferation which creates a pressure gradient for expansion. Since many cells respond to soluble factors by directed migration (chemotaxis), we tested the responsiveness of human and mouse keratinocytes to numerous growth factors. We also characterized the matrix components most conducive to human keratinocyte adhesion.

The Boyden chamber (cells and chemoattractant separated by a micropore filter) was used for chemotaxis assays. In separate experiments keratinocytes were added to 24-well plates individually coated with the matrix components (collagen I and IV and laminin @ 50 μg/ml, matrigel @ 1 mg/ml dilution and fibronectin @ 500 μg/ml). The attached cells in the two assays were then counted with an Optomax image analyzer linked to an IBM computer.

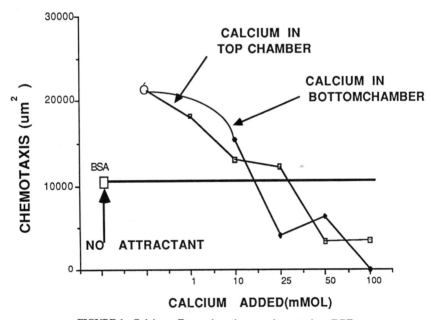

FIGURE 1. Calcium effect on keratinocyte chemotaxis to EGF.

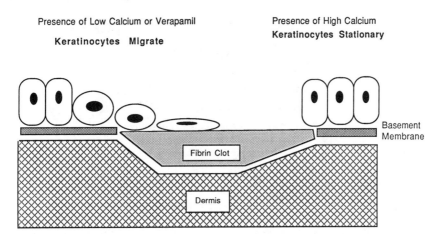

FIGURE 2. Possible mechanism of effect of calcium on keratinocyte migration in wound healing.

A number of growth factors were chemotactic for these keratinocytes, most notably PDGF, bFGF, EGF and TGF-β. The positive chemotactic responses of mouse and human keratinocytes were not identical. Adhesion of keratinocytes proceeded best in the presence of collagen I and IV, laminin and matrigel. The addition of calcium inhibited both chemotaxis and adhesion (FIGS. 1 and 2). This effect of calcium could be reversed with verapamil. The blocking of keratinocyte migration by calcium may be relevant to the healing of certain chronic wounds.

Presence of Interleukin-1 Receptors on Keratinocytes

K. PAGANELLI PARKER,[a] D. N. SAUDER,[b]

AND PATRICIA L. KILIAN[a]

[a]Department of Immunopharmacology
Roche Research Center
Hoffmann-La Roche, Inc.
340 Kingsland Street
Nutley, New Jersey 07110
and
[b]Division of Dermatology
McMaster University
Hamilton, Ontario L8N 3Z5

Two major types of interleukin-1 (IL-1) proteins designed IL-1 alpha and IL-1 beta have been described.[1,2] Both IL-1 proteins are produced by activated macrophages and other cell types and exhibit a wide range of biological activities associated with immune regulation and the inflammatory response.

The skin may be one of the major target organs in which IL-1 has important metabolic, biochemical, and physiological effects. IL-1-like activity is found in stratum corneum and epidermis[3,4] and keratinocytes have been shown to produce IL-1[5] and to contain mRNA for both IL-1 species.[6] The precise role that keratinocyte-derived IL-1 plays in regulating normal skin function is largely unknown. One recent study suggests,

TABLE 1. Binding Characteristics of IL-1 Receptors on Various Cell Types

	Binding Constants		
Cell Type	$K_D{}^a$	Receptor Number Sites/Cell	Reference
Human keratinocytes	2.5×10^{-11}	1930	this paper
Mouse EL-4 thymoma	2.0×10^{-11}	1500	1, 8
Mouse 3T3 fibroblast	4.0×10^{-11}	4500	9
Human CRL 1445 dermal fibroblast	2.0×10^{-11}	1600	1

^{a}Apparent dissociation constant.

however, that IL-1 may act in an autocrine fashion to stimulate keratinocyte proliferation.[7]

The objective of this study was to determine if keratinocytes possess IL-1 receptors. In order to meet this objective, normal human keratinocytes were grown in culture and the radioreceptor assay using [125-I]-labeled human recombinant IL-1 alpha was performed as described.[8,9] These experiments demonstrate that high-affinity IL-1 receptors are present on keratinocytes. Scatchard plot analysis for binding studies carried out at 4°C reveals a single type of high-affinity binding site on keratinocytes with an apparent dissociation constant of approximately 2.5×10^{-11} M and the presence of approximately 1900 receptors per cell. These values are similar to those obtained for other cell types as shown in TABLE 1. Characterization of the IL-1 receptor on keratinocytes by chemical cross-linking experiments is in progress as are other studies aimed at identifying agents which modulate IL-1 receptor expression by these cells.

The presence of IL-1 receptors on keratinocytes supports an autocrine role for IL-1 in keratinocyte function. This may have important ramifications for understanding normal skin physiology and pathophysiological conditions.

REFERENCES

1. LOMEDICO, P. T., P. L. KILIAN, U. GUBLER, A. STERN & R. CHIZZONITE. 1986. Molecular biology of interleukin-1. Cold Spring Harbor Symp. Quant. Biol. **51:** 631–639.
2. OPPENHEIM, J. J., E. J. KOVACS, K. MATSUSHIMA & S. K. DURUM. 1986. There is more than one interleukin-1. *Immunol. Today* **1:** 45–56.
3. GAHRING, L. C., A. BUCKLEY & R. A. DAYNES. 1985. Presence of epidermal-derived thymocyte activating factor/interleukin-1 in normal human stratum corneum. J. Clin. Invest. **76:** 1585–1591.
4. HAUSER, C., J. H. SAURAT, A. SCHMITT, F. JAUNIN & J.-M. DAYER. 1986. Interleukin-1 is present in normal human epidermis. J. Immunol **136:** 3317–3323.
5. SAUDER, D. N., C. S. CARTER, S. I. KATZ & J. J. OPPENHEIM. 1982. Epidermal cell production of thymocyte activating factor (ETAF). J. Invest. Dermatol. **79:** 34–39.
6. KUPPER, T. S., D. W. BALLARD, A. O. CHUA, J. S. MCGUIRE, P. M. FLOOD, M. C. HOROWITZ, R. LANDON, J. LIGHTFOOT & U. GUBLER. 1986. Human keratinocytes contain mRNA indistinguishable from monocyte interleukin 1α and β mRNA. J. Exp. Med. **164:** 2095–2100.
7. RISTOW, H. J. 1987. A major factor contributing to epidermal proliferation in inflammatory skin diseases appears to be interleukin-1 or a related protein. Proc. Natl. Acad. Sci. USA **84:** 1940–1944.
8. KILIAN, P. L., K. L. KAFFKA, A. S. STERN, D. WOEHLE, W. R. BENJAMIN, T. M. DECHIARA, U. GUBLER, J. J. FARRAR, S. B. MIZEL & P. T. LOMEDICO. 1986. Interleukin-1 alpha and interleukin-1 beta bind to the same receptor on T cells. J. Immunol. **136:** 4509–4514.
9. MIZEL, S. B., P. L. KILIAN, J. C. LEWIS, K. A. PAGANELLI & R. A. CHIZZONITE. 1987. The interleukin-1 receptor. Dynamics of interleukin-1 binding and internalization in T cells and fibroblasts. J. Immunol. **138:** 2906–2912.

Epidermal-Derived Lymphokines and Their Presence in Allergic and Irritant Skin Reactions

KRISTIAN THESTRUP-PEDERSEN,
CHRISTIAN GRØNHØJ LARSEN, THOMAS TERNOWITZ,
AND CLAUS ZACHARIAE

University of Aarhus
Department of Dermatology
Marselisborg Hospital
8000 Aarhus C, Denmark

We have studied epidermal-derived lymphokines in 32 patients with allergic or toxic eczema and persons without skin disease.

On the one hand, we performed patch testing in patients with known type IV cutaneous allergy, and on the other hand we tested the skin using 3% sodium lauryl sulphate in petrolatum in order to induce an irritant reaction. After 48 hours we used the suction blister technique to secure epidermal tissue and studied its content of interleukin-1 (IL-1) using the C3H thymocyte assay, and the presence of a newly described epidermal lymphocyte chemotactic factor (ELCF).[1]

IL-1 was found in normal skin. During the evolution of an allergic patch test, IL-1 increased 2.8-fold in the test area and 1.9-fold in a nontest area compared with pretest values (TABLE 1).[2] However, IL-1 was not increased in epidermis overlying an irritant skin patch test (TABLE 2).

ELCF was not present in normal skin. However, ELCF activity could be measured in epidermis overlying a positive allergic patch test. Similar studies of a toxic-irritant patch test using sodium lauryl sulphate showed that ELCF was increased contrary to IL-1.

IL-1 and ELCF may thus be important lymphokines in immune-mediated skin inflammation. Preliminary studies of the time course in patch tests have shown that

TABLE 1.[a]

	ETAF/IL-1 (Units/cm^2)			ELCF (CI)[b]		
	0 h	48 h−	48 h+	0 h	48 h−	48 h+
Mean	269	571	836	1.26	1.51	2.31
SD	162	467	554	0.19	0.31	0.68
No.	7	13	13	7	13	13

[a]Thirteen patients were tested for allergic contact dermatitis using patch testing. The readings were done after 48 hours. Epidermal sheets were isolated after 48 hours using a suction blister technique. Homogenates of epidermal sheets from the test area (+ test) and nontest area (− test) were measured for ETAF/IL-1-like activity and presence of ELCF. Seven of the patients were additionally measured in the nontest area immediately before applying the patch test.

[b]CI: chemotactic index.

TABLE 2.[a]

	ETAF/IL-1 (Units/cm^2)			ELCF (CI)		
	0 h	48 h−	48 h+	0 h	48 h−	48 h+
Mean	218	220	244	1.14	1.22	2.09
SD	120	149	169	0.11	0.22	0.68
No.	15	15	15	15	15	15

[a]The table shows the mean values of ETAF/IL-1 and ELCF activity in epidermis before and 48 hours after applying a sodium lauryl sulphate 3% patch test.

both IL-1 and ELCF exist in epidermis before the clinical appearance of a positive allergic patch test.

REFERENCES

1. TERNOWITZ, T. & K. THESTRUP-PEDERSEN. 1986. Epidermis and lymphocyte interaction during a tuberculin skin reaction. II. Epidermis contains specific lymphocyte chemotactic factors. J. Invest. Dermatol. **87:** 613–617.
2. LARSEN, C. G., T. TERNOWITZ, F. G. LARSEN & K. THESTRUP-PEDERSEN. 1988. Epidermis and lymphocyte interactions during an allergic patch test reaction. Increased activity of ETAF/IL-1, epidermal derived lymphocyte chemotactic factor and mixed skin lymphocyte reactivity in persons with type IV allergy. J. Invest. Dermatol. In press.

Description of an Epidermal Lymphocyte Chemotactic Factor Which Specifically Attracts OKT4-Positive Lymphocytes

CLAUS ZACHARIAE, THOMAS TERNOWITZ,
CHRISTIAN GRØNHØJ LARSEN,
AND KRISTIAN THESTRUP-PEDERSEN

Department of Dermatology
Marselisborg Hospital
University of Aarhus
DK-8000 Aarhus C, Denmark

INTRODUCTION

A cell-mediated immune reaction in the skin is histologically characterized by accumulation of T-lymphocytes. We have recently described how epidermis overlying a positive tuberculin reaction contains epidermal lymphocyte chemotactic factor(s) (ELCF), which preferentially activates T-lymphocytes for migration.[1] However, the affinity of ELCF for different T-cell subsets is not known.

MATERIAL AND METHODS

Defibrinated venous blood was separated on Isopaque Ficoll gradients. The cells were washed and mixed with anti-CD4 (OKT4) and anti-CD8 (OKT8) monoclonal antibodies. Separation was done using fluorescence-activated cell sorting following which the cells were washed, resuspended in RPMI-1640 with 10% human A serum, and left unstimulated at 37°C for 2 days. They were then labelled with chromium-51 and adjusted to a final concentration of 3.5×10^6 cells/ml.

Chemoattractants

ELCF was obtained from skin exhibiting a 48-hour positive tuberculin reaction using the suction blister technique. The epidermal sheets were homogenized, centrifuged, dialyzed and ultrafiltrated. We pooled ELCF from several persons before studies of chemotaxis. The standard dilution was 1 epidermal sheet per 300 μl of medium.

LTB$_4$ and FMLP (2.5×10^{-8} M) was also included. Lymphocyte chemotaxis was performed by a ^{51}Cr Boyden assay using a polycarbonate filter with pore size 5 μm (Nuclepore Corp., Pleasanton, CA). The chemotactic factors were placed in the lower compartment. The migration time was 60 min. After the incubation period the radioactivity of the migrated cells and the cells still adherent to the filter was determined. The chemotaxis was expressed as chemotactic index (CI), being the ratio of active migration in the presence of chemotactic factors to random migration in the presence of medium alone.

RESULTS

FIGURE 1 shows the individual experiments. The migration indices of OKT4[+] and OKT8[+] cells was significantly different ($p < 0.05$), when ELCF was used as chemoattractant. LTB₄, but not FMLP, induced significant migration in both OKT4[+] and OKT8[+]lymphocytes.

DISCUSSION

This study confirms that T-lymphocytes show active migration towards the chemoattractants ELCF and LTB₄, but not to FMLP. There seems to be a discrimina-

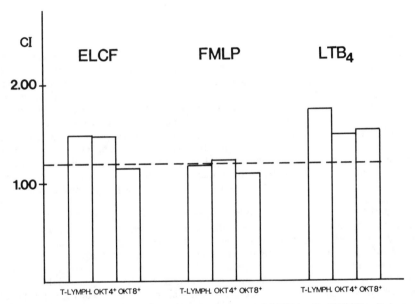

FIGURE 1. Chemotaxis of T-lymphocytes and OKT4[+] and OKT8[+] cells. OKT4[+] cells versus OKT8[+] cells: $p < 0.05$ for ELCF. CI: chemotactic index; ELCF: epidermal lymphocyte chemotactic factor.

tory capacity of OKT4[+] and OKT8[+] T-cells in their response to the newly described ELCF. Our observations may be of significance in order to understand the predominance of OKT4[+] cells in a variety of skin diseases such as contact allergy of mycosis fungoides.

REFERENCE

1. TERNOWITZ, T. & K. THESTRUP-PEDERSEN. 1986. Epidermis and lymphocyte interactions during a tuberculin skin reaction. II. Epidermis contains specific lymphocyte chemotactic factors. J. Invest. Dermatol. **87:** 613–616.

Protein Kinase C-Mediated Expression of Transforming Growth Factor-α in Normal Human Keratinocytes

MARK R. PITTELKOW[a] AND ROBERT J. COFFEY, JR.[b]

[a]Department of Dermatology
Mayo Clinic/Foundation
Rochester, Minnesota 55905
and
[b]Department of Medicine and Cell Biology
Vanderbilt University
Nashville, Tennessee 37232

Transforming growth factor-α (TGF-α) was initially isolated from virally transformed mouse mesenchymal cells and is a potent mitogen for a variety of cell types in culture. TGF-α shares with epidermal growth factor (EGF) significantly sequence homology and biological activity and appears to activate cellular responses via a shared membrane receptor, the EGF receptor.[1] Recently, TGF-α has been shown to be expressed in human epidermis and produced by normal human keratinocytes (HK) *in vitro*.[2] TGF-α and EGF induce TGF-α mRNA and protein expression in cultured HK. TGF-α, like EGF, is a potent mitogen for HK proliferation. TGF-α may serve, therefore, to regulate the growth of epidermis. In this fashion, epidermal growth control could be locally mediated by autocrine/paracrine production of TGF-α and receptor-ligand activation events that would regulate keratinocyte proliferation and differentiation.

Phorbol esters, such as 12-0-tetradecanoyl phorbol-13-acetate (TPA), are well known hyperplastic agents that markedly increase germinative keratinocyte replication when applied to skin. These studies were designed to determine whether phorbol ester-induced epidermal hyperplasia is associated with enhanced expression of TGF-α transcript and protein. In this regard, TGF-α may be an endogenously controlled positive signal for keratinocyte proliferation that can be upregulated by specific exogenous chemical compounds. Certain of these chemicals have been recognized to be potent tumor promoters as well as inducers of epidermal hyperplasia. Therefore, TGF-α may mediate both normal and pathologic states of epidermal growth and differentiation.

HK were obtained from neonatal foreskin or adult skin and propagated in serum-free medium, MCDB 153, supplemented with growth factors, hormones and additives.[3] Confluent cultures of HK deprived of exogenous growth factors for 48 hr express low or undetectable levels of TGF-α mRNA and protein. TPA and other phorbol esters that activate protein kinase C (PKC) were shown to dramatically induce TGF-α mRNA accumulation in treated cultures of HK. TPA exposure increases TGF-α mRNA levels within 1 hr and maximum induction occurs by 5 hr of treatment. EGF also induces TGF-α mRNA expression, but significantly less than TPA treatment (FIG. 1). 1,2-sn-dioctanoylglycerol (DiC8), a synthetic diacylglycerol that activates PKC in intact cells, also induces TGF-α mRNA accumulation, but less markedly than TPA. TPA increases production and secretion of TGF-α protein by 20-fold at 24 hr. Greater than 1 ng/ml of TGF-α protein is secreted into culture medium by HK treated with TPA (10 ng/ml). This concentration is proportional to amounts of TGF-α secreted by various carcinoma cell lines that express high levels of this growth factor.

Synthetic antagonists of PKC, such as 1-(5-isoquinolinylsulfonyl)-2-methyl-piperazine (H7) specifically inhibited DiC8-mediated accumulation of TGF-α mRNA. Cycloheximide failed to inhibit TGF-α mRNA expression. In fact, cyclohex-imide alone, enhanced TGF-α mRNA acccumulation. Actinomycin D, an inhibitor of DNA synthesis, completely inhibited transcriptional activation of TGF-α mRNA by TPA. Activation of PKC by active phorbol esters or diacylglycerols appears to regulate, at least in part, TGF-α gene expression.

Autocrine/paracrine growth of epidermis may be mediated by PKC activation and expression of the TGF-α gene. Epidermal hyperplasia resulting from active phorbol ester treatment of skin may be dependent on dramatic upregulation of TGF-α mRNA

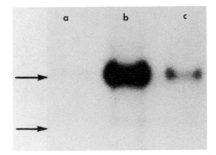

FIGURE 1. Induction of TGF-α mRNA accumulation by TPA and EGF. Northern blot of poly A$^+$ RNA (2 μg per lane) isolated from HK cultured in the absence of growth factors for 2 days and refed fresh medium without growth factors for 4 hr (control) (a), containing TPA (10 ng/ml) (b), or EGF (10 ng/ml) (c). Hybridization performed with a probe complementary to TGF-α transcripts. ($\rightarrow$) indicate 28s and 18s rRNA markers.

and protein production. Furthermore, other hyperproliferative epidermal diseases, such as psoriasis, also may be perpetuated by misregulated TGF-α gene expression.

REFERENCES

1. PITTELKOW, M. R., R. J. COFFEY, JR. & H. L. MOSES. 1988. Keratinocytes produce and are regulated by transforming growth factors. Ann. N.Y. Acad. Sci. This volume.
2. COFFEY, R. J., JR., R. DERYNCK, J. N. WILCOX, T. S. BRINGMAN, A. S. GOUSTIN, H. L. MOSES & M. R. PITTELKOW. 1987. Production and auto-induction of transforming growth factor-α in human keratinocytes. Nature **328**: 817–820.
3. WILLE, J. J., JR., M. R. PITTELKOW, G. D. SHIPLEY & R. E. SCOTT. 1984. Integrated control of growth and differentiation of normal human prokeratinocytes cultured in serum-free medium: Clonal analyses, growth kinetics, and cell cycle studies. J. Cell. Physiol. **121**: 31–44.

Subject Index

Actinomycin D, effects of, on TGF-α mRNA activity, 353
Adenylate cyclase, and malignant cell lines, 147, 149–151
Adenylate cyclase, stimulation of, 141, 147–153, 156
Adhesion, of keratinocytes, 344–345
AHH, *see* Aryl hydrocarbon hydroxylase
Allergic eczema, and lymphokines, 348
Allergy, hemopoietin role in, 291
Alopecia, and vitamin D, 24
Alpha-beta T-cells, 227–228
AMP, cyclic (cAMP)
  accumulation of, in ROS cells, 149–153
  and catecholamine effects, 208
  in HHM, 146
  inhibitory effect of, on collagenase release, 168
  and intracellular messenger systems, 201
  in melanocyte proliferation, 189
  and mitogen activity, 183, 186, 189
  and Thy-1 expression, 332
Anagen
  and melanocyte activity, 188
  TGF-α role in, 218
Angiogenesis, growth factor stimulation of, 214
Antibiotic(s), effect of, on TGF-α mRNA activity, 353
Antigen presentation
  by LC and T-cells, 242
  IL-1 mediation of, 242
Antigen receptors, on T-cells, 226
Antigen(s), processing of, in skin, 225
Antigen(s), processing of, in skin, 225
Antigen-presenting cells, 228–229
apo E, *see* Apolipoprotein E
Apolipoprotein E (apo E)
  in cholesterol transport, 164–165
  distribution and functions of, 160
  in lipid metabolism, 165
  production of, 160–168
Aryl hydrocarbon hydroxylase (AHH), induction of, by coal tar, 103
Athymia, hereditary, and nude mice, 238
Athymic mouse epidermis, response of, to TPA, 273–274
Autocrine/paracrine interaction, effects of, 299
Autocrine/paracrine mechanisms, in TGF-α production, 352–353
Autocrine/paracrine regulation, of keratinocytes, 211–212
Autostimulatory factors, in oncogenesis, 297

Basal cell carcinomas (BCC)
  and IL-1 activity, 178
  collagenase expression in, 174, 176
Basement membrane zone (BMZ)
  in morphogenesis, 174
  remodeling of, in wound healing, 174
Basic fibroblast growth factor (bFGF), 345
  cell-to-cell transfer of, 188–189
  keratinocyte synthesis of, 180, 185–186
  in melanomas, 180, 185–187
  mitogenic activity of, 185–186
  relation to TPA, 180, 185–187, 189
BCC, *see* Basal cell carcinomas
bFGF, *see* Basic fibroblast growth factor
Birbeck granules
  and T-cell lymphoma, 228
  in Langerhans cells (LC), 228–229
BMZ, *see* Basement membrane zone
Bone
  as hHCF target organ, 140–141
  resorption of
    and HHM, 146
    and PTH-like factor, 153–154, 156
BSF-1 (IL-4), 262

Caᵢ, *see* Intracellular calcium
Calcium
  effects of
    on keratinocyte adhesion and chemotaxis, 345
    on PTH-like factor production, 150–151
  and IFNγ-induced MHC expression, 323–324
  in MHC antigen expression, 332
  and Thy-1 expression, 332
Calcium gradients, in epidermal differentiation and homeostasis, 194–195
Calcium ionophores
  and PI metabolism, 195
  and phosphatidylinositol (PI) metabolism, 193–195
Calcium resorption, TGF-α in, 213
Carcinogen(s), metabolism of, by keratinocytes, 102–106
Catabolin, collagenase induction by, 168
Catecholamines, antimitotic effect of, cAMP mediation of, 208
CD antigens, 227–230
CE, *see* Cornified envelope
Cell cycle phase(s), inhibition of, by chalones and TGF, 206, 208–209
Cell-cell communication, mechanisms of, 167–168
Chalone(s), 299, 309–310

Index of Contributors